T0262895

THE CHEMISTRY AND PHYSICS OF ENGINEERING MATERIALS

Volume 2: Limitations, Properties, and Models

THE CHEMISTRY AND PHYSICS OF ENGINEERING MATERIALS

Volume 2: Limitations, Properties, and Models

Edited by
Alexandr A. Berlin
Roman Joswik
Nikolai I. Vatin

Reviewers and Advisory Members
A. K. Haghi
Gennady E. Zaikov

Apple Academic Press Inc.
3333 Mistwell Crescent
Oakville, ON L6L 0A2 Canada

Apple Academic Press Inc.
9 Spinnaker Way
Waretown, NJ 08758 USA

© 2016 by Apple Academic Press, Inc.

First issued in paperback 2021

Exclusive worldwide distribution by CRC Press, a member of Taylor & Francis Group
No claim to original U.S. Government works

ISBN 13: 978-1-77463-129-4 (pbk)
ISBN 13: 978-1-77188-736-6 (hbk)

The Chemistry and Physics of Engineering Materials (2-volume set)
ISBN 13: 978-1-77188-742-7 (hbk)

Library and Archives Canada Cataloguing in Publication

The chemistry and physics of engineering materials / edited by Alexandr A. Berlin, Roman Joswik, Nikolai I. Vatin ; reviewers and advisory members, A.K. Haghi, Gennady E. Zaikov. Reissued with one fewer chapter from each volume with revised indexes.

Includes bibliographical references and indexes.
Contents: Volume 2. Limitations, properties, and models.
ISBN 978-1-77188-736-6 (v. 2 : hardcover) --ISBN 978-0-42945-358-8 (PDF)

1. Materials--Mechanical properties. 2. Strength of materials. 3. Metallurgical analysis. 4. Polymers--Structure. 5. Chemistry, Technical.I. Berlin, Al. Al., 1940-, editor II. Joswik, Roman, editor III. Vatin, Nikolai I. (Nikolai Ivanovich), editor

| TA405.C54 2018 | 620.1'1292 | C2018-901927-1 |

CIP data on file with US Library of Congress

Apple Academic Press also publishes its books in a variety of electronic formats. Some content that appears in print may not be available in electronic format. For information about Apple Academic Press products, visit our website at **www.appleacademicpress.com** and the CRC Press website at **www.crcpress.com**

CONTENTS

LIST OF CONTRIBUTORS

I. B. Abdrakhmanov
Establishment of Russian Science Academy, Institute of Organic Chemistry of Ufa Science Center of RSA, 450054, 71 Prospect Octyabrya str, Ufa, Russia, Tel.: +(347)235-38-15; E-mail: Chemhet@anrb.ru

V. A. Babkin
Sebryakov Department, Volgograd State Architect-Build University, 400074, Russia

O. A. Baulin
Ufa State Petroleum Technical University, 1 Kosmonavtov St. Ufa, 45006, Russia

D. A. Chuvashov
Bashkir State University, 32, Validy Str., Ufa, 450076, Ufa, Russia

R. Ya. Deberdeev
Kazan National Research Technological University, Kazan, Tatarstan, Russia

Mandana Dilamian
University of Guilan, Rasht, Iran

V. Yu. Dmitriev
Sebryakov Department, Volgograd State Architect-Build University, 400074, Russia

A. K. Haghi
University of Guilan, Rasht, Iran

B. A. Howell
Central Michigan University, Mount Pleasent, Michigan, USA

A. V. Ignatov
Sebryakov Department, Volgograd State Architect-Build University, 400074, Russia

V. F. Kablov
Volzhsky Polytechnical Institute (branch), Volgograd State Technical University, Russia; E-mail: kablov@volpi.ru

S. L. Khursan
Establishment of Russian Science Academy, Institute of Organic Chemistry of Ufa Science Center of RSA, 450054, 71 Prospect Octyabrya str, Ufa, Russia, Tel.: +(347)235-38-15; E-mail: chemhet@anrb.ru

K. R. Khusnitdinov
Establishment of Russian Science Academy, Institute of Organic Chemistry of Ufa Science Center of RSA, 450054,71 Prospect Octyabrya str, Ufa, Russia, Tel.: +(347)235-38-15; E-mail: chemhet@anrb.ru

R. N. Khusnitdinov
Establishment of Russian Science Academy, Institute of Organic Chemistry of Ufa Science Center of RSA, 450054, 71 Prospect Octyabrya str, Ufa, Russia

V. I. Kodolov
Kalashnikov Izhevsk State Technical University, Russian Federation

Gennady G. Komissarov
N.N. Semenov Institute for Chemical Physics, Russian Academy of Sciences, Kosygin St. 4, Moscow
119991, Russia, E-mail: komiss@chph.ras.ru; gkomiss@yandex.ru

G. A. Korablev
Izhevsk State Agricultural Academy, Izhevsk 426000, Russia; E-mail: korablev@udm.net

R. G. Korablev
Izhevsk State Agricultural Academy, Izhevsk 426000, Russia

G. V. Kozlov
FSBEI HPE "Kh.M. Berbekov Kabardino-Balkarian State University," Chernyshevsky st., 173,
Nal'chik-360004, Russian Federation

G. M. Magomedov
FSBEI HPE "Daghestan State Pedagogical University," M. Yaragskii st., 57, Makhachkala-367003,
Russian Federation

A. K. Mikitaev
Kh.M. Berbekov Kabardino-Balkarian State University, Chernyshevsky st., 173, Nal'chik-360004,
Russian Federation

M. S. Mohyeldin
Polymeric Materials Department, Advanced Technologies and New Materials Research Institute, City
of Scientific Research and Technological Applications, Alexandria, Egypt

A. G. Moustafin
Establishment of Russian Science Academy, Institute of Organic Chemistry of Ufa Science Center
of RSA, 450054, 71 Prospect Octyabrya str, Ufa, Russia, Tel.: +(347)235-38-15; E-mail: chemhet@
anrb.ru

A. K. Osipov
Izhevsk State Agricultural Academy, Izhevsk 426000, Russia

N. G. Petrova
Ministry of Informatization and Communication of the Udmurt Republic, Russian Federation

A. I. Rakhimov
Volgograd State Technical University, Lenin Avenue, 28, Volgograd, 400005, Russia

N. A. Rakhimova
Volgograd State Technical University, Lenin Avenue, 28, Volgograd, 400005, Russia

F. B. Shevlyakov
Ufa State Petroleum Technical University, 1 Kosmonavtov St. Ufa, 45006, Russia

E. S. Titova
Volgograd State Technical University, Lenin Avenue, 28, Volgograd, 400005, Russia

V. V. Trifonov
Sebryakov Department, Volgograd State Architect-Build University, Russian Federation

T. G. Umergalin
Ufa State Petroleum Technical University, 1 Kosmonavtov St. Ufa, 45006, Russia

N. I. Vatin
Civil Engineering Institute, Saint-Petersburg State Polytechnical University, 29 Polytechnicheskaya Street, Saint-Petersburg, 195251, Russia; E-mail: vatin@mail.ru

Kh. Sh. Yakh'yaeva
FSBEI HPE "M.M. Dzhambulatov Daghestan State Agrarian University," M. Gadzhiev st, 180, Makhachkala, 367032, Russian Federation

G. E. Zaikov
Institute of Biochemical Physics, Russian Academy of Sciences, 4 Kosygin str., Moscow 119334, Russian Federation; E-mail: chembio@sky.chph.ras.ru

V. P. Zakharov
Bashkir State University, 32, Validy Str., Ufa, 450076, Ufa, Russia; E-mail: ZaharovVP@mail.ru

LIST OF ABBREVIATIONS

ADH	adipic dihydrazide
ART	activation-relaxation technique
ATNMRI	Advanced Technologies and New Materials Research Institute
BET	Brunauer–Emmett–Teller
BGK	Bhatnagar-Gross-Krook
BKS	van Beest, Kramer and van Santen
BMH	Born–Mayer–Huggins
BSA	bovine serum albumin
CA	chitosan
CNT	classical nucleation theory
DLCA	diffusion-limited cluster–cluster aggregation
DMC	dynamic Monte Carlo
ECM	extracellular matrix
EDP	electric desalting plant
FENE	finitely extensible nonlinear elastic
FSSE	first shell substitution effect
GAG	glycosaminoglycan
GNC	globular nanocarbon
HA	hyaluronic acid
HCl	hydrochloridric acid
HMDS	hexamethyldisilazane
HRTEM	high resolution transmission electron microscopy
HYAL	hyaluronidase
LAAO	limited access of atmospheric oxygen
LDPE	low density polyethylene
MBA	1-methyl-2-butenyl-aniline
MC	Monte Carlo
MD	molecular dynamics
MFI	melt flow index

PCCM	polycondensation capable monomers
PCT	physical and chemical transformations
PDF	pair distribution function
PEI	polyethyleneimine
PLL	poly-L-lysine
PP	polypropylene
RCP	random close packing
ROS	reactive oxygen species
SAIA	Slovak Academic Information Agency
SANS	small-angle neutron scattering data
SEM	scanning electron microscopy
SF	synovial fluid
SR	sustained release
THz	terahertz
TM	technical materials
TMOS	tetramethoxysilane
UHMPE	ultrahigh molecular polyethylene
VACF	velocity autocorrelation function

LIST OF SYMBOLS

a	Acceleration of particle
A	Atom-type dependent constants
B	Strength of the three-body interaction
b_0	Ideal bond length
C_{ik}	Constants
D	Effective dielectric function
d_f	Fractal dimension, df, of porous systems
E	Potential energy
E_{kin}	Kinetic energy
F	Force
H	Hamiltonian
H	Strength
H_{ij}	Strength
k_B	Boltzmann constant, $1.3806 \times 10\text{--}23$ J/K
K_b	Force constant for bound atoms
K_θ	Force constant for bond angels theta
m	Mass of particle
M	Total sampling number
N	Number of atoms
P	Momentum
q	Atomic charges
r	Distance between two atoms
S	Entropy
T	Temperature
t	Time
$u(r^N)$	Potential energy
V	Potential
v	Velocity
x	Direction
Zi	Formal ionic charges
Δt	Time interval

Greek Symbols

$\varnothing_{ijkl}$	Torsion angles
θ_{ijk}	Band angel
φ_{LJ}	Lennard-Jones potential
α	Electric polarizability
ε	Energy
ε_0	Permittivity of free space
η	Exponent of steric repulsion
η_{ij}	Exponent of steric repulsion
θ	Angle between the vector position of the atoms
ρ	Density
σ	Length parameters
τ	Relaxation time

PREFACE

The collection of topics in the two-volume publication reflects the diversity of recent advances in this field with a broad perspective which may be useful for scientists as well as for graduate students and engineers. This new book presents leading-edge research from around the world in this dynamic field.

Diverse topics published in this book are the original works of some of the brightest and most well-known international scientists in two separate volumes.

In the first volume, modern analytical methodologies are presented here.

The first volume offers scope for academics, researchers, and engineering professionals to present their research and development works that have potential for applications in several disciplines of engineering and science. Contributions range from new methods to novel applications of existing methods to provide an understanding of the material and/or structural behavior of new and advanced systems.

In the second volume, limitations, properties and models are presented. These two volumes:

- are collections of articles that highlight some important areas of current interest in recent advances in chemistry and physics of engineering materials
- give an up-to-date and thorough exposition of the present state-of-the-art of chemical physics
- describe the types of techniques now available to the chemist and technician, and discuss their capabilities, limitations and applications.
- provide a balance between chemical and material engineering, basic and applied research.

We would like to express our deep appreciation to all the authors for their outstanding contributions to this book and to express our sincere gratitude for their generosity. All the authors eagerly shared their experiences and expertise in this new book. Special thanks go to the referees for their valuable work.

ABOUT THE EDITORS

Alexandr A. Berlin, DSc
Director, Institute of Chemical Physics, Russian Academy of Sciences, Moscow, Russia
Professor Alexandr A. Berlin, DSc, is the Director of the N. N. Semenov Institute of Chemical Physics at the Russian Academy of Sciences, Moscow, Russia. He is a member of the Russian Academy of Sciences and many national and international associations. Dr. Berlin is world-renowned scientist in the field of chemical kinetics (combustion and flame), chemical physics (thermodynamics), chemistry and physics of oligomers, polymers, and composites and nanocomposites. He is a contributor to 100 books and volumes and has authored over 1000 original papers and reviews.

Roman Joswik, PhD
Director, Military Institute of Chemistry and Radiometry, Warsaw, Poland
Roman Joswik, PhD, is the Director of the Military Institute of Chemistry and Radiometry in Warsaw, Poland. He is a specialist in the field of physical chemistry, chemical physics, radiochemistry, organic chemistry, and applied chemistry. He has published several hundred original scientific papers as well as reviews in the field of radiochemistry and applied chemistry.

Nikolai I. Vatin, DSc
Director of Civil Engineering Institute, Saint-Petersburg State Polytechnical University, Chief of Construction of Unique Buildings and Structures Department
Nikolai I. Vatin, DSc, is the Chief scientific editor of *Magazine of Civil Engineering* and Editor of *Construction of Unique Buildings and Structures*. He is a specialist in the field of chemistry and chemical technology. He published several hundred scientific papers (original and review) and several volumes and books.

A. K. Haghi, PhD

Member of the Canadian Research and Development Center of Sciences and Cultures (CRDCSC), Montreal, Quebec, Canada; Editor-in-Chief, International Journal of Chemoinformatics and Chemical Engineering; Editor-in-Chief, Polymers Research Journal

A. K. Haghi, PhD, holds a BSc in urban and environmental engineering from University of North Carolina (USA); a MSc in mechanical engineering from North Carolina A&T State University (USA); a DEA in applied mechanics, acoustics and materials from Université de Technologie de Compiègne (France); and a PhD in engineering sciences from Université de Franche-Comté (France). He is the author and editor of 165 books as well as 1000 published papers in various journals and conference proceedings. Dr. Haghi has received several grants, consulted for a number of major corporations, and is a frequent speaker to national and international audiences. Since 1983, he served as a professor at several universities. He is currently Editor-in-Chief of the *International Journal of Chemoinformatics and Chemical Engineering* and *Polymers Research Journal* and on the editorial boards of many international journals. He is a member of the Canadian Research and Development Center of Sciences and Cultures (CRDCSC), Montreal, Quebec, Canada.

Gennady E. Zaikov, DSc

Head of the Polymer Division, N. M. Emanuel Institute of Biochemical Physics, Russian Academy of Sciences, Moscow; Professor, Moscow State Academy of Fine Chemical Technology and Kazan National Research Technological University, Russia

Gennady E. Zaikov, DSc, is the Head of the Polymer Division at the N. M. Emanuel Institute of Biochemical Physics, Russian Academy of Sciences, Moscow, Russia, and Professor at Moscow State Academy of Fine Chemical Technology, Russia, as well as Professor at Kazan National Research Technological University, Kazan, Russia. He is also a prolific author, researcher, and lecturer. He has received several awards for his work, including the Russian Federation Scholarship for Outstanding Scientists. He has been a member of many professional organizations and on the editorial boards of many international science journals.

CHAPTER 1

UPDATE ON AEROGELS MATERIAL AND TECHNOLOGY

MANDANA DILAMIAN and A. K. HAGHI

University of Guilan, Rasht, Iran

CONTENTS

ABSTRACT

In this chapter, different simulation methods for modeling the porous structure of silica aerogels and evaluating its structure and properties have been updated and reviewed in detail. This review has been divided in two sections. In section one, the *"basic concepts"* has been reviewed and in the second parts of this chapter the *"research methodology"* has been updated in detail.

1.1 PART 1: INTRODUCTION ON CONCEPTS

A deeper understanding of phenomena on the microscopic scale may lead to completely new fields of application. As a tool for microscopic analysis, molecular simulation methods such as the molecular dynamics (MD), Monte Carlo (MC) methods have currently been playing an extremely important role in numerous fields, ranging from pure science and engineering to the medical, pharmaceutical, and agricultural sciences. MC methods exhibit a powerful ability to analyze thermodynamic equilibrium, but are unsuitable for investigating dynamic phenomena. MD methods are useful for thermodynamic equilibrium but are more advantageous for investigating the dynamic properties of a system in a nonequilibrium situation. The importance of these methods is expected to increase significantly with the advance of science and technology. The purpose of this study is to consider the most suitable method for modeling and characterization of aerogels. Initially, giving an introduction to the Molecular Simulations and its methods help us to have a clear vision of simulating a molecular structure and to understand and predict properties of the systems even at extreme conditions. Considerably, molecular modeling is concerned with the description of the atomic and molecular interactions that govern microscopic and macroscopic behaviors of physical systems. The connection between the macroscopic world and the microscopic world provided by the theory of statistical mechanics. This is the basic of molecular simulations. There are numerous studies mentioned the structure and properties of aerogels and xerogels via experiments and computer simulations.

Computational methods can be used to address a number of the outstanding questions concerning aerogel structure, preparation, and properties. In a computational model, the material structure is known exactly and completely, and so structure/property relationships can be determined and understood directly. Techniques applied in the case of aerogels include both "mimetic" simulations, in which the experimental preparation of an aerogel is imitated using dynamical simulations, and reconstructions, in which available experimental data is used to generate a statistically representative structure. The idea of using molecular dynamics for understanding physical phenomena goes back centuries.

Computer simulations are hopefully used to understand the properties of assemblies of molecules in terms of their structure and the microscopic interactions between them. This serves as a complement to conventional experiments, enabling us to learn something new, something that cannot be found out in other ways.

The main concept of molecular simulations for a given intermolecular "exactly" predict the thermodynamic (pressure, heat capacity, heat of adsorption, structure) and transport (diffusion coefficient, viscosity) properties of the system. In some cases, experiment is impossible (inside of stars weather forecast), too dangerous (flight simulation explosion simulation), expensive (high pressure simulation wind channel simulation), blind (Some properties cannot be observed on very short time-scales and very small space-scales).

The two main families of simulation technique are MD and MC; additionally, there is a whole range of hybrid techniques which combine features from both. In this lecture we shall concentrate on MD. The obvious advantage of MD over MC is that it gives a route to dynamical properties of the system: transport coefficients, time-dependent responses to perturbations, rheological properties and spectra. Computer simulations act as a bridge (Fig. 1.1) between microscopic length and time scales and the macroscopic world of the laboratory: we provide a guess at the interactions between molecules, and obtain 'exact' predictions of bulk properties. The predictions are 'exact' in the sense that they can be made as accurate as we like, subject to the limitations imposed by our computer budget. At the same time, the hidden detail behind bulk measurements can be revealed. An example is the link between the diffusion coefficient and velocity auto-correlation function (the former easy to measure experimentally, the latter much harder). Simulations act as a bridge in another sense: between theory and experiment. We may test a theory by conducting a simulation using the same model, and we may test the model by comparing with experimental results. We may also carry out simulations on the computer that are difficult or impossible in the laboratory (for example, working at extremes of temperature or pressure) [1].

The purpose of molecular simulations is described as:
a. mimic the real world:
   • predicting properties of (new) materials;

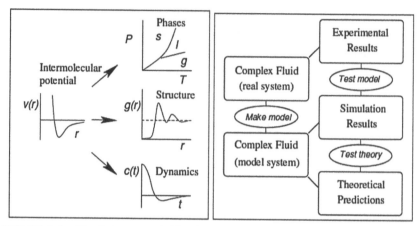

FIGURE 1.1   Simulations as a bridge between (a) microscopic and macroscopic; (b) theory and experiments.

- computer 'experiments' at extreme conditions (carbon phase behavior at very high pressure and temperature);
- understanding phenomena on a molecular scale (protein conformational change with molecular dynamics, empirical potential, including bonds, angles dihedrals).

b. model systems:
- test theory using same simple model;
- explore consequences of model;
- explain poorly understood phenomena in terms of essential physics.

Molecular scale simulations are usually accomplished in three stages by developing a molecular model, calculating the molecular positions, velocities and trajectories, and finally collecting the desired properties from the molecular trajectories. It is the second stage of this process that characterizes the simulation method. For MD the molecular positions are deterministically generated from the Newtonian equations of motion. In other methods, for instance the MC method, the molecular positions are generated randomly by stochastic methods. Some methods have a combination of deterministic and stochastic features. It is the degree of this determinism that distinguishes between different simulation methods [43].

In other words, MD simulations are in many respects very similar to real experiments. When we perform a real experiment, we proceed as follows. We prepare a sample of the material that we wish to study. We connect this sample to a measuring instrument (e.g., a thermometer, nometer, or viscosimeter), and we measure the property of interest during a certain time interval. If our measurements are subject to statistical noise (like most of the measurements), then the longer we average, the more accurate our measurement becomes. In a Molecular Dynamics simulation, we follow exactly the same approach. First, we prepare a sample: we select a model system consisting of N-particles and we solve Newton's equations of motion for this system until the properties of the system no longer change with time (we equilibrate the system). After equilibration, we perform the actual measurement. In fact, some of the most common mistakes that can be made when performing a computer experiment are very similar to the mistakes that can be made in real experiments (e.g., the sample is not prepared correctly, the measurement is too short, the system undergoes an irreversible change during the experiment, or we do not measure what we think) [5].

1.1.1   HISTORICAL BACKGROUND

Before computer simulation appeared on the scene, there was only one way to predict the properties of a molecular substance, namely by making use of a theory that provided an approximate description of that material. Such approximations are inevitable precisely because there are very few systems for which the equilibrium properties can be computed exactly (examples are the ideal gas, the harmonic crystal, and a number of lattice models, such as the two-dimensional Ising model for ferromagnetism). As a result, most properties of real materials were predicted on the basis of approximate theories (examples are the van der Waals equation for dense gases, the Debye-Huckel theory for electrolytes, and the Boltzmann equation to describe the transport properties of dilute gases).

Given sufficient information about the intermolecular interactions, these theories will provide us with an estimate of the properties of interest. Unfortunately, our knowledge of the intermolecular interactions of all but the simplest molecules is also quite limited. This leads to a problem if we wish to test the validity of a particular theory by comparing directly to experiment. If we find that theory and experiment disagree, it may mean that our theory is wrong, or that we have an incorrect estimate of the intermolecular interactions, or both. Clearly, it would be very nice if we could obtain essentially exact results for a given model system without having to rely on approximate theories. Computer simulations allow us to do precisely that. On the one hand, we can now compare the calculated properties of a model system with those of an experimental system: if the two disagree, our model is inadequate; that is, we have to improve on our estimate of the intermolecular interactions. On the other hand, we can compare the result of a simulation of a given model system with the predictions of an approximate analytical theory applied to the same model. If we now find that theory and simulation disagree, we know that the *theory* is flawed. So, in this case, the computer simulation plays the role of the experiment designed to test the theory. This method of screening theories before we apply them to the real world is called a *computer experiment.* This application of computer simulation is of tremendous importance. It has led to the revision of some very respectable theories, some of them dating back to Boltzmann. And it has changed the way in which we construct new theories. Nowadays it is becoming increasingly rare that a theory is applied to the real world before being tested by computer simulation. But note that the computer as such offers us no understanding, only numbers. And, as in a real experiment, these numbers have statistical errors. So what we get out of a simulation is never directly a theoretical relation. As in a real experiment, we still have to extract the useful information [29].

The early history of computer simulation illustrates this role of computer simulation. Some areas of physics appeared to have little need for simulation because very good analytical theories were available (e.g., to predict the properties of dilute gases or of nearly harmonic crystalline solids). However, in other areas, few if any exact theoretical results were known, and progress was much hindered by the lack of unambiguous tests to assess the quality of approximate theories. A case in point was the theory of dense liquids. Before the advent of computer simulations, the only way to model liquids was by mechanical simulation [3–5] of large assemblies of macroscopic spheres (e.g., ball bearings). Then the main problem becomes how to arrange these balls in the same way as atoms in a liquid. Much work on this topic was done by the famous British scientist J.D. Bernal, who built and analyzed such mechanical models for liquids.

The first proper MD simulations were reported in 1956 by Alder and Wainwright at Livermore, who studied the dynamics of an assembly of hard spheres. The first MD simulation of a model for a "real" material was reported in 1959 (and published in 1960) by the group led by Vineyard at Brookhaven, who simulated radiation damage in crystalline Cu. The first MD simulation of a real liquid (argon) was reported in 1964 by Rahman at Argonne. After that, computers were increasingly becoming available to scientists outside the US government labs, and the practice of simulation started spreading to other continents. Much of the methodology of computer simulations has been developed since then, although it is fair to say that the basic algorithms for MC and MD have hardly changed since the 1950s. The most common application of computer simulations is to predict the properties of materials. The need for such simulations may not be immediately obvious [29].

1.1.2 MOLECULAR DYNAMIC: INTERACTIONS AND POTENTIALS

The concept of the MD method is rather straightforward and logical. The motion of molecules is generally governed by Newton's equations of motion in classical theory. In MD simulations, particle motion is simulated on a computer according to the equations of motion. If one molecule moves solely on a classical mechanics level, a computer is unnecessary because mathematical calculation with pencil and paper is sufficient to solve the motion of the molecule. However, since molecules in a real system are numerous and interact with each other, such mathematical analysis is impracticable. In this situation, therefore, computer simulations become a powerful tool for a microscopic analysis [76]. Molecular dynamics simulation consists of the numerical, step-by-step, solution of the classical equations of motion, which for a simple atomic system may be written [1, 2],

$$m_i \ddot{r}_i = f_i \tag{1}$$

$$f_i = -\frac{\partial}{\partial r_i} v \tag{2}$$

For this purpose we need to be able to calculate the forces f_i acting on the atoms, and these are usually derived from a potential energy $U(r^N)$, where $r^N = (r_1; r_2; \ldots r_N)$ represents the complete set of 3N atomic coordinates [5].

The energy, E, is a function of the atomic positions, R, of all the atoms in the system, these are usually expressed in term of Cartesian coordinates. The value of the energy is calculated as a sum of internal, or bonded, terms E-bonded, which describe the bonds, angles and bond rotations in a molecule, and a sum of external or nonbonded terms, $E_{nonbonded}$. These terms account for interactions between nonbonded atoms or atoms separated by 3 or more covalent bonds.

$$V(R) = E_{bonded} + E_{non-bonded} \tag{3}$$

1.1.2.1   Non-Bonded Interactions

There are two potential functions we need to be concerned about between nonbonded atoms:
- Van der Waals Potential
- Electrostatic Potential

The energy term representing the contribution of nonbonded interactions in the CHARMM potential function has two components, the Van der Waals interaction energy and the electrostatic interaction energy. Some other potential functions also include an additional term to account for hydrogen bonds. In the CHARMM potential energy function, these interactions are account for by the electrostatic and Van der Waals interactions:

$$E_{non-bonded} = E_{van-dar-waals} + E_{electrostactic} \tag{4}$$

The van der Waals interaction between two atoms arises from a balance between repulsive and attractive forces. The repulsive force arises at short distances where the electron–electron interaction is strong. The attractive force, also referred to as the dispersion force, arises from fluctuations in the charge distribution in the electron clouds. The fluctuation in the electron distribution on one atom or molecules gives rise to an instantaneous dipole which, in turn, induces a dipole in a second atom or molecule giving rise to an attractive interaction. Each of these two effects is equal to zero at infinite atomic separation r and become significant as the distance decreases. The attractive interaction is longer range than the repulsion but as the distance become short, the repulsive interaction becomes dominant. This gives rise to a minimum in the energy. Positioning of the atoms at the optimal distances stabilizes the system. Both value of energy at the minimum E* and the optimal separation of atoms r* (which is roughly equal to the sum of Van der Waals radii of the atoms) depend on chemical type of these atoms (Fig. 1.2).

The van der Waals interaction is most often modeled using the Lennard-Jones 6–12 potential which expresses the interaction energy using

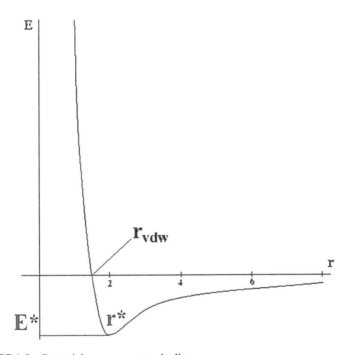

FIGURE 1.2 Potential energy vs. atomic distances.

the atom-type dependent constants A and C [Eq (5)]. Values of A and C may be determined by a variety of methods, like nonbonding distances in crystals and gas-phase scattering measurements.

$$E_{van-der-waals} = \Sigma_{nonbonded\ pairs} \left(\frac{A_{ik}}{r_{ik}^{12}} - \frac{C_{ik}}{r_{ik}^{6}} \right) \tag{5}$$

The part of the potential energy $U_{nonbonded}$ representing nonbonded interactions between atoms is traditionally split into: 1-body, 2-body, 3-body terms):

$$u_{non-bonded} (r^N) = \Sigma_i u(r_i) + \Sigma_i \Sigma_{j>i} v(r_i r_j) \tag{6}$$

The $u(r)$ term represents an externally applied potential field or the effects of the container walls; it is usually dropped for fully periodic simulations of bulk systems. Also, it is usual to concentrate on the pair potential v $(r_i; r_j) = v (r_{ij})$ and neglect three-body (and higher order) interactions. There is an extensive literature on the way these potentials are determined experimentally, or modeled theoretically [1–4]. In some simulations of complex fluids, it is sufficient to use the simplest models that faithfully represent the essential physics. In this chapter we shall concentrate on continuous, differentiable pair-potentials (although discontinuous potentials such as hard spheres and spheroids have also played a role. [5]

The most common model that describes matter in its different forms is a collection of spheres that we call "atoms" for brevity. These "atoms" can be a single atom such as Carbon (C) or Hydrogen (H) or they can represent a group of atoms such as CH_2 or CS_2. These spheres can be connected together to form larger molecules. The interactions between these atoms are governed by a force potential that maintains the integrity of the matter and prevents the atoms from collapsing. The most commonly used potential, that was first used for liquid argon [57], is the Lennard-Jones potential [44]. This potential has the following general form:

$$\emptyset_{LJ} (r) = A\varepsilon \left[\left(\frac{\sigma}{r_{ij}} \right)^m - \left(\frac{\sigma}{r_{ij}} \right)^n \right] \tag{7}$$

where $r_{ij} = r_i - r_j$ is the distance between a pair of atoms i and j. This potential has a short range repulsive force that prevents the atoms from collapsing into each other and also a longer range attractive tail that prevents the disintegration of the atomic system. Parameters m and n determine the range and the strength of the attractive and repulsive forces applied by the potential where normally m is larger than n. The common values used for these parameters are $m = 12$ and $n = 6$. The constant A depends on m and n and with the mentioned values, it will be $A = 4$. The result is the well-known 6–12 Lennard-Jones potential. Two other terms in the potential, namely ε and σ, are the energy and length parameters. In the case of a molecular system several interaction sites or atoms are connected together to form a long chain, ring or a molecule in other forms. In this case, in addition to the mentioned Lennard-Jones potential, which governs the Intermolecular potential, other potentials should be employed. An example of an intermolecular potential is the torsional

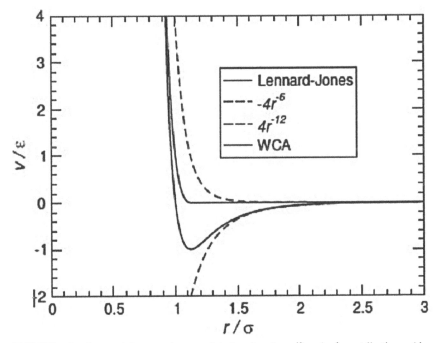

FIGURE 1.3 Lennard-Jones pair potential showing the r^{-12} and r^{-6} contributions. Also shown is the WCA shifted repulsive part of the potential.

potential, which was first introduced by Ryckaert and Bellemans [43, 74]. The electrostatic interaction between a pair of atoms is represented by Coulomb potential; D is the effective dielectric function for the medium and r is the distance between two atoms having charges q_i and q_k.

$$E_{electrostatic} = \Sigma_{nonbonded\ pairs} \frac{q_i q_k}{D r_{ik}} \qquad (8)$$

For applications in which attractive interactions are of less concern than the excluded volume effects which dictate molecular packing, the potential may be truncated at the position of its minimum, and shifted upwards to give what is usually termed the WCA model. If electrostatic charges are present, we add the appropriate Coulomb potentials:

$$v^{coulomb}(r) = \frac{Q_1 Q_2}{4\pi\varepsilon_0 r} \qquad (9)$$

where Q_1, Q_2 are the charges and ε_0 is the permittivity of free space. The correct handling of long-range forces in a simulation is an essential aspect of polyelectrolyte simulations.

1.1.2.2   Bonding Potentials

For molecular systems, we simply build the molecules out of site-site potentials of the form of Eq. (7) or similar. Typically, a single-molecule quantum-chemical calculation may be used to estimate the electron density throughout the molecule, which may then be modeled by a distribution of partial charges via Eq. (9), or more accurately by a distribution of electrostatic multipoles [3, 81]. For molecules we must also consider the intermolecular bonding interactions. The simplest molecular model will include terms of the following kind:

The E-bonded term is a sum of three terms:

$$E_{bonded} = E_{bond-strech} + E_{angel-bend} + E_{rotate-along-bond} \qquad (10)$$

Which correspond to three types of atom movement:
- Stretching along the bond
- Bending between bonds
- Rotating around bonds

The first term in the above equation is a harmonic potential representing the interaction between atomic pairs where atoms are separated by one covalent bond, that is, 1,2-pairs. This is the approximation to the energy of a bond as a function of displacement from the ideal bond length, b_0. The force constant, K_b, determines the strength of the bond. Both ideal bond lengths b_0 and force constants K_b are specific for each pair of bound atoms, that is, depend on chemical type of atoms-constituents.

$$E_{bond-strach} = \Sigma_{1,2\,pairs}\, K_b\, (b - b_0)^2 \tag{11}$$

Values of force constant are often evaluated from experimental data such as infrared stretching frequencies or from quantum mechanical calculations. Values of bond length can be inferred from high resolution crystal structures or microwave spectroscopy data. The second term in above equation is associated with alteration of bond angles theta from ideal values q_0, which is also represented by a harmonic potential. Values of q_0 and K_q depend on chemical type of atoms constituting the angle. These two terms describe the deviation from an ideal geometry; effectively, they are penalty functions and that in a perfectly optimized structure, the sum of them should be close to zero.

$$E_{bond-bend} = \Sigma_{angles}\, K_\theta\, (\theta - \theta_0)^2 \tag{12}$$

The third term represents the torsion angle potential function which models the presence of steric barriers between atoms separated by three covalent bonds (1,4 pairs). The motion associated with this term is a rotation, described by a dihedral angle and coefficient of symmetry n = 1, 2, 3), around the middle bond. This potential is assumed to be periodic and is often expressed as a cosine function. In addition to these term, the CHARMM force field has two additional terms; one is the Urey-Bradley term, which is an interaction based on the distance between atoms separated by two bond (1,3 interaction). The second additional term is the improper dihedral term (see the section on CHARMM) which is used to maintain chirality and planarity.

$$E_{rotate-along-bond} = \Sigma_{1,4\,pairs}\, K_\emptyset\, (1 - \cos(n\emptyset)) \tag{13}$$

The parameters for these terms, K_b, K_q, K_p, are obtained from studies of small model compounds and comparisons to the geometry and vibrational spectra in the gas phase (IR and Raman spectroscopy), supplemented with ab initio quantum calculations [5].

$$U_{intramolecular} = \frac{1}{2}\Sigma_{bonds} k_{ij}^r \left(r_{ij} - r_{eq}\right)^2 \tag{14a}$$

$$+ \frac{1}{2}\Sigma \ k_{ijk}^\theta \left(\theta_{ijk} - \theta_{eq}\right)^2 \tag{14b}$$

$$+ \frac{1}{2}\Sigma_{\substack{torsion \\ angles}} \Sigma_m k_{ijkl}^{\emptyset,m}(1 + \cos(m\emptyset_{ijkl} - \gamma_m)) \tag{14c}$$

The geometry is illustrated in Fig. 1.4. The "bonds" will typically involve the separation $r_{ij} = |r_i - r_j|$ between adjacent pairs of atoms in a molecular framework, and we assume in Eq (14a), a harmonic form with specified equilibrium separation, although this is not the only possibility.

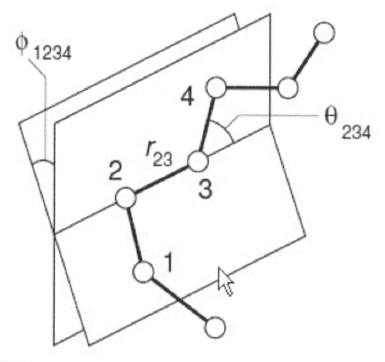

FIGURE 1.4 Geometry of a simple chain molecule, illustrating the definition of interatomic distance r23, bend angle θ234, and torsion angle 1234 [5].

The "bend angles" θ_{ijk} are between successive bond vectors such as $r_i - r_j$ and $r_j - r_k$, and therefore involve three atom coordinates:

$$\cos \theta_{ijk} = \hat{r}_{ij} \cdot \hat{r}_{jk} = \left(r_{ij} \cdot r_{ij} \right)^{-1/2} \left(r_{jk} \cdot r_{jk} \right)^{-1/2} \left(r_{ij} \cdot r_{jk} \right) \quad (15)$$

where $\check{r} = \mathbf{r}/r$. Usually this bending term is taken to be quadratic in the angular displacement from the equilibrium value, as in Eq (14b), although periodic functions are also used. The "torsion angles" are defined in terms of three connected bonds, hence four atomic coordinates [5]:

$$\cos \varnothing_{ijkl} = -\hat{n}_{ijk} \cdot \hat{n}_{jkl}, \; where \; n_{ijk} = r_{ij} \times r_{jk}, \; n_{jkl} = r_{jk} \times r_{kl} \quad (16)$$

where $\check{n} = \mathbf{n}/n$, the unit normal to the plane denied by each pair of bonds. Usually the torsional potential involves an expansion in periodic functions of order m = 1, 2,... Eq (14c). A simulation package force-field will specify the precise form of Eq. (14), and the various strength parameters k and other constants therein. Actually, Eq. (14) is a considerable oversimplification. Molecular mechanics force-fields, aimed at accurately predicting structures and properties, will include many cross-terms (e.g., stretch-bend): MM3 [7, 53] and MM4 [7] are examples. Quantum mechanical calculations may give a guide to the "best" molecular force-field; also comparison of simulation results with thermo physical properties and vibration frequencies is invaluable in force-field development and refinement. A separate family of force fields, such as AMBER [8], CHARMM [15] and OPLS [45] are geared more to larger molecules (proteins, polymers) in condensed phases; their functional form is simpler, closer to that of Eq. (14), and their parameters are typically determined by quantum chemical calculations combined with thermo physical and phase coexistence data. This field is too broad to be reviewed here; several molecular modeling texts (albeit targeted at biological applications) should be consulted by the interested reader. The modeling of long chain molecules will be of particular interest to us, especially as an illustration of the scope for progressively simplifying and "coarse-graining" the potential model. Various explicit-atom potentials have been devised for the n-alkanes. More approximate potentials have also been constructed [26, 28] in which the CH_2 and CH_3 units are represented by single "united atoms." These potentials are typically less accurate and less transferable than the explicit-atom potentials, but significantly less expensive; comparisons have been made between the two approaches [29]. For more complicated molecules this approach may need to be modified. In the liquid crystal field, for instance, a compromise has been suggested [32]: use the united-atom approach for hydrocarbon chains, but model phenyl ring hydrogen's explicitly.

In polymer simulations, there is frequently a need to economize further and coarse-grain the interactions more dramatically: significant progress has been made in recent years in approaching this problem systematically [85]. Finally, the most fundamental properties, such as the entanglement length in a polymer melt [72], may be investigated using a simple chain of pseudoatoms or beads (modeled using the WCA potential of Fig. 1.3, and each representing several monomers), joined by an attractive finitely extensible nonlinear elastic (FENE) potential 24 which is illustrated in Fig. 1.4.

$$v^{FENE}(r) = \begin{cases} -\frac{1}{2}kR_0^2 \ln(1 - (r/R_0)^2) & r < R_0 \\ \infty & r \geq R_0 \end{cases} \qquad (17)$$

The key feature of this potential is that it cannot be extended beyond $r = R_0$, ensuring (for suitable choices of the parameters k and R_0) that polymer chains cannot move through one another. The empirical potential energy function is differentiable with respect to the atomic coordinates; this gives the value and the direction of the force acting on an atom and thus it can be used in a molecular dynamics simulation. The empirical potential function has several limitations, which result in inaccuracies in the calculated potential energy. One limitation is due to the fixed set of atom types employed when determining the parameters for the force field. Atom types are used to define an atom in a particular bonding situation, for example an aliphatic carbon atom in an sp3 bonding situation has different properties than a carbon atom found in the His ring. Instead of presenting each atom in the molecule as a unique one described by unique set of parameters, there is a certain amount of grouping in order minimizes the number of atom types. This can lead to type-specific errors. The properties of certain atoms, like aliphatic carbon or hydrogen atoms, are less sensitive to their surroundings and a single set of parameters may work quite well, while other atoms like oxygen and nitrogen are much more influenced by their neighboring atoms. These atoms require more types and parameters to account for the different bonding environments.

Another important point to take into consideration is that the potential energy function does not include entropic effects. Thus, a minimum value of E calculated as a sum of potential functions does not necessarily correspond to the equilibrium, or the most probable state; this corresponds to the minimum of free energy. Because of the fact that experiments are generally carried out under isothermal-isobaric conditions (constant pressure, constant system size and constant temperature) the equilibrium state corresponds to the minimum of Gibb's Free Energy, G. While just an energy calculation ignores entropic effects, these are included in molecular dynamics simulations.

1.1.2.3   Statistical Mechanics

Molecular simulations are based on the framework of statistical mechanics/thermodynamics. Molecular dynamics simulations generate information at the microscopic level, including atomic positions and velocities. The conversion of this microscopic information to macroscopic observables such as pressure, energy, heat capacities, etc., requires statistical mechanics. In a molecular dynamics simulation, one often wishes to explore the macroscopic properties of a system through microscopic simulations, for example, to calculate changes in the binding free energy of a particular drug candidate, or to examine the energetics and mechanisms of conformational change. The connection between microscopic simulations and macroscopic properties is made via statistical mechanics which provides the rigorous mathematical expressions that relate macroscopic properties to the distribution and motion of the atoms and molecules of the N-body system; molecular dynamics simulations provide the means to solve the equation of motion of the particles and evaluate these mathematical formulas. With molecular dynamics simulations, one can study both thermodynamic properties and/or time dependent (kinetic) phenomenon [10].

Statistical mechanics is the branch of physical sciences that studies macroscopic systems from a molecular point of view. The goal is to understand and to predict macroscopic phenomena from the properties of individual molecules making up the system. The system could range from a collection of solvent molecules to a solvated protein-DNA complex. In order to connect the macroscopic system to the microscopic system, time independent statistical averages are often introduced. The thermodynamic state of a system is usually defined by a small set of parameters, for example, the temperature, T, the pressure, P, and the number of particles, N. Other thermodynamic properties may be derived from the equations of state and other fundamental thermodynamic equations. The mechanical or microscopic state of a system is defined by the atomic positions, q, and momenta, p; these can also be considered as coordinates in a multidimensional space called phase space. For a system of N particles, this space has 6N dimensions. A single point in phase space, denoted by G, describes the state of the system. An ensemble is a collection of points in phase space satisfying the conditions of a particular thermodynamic state. A molecular dynamics simulation generates a sequence of points in phase space as a function of time; these points belong to the same ensemble, and they correspond to the different conformations of the system and their respective momenta. Several different ensembles are described below [11]. An ensemble is a collection of all possible systems which have different microscopic states but have an identical macroscopic or thermodynamic state.

1.1.2.4   Newton's Equation of Motion

We view materials as a collection of discrete atoms. The atoms interact by exerting forces on each other. Force Field is defined as a mathematical expression that describes the dependence of the energy of a molecule on the coordinates of the atoms in the molecule. The molecular dynamics simulation method is based on Newton's second law or the equation of motion, $F = ma$, where F is the force exerted on the particle, m is its mass and "a" is its acceleration. From knowledge of the force on each atom, it is possible to determine the acceleration of each atom in the system. Integration of the equations of motion then yields a trajectory that describes the positions, velocities and accelerations of the particles as they vary with time. From this trajectory, the average values of properties can be determined. The method is deterministic; once the positions and velocities of each atom are known, the state of the system can be predicted at any time in the future or the past. Molecular dynamics simulations can be time consuming and computationally expensive. However, computers are getting faster and cheaper. Based on the interaction model, a simulation computes the atoms' trajectories numerically. Thus a molecular simulation necessarily contains the following ingredients [5, 17, 43, 76].

- The model that describes the interaction between atoms. This is usually called the interatomic potential: $V(\{rib\})$, where $\{ri\}$ represent the position of all atoms.
- Numerical integrator that follows the atoms equation of motion. This is the heart of the simulation. Usually, we also need auxiliary algorithms to set up the initial and boundary conditions and to monitor and control the systems state (such as temperature) during the simulation.
- Extract useful data from the raw atomic trajectory information. Compute materials properties of interest. Visualization [17].

The dynamics of classical objects follow the three laws of Newton. Here, we review the Newton's laws (Fig. 1.5):

First law: Every object in a state of uniform motion tends to remain in that state of motion unless an external force is applied to it. This is also called the Law of inertia.

FIGURE 1.5   Sir Isaac Newton (1643–1727 England).

An object's mass m, its acceleration a and the applied force F are related by

$$F_i = m_i a_i \tag{18}$$

where F_i is the force exerted on particle i, m_i is the mass of particle i and a_i is the acceleration of particle i.

For every action there is an equal and opposite reaction.

The *second law* gives us the equation of motion for classical particles. Consider a particle its position is described by a vector r = (x, y, z). The velocity is how fast r changes with time and is also a vector: v = dr/dt = (v_x, v_y, v_z). In component form, v_x = dx/dt, v_y = dy/dt, v_z = dz/dt. The acceleration vector is then time derivative of velocity, that is, a = dv/dt. The force can also be expressed as the gradient of the potential energy,

$$F_i = -\nabla_i V \tag{19}$$

Combining of equation of motion and gradient of the potential energy yields:

$$-\frac{dV}{dr_i} = m_i \frac{d^2 r_i}{dt^2} \tag{20}$$

where V is the potential energy of the system. Newton's equation of motion can then relate the derivative of the potential energy to the changes in position as a function of time. The kinetic energy is given by the velocity,

$$E_{kin} = \frac{1}{2} m |v|^2 \tag{21}$$

The total energy is the sum of potential and kinetic energy contributions,

$$E_{tot} = E_{kin} + V \tag{22}$$

When we express the total energy as a function of particle position r and momentum p = mv, it is called the Hamiltonian of the system,

$$H(r,p) = \frac{|P|^2}{2m} + V(r) \tag{23}$$

The Hamiltonian (i.e., the total energy) (Fig. 1.6) is a conserved quantity as the particle moves. To see this, let us compute its time derivative,

$$\frac{dE_{tot}}{dt} = mv \cdot \frac{dv}{dt} + \frac{dV(r)}{dr} \cdot \frac{dr}{dt}$$

$$= m\frac{dr}{dt} \cdot \left[-\frac{1}{m}\frac{dV(r)}{dr} \right] + \frac{dV(r)}{dr} \cdot \frac{dr}{dt} \tag{24}$$

$$= 0$$

Therefore, the total energy is conserved if the particle follows the Newton's equation of motion in a conservative force field (when force can be written in terms of spatial derivative of a potential field), while the kinetic energy and potential energy can interchange with each other. The Newton's equation of motion can also be written in the Hamiltonian form, [17]

$$\frac{dr}{dt} = \frac{\partial H(r,p)}{\partial p} \tag{25}$$
$$\frac{dp}{dt} = -\frac{\partial H(r,p)}{\partial r}$$

Hence,

$$\frac{dH}{dt} = \frac{\partial H(r,p)}{\partial r} \cdot \frac{dr}{dt} + \frac{\partial H(r,p)}{\partial p} \cdot \frac{dp}{dt} = 0 \tag{26}$$

FIGURE 1.6 Sir William Rowan Hamilton (1805–1865 Ireland).

Newton's Second Law of motion: a simple application

$$F = ma = m\frac{dv}{dt} = m\frac{d^2x}{dt^2} \qquad (27)$$

$$a = \frac{dv}{dt}$$

Taking the simple case where the acceleration is constant. We obtain an expression for the velocity after integration:

$$v = at + v_0 \qquad (28)$$

And since

$$v = \frac{dx}{dt} \qquad (29)$$

We can once again integrate to obtain

$$x = vt + x_0 \qquad (29a)$$

Combining this equation with the expression for the velocity, we obtain the following relation which gives the value of x at time t as a function of the acceleration, a, the initial position, x_0, and the initial velocity, v_0.

$$x = vt + x_0$$

$$x = at^2 + v_0 t + x_0 \qquad (30)$$

The acceleration is given as the derivative of the potential energy with respect to the position, r,

$$a = -\frac{1}{m}\frac{dE}{dr} \qquad (31)$$

Therefore, to calculate a trajectory, one only needs the initial positions of the atoms, an initial distribution of velocities and the acceleration, which is determined by the gradient of the potential energy function. The equations of motion are deterministic, for example, the positions and the velocities at time zero determine the positions and velocities at all other times, t. The initial positions can be obtained from experimental structures, such as the x-ray crystal structure of the protein or the solution structure determined by NMR spectroscopy. The initial distribution of velocities are usually determined from a random distribution with the magnitudes conforming to the required temperature and corrected so there is no overall momentum, i.e.,

$$P = \sum_{i=1}^{N} m_i v_i = 0 \tag{32}$$

The velocities, v_i, are often chosen randomly from a Maxwell-Boltzmann or Gaussian distribution at a given temperature, which gives the probability that an atom i has a velocity v_x in the x direction at a temperature T.

$$p(v_{ix}) = (\frac{m_i}{2\pi k_B T})^{1/2} exp \left[-\frac{1}{2} \frac{m_i v_{ix}^2}{k_B T} \right] \tag{33}$$

The temperature can be calculated from the velocities using the relation

$$T = \frac{1}{(3N)} \sum_{i=1}^{N} \frac{|p_i|}{2m_i} \tag{34}$$

where N is the number of atoms in the system.

1.1.2.5   Integration Algorithms

Solving Newton's equations of motion does not immediately suggest activity at the cutting edge of research. The molecular dynamics algorithm in most common use today may even have been known to Newton [41]. Nonetheless, the last decade has seen a rapid development in our understanding of numerical algorithms; a forthcoming review [21] and a book summarize the present state of the field.

Continuing to discuss, for simplicity, a system composed of atoms with coordinates $r^N = (r_1, r_2, ... r_N)$ and potential energy $u(r^N)$, we introduce the atomic momenta $p^N = (p_1, p_2, ... p_N)$, in terms of which the kinetic energy may be written $\kappa(p^N) = \sum_{i=1}^{N} |p_i|^2 / 2m_i$. Then the energy, or Hamiltonian, may be written as a sum of kinetic and potential terms $H = K + U$. Write the classical equations of motion as

$$\dot{r_i} = p_i / m_i$$
$$\dot{p_i} = f_i \tag{35}$$

This is a system of coupled ordinary differential equations. Many methods exist to perform step-by-step numerical integration of them. Characteristics of these equations are: (a) they are 'stiff,' that is, there may be short and long timescales, and the algorithm must cope with both; (b) calculating the forces is expensive, typically involving a sum over pairs of atoms, and should be performed as infrequently as possible. Also we must bear in mind that the advancement of the coordinates fulfills two functions: (i) accurate calculation of dynamical properties, especially over times as long as typical correlation times τ_a of properties a of interest (we shall define this later); (ii) accurately staying on the constant-energy hyper surface, for much longer times $\tau_{run} \gg \tau_a$, in order to sample the correct ensemble [5].

To ensure rapid sampling of phase space, we wish to make the time step as large as possible consistent with these requirements. For these reasons, simulation algorithms have tended to be of *low order* (i.e., they do not involve storing high derivatives of positions, velocities, etc.): this allows the time step to be increased as much as possible without jeopardizing energy conservation. It is unrealistic to expect the numerical method to accurately follow the true trajectory for very long times τ_{run}. The 'ergodic' and 'mixing' properties of classical trajectories, that is, the fact that nearby trajectories diverge from each other exponentially quickly, make this impossible to achieve. All these observations tend to favor the Verlet algorithm in one form or another, and we look closely at this in the following section. For historical reasons only, we mention the more general class of predictor-corrector methods which have been optimized for classical mechanical equations [35, 52].

The potential energy is a function of the atomic positions (3N) of all the atoms in the system. Due to the complicated nature of this function, there is no analytical solution to the equations of motion; they must be solved numerically. Numerous numerical algorithms have been developed for integrating the equations of motion. Different algorithms have been listed below:

- Verlet algorithm
- Leap-frog algorithm
- Velocity Verlet
- Beeman's algorithm

It is important to attention which algorithm to use, one should consider the following criteria:

- The algorithm should conserve energy and momentum.
- It should be computationally efficient
- It should permit a long time step for integration.

All the integration algorithms assume the positions, velocities and accelerations can be approximated by a Taylor series expansion:

$$y(t_{i+1}) = y(t_i) + hy'(t_i) + \frac{h^2}{2}y''(t_i) + \cdots + \frac{h^n}{n!}y^{(n)}(t_i) + \frac{h^{n+1}}{(n+1)!}y^{(n+1)}(\xi_i)$$

$$r(t + \delta t) = r(t) + v(t)\delta t + \frac{1}{2}a(t)\delta t^2 + \cdots$$

$$v(t + \delta t) = v(t) + a(t)\delta t + \frac{1}{2}b(t)\delta t^2 + \cdots$$

$$a(t + \delta t) = a(t) + b(t)\delta t + \cdots \tag{36}$$

where r is the position, v is the velocity (the first derivative with respect to time), a is the acceleration (the second derivative with respect to time), etc. To derive the Verlet algorithm one can write

$$r(t + \delta t) = r(t) + v(t)\delta t + \frac{1}{2}a(t)\delta t^2$$

$$r(t - \delta t) = r(t) - v(t)\delta t + \frac{1}{2}a(t)\delta t^2 \tag{37}$$

Summing these two equations, one obtains

$$r(t + \delta t) = 2r(t) - r(t - \delta t) + a(t)\delta t^2 \tag{38}$$

1.1.2.5.1   Verlet Algorithms

The Verlet algorithm uses positions and accelerations at time t and the positions from time t-dt to calculate new positions at time t + dt. The Verlet algorithm uses no explicit velocities. The advantages of the Verlet algorithm are: (i) it is straightforward, and (ii) the storage requirements are

TABLE 1.1 Verlet's Algorithm Characteristics

Pros	Cons
• Simple and Effective	• Not as accurate as RK
• Low Memory and CPU Requirements (don't need to store velocities or perform multiple force calculations)	• We never calculate velocities
• Time Reversible	
• Very stable even with large numbers of interacting particles	

modest. The disadvantage is that the algorithm is of moderate precision. The advantages and disadvantages of this algorithm have been summarized in Table 1.1 [5, 76].

We can estimate the velocities using a finite difference:

$$v(t) = \frac{1}{2\Delta t}[r(t + \Delta t) - r(t - \Delta t)] + O(\Delta t^2) \qquad (39)$$

There are variations of the Verlet algorithm, such as the leapfrog algorithm, which seek to improve velocity estimations.

1.1.2.5.2   The Leap-Frog Algorithm

In this algorithm, the velocities are first calculated at time $t + 1/2dt$; these are used to calculate the positions, r, at time $t + dt$. In this way, the velocities leap over the positions, then the positions leap over the velocities. The advantage of this algorithm is that the velocities are explicitly calculated, however, the disadvantage is that they are not calculated at the same time as the positions. The velocities at time t can be approximated by the relationship: [76].

$$v(t) = \frac{1}{2}\left[v\left(t - \frac{1}{2}\delta t\right) + v\left(t + \frac{1}{2}\delta t\right)\right] \qquad (40)$$

1.1.2.5.3   The Velocity Verlet Algorithm

This algorithm yields positions, velocities and accelerations at time t. There is no compromise on precision.

$$r(t + \delta t) = r(t) + v(t)\delta t + \frac{1}{2}a(t)\delta t^2$$

$$r(t + \delta t) = v(t) + \frac{1}{2}[a(t) + a(t + \delta t)]\delta t \tag{41}$$

The MD method is applicable to both equilibrium and nonequilibrium physical phenomena, which makes it a powerful computational tool that can be used to simulate many physical phenomena (if computing power is sufficient). The main procedure for conducting the MD simulation using the velocity Verlet method is shown in the following steps (see appendix 1):

1. Specify the initial position and velocity of all molecules.
2. Calculate the forces acting on molecules.
3. Evaluate the positions of all molecules at the next time.
4. Evaluate the velocities of all molecules at the next time step.
5. Repeat the procedures from step 2.

In the above procedure, the positions and velocities will be evaluated at every time interval h in the MD simulation. The method of evaluating the system averages is necessary to make a comparison with experimental or theoretical values. Since microscopic quantities such as positions and velocities are evaluated at every time interval in MD simulations, a quantity evaluated from such microscopic values for example, the pressure will differ from that measured experimentally. In order to compare with experimental data, instant pressure is sampled at each time step, and these values are averaged during a short sampling time to yield a macroscopic pressure. This average can be expressed as

$$\overline{A} = \sum_{n=1}^{N} A_n / N \tag{42}$$

where A_n is the n-th sampled value of an arbitrary physical quantity A, and $\ddot{A}$, called the "time average," is the mathematical average of N sampling data [76].

This algorithm is closely related to the Verlet algorithm:

$$r(t + \delta t) = r(t) + v(t)\delta t + \frac{2}{3}a(t)\delta t^2 - \frac{1}{6}a(t - \delta t)\delta t^2$$

$$v(t + \delta t) = v(t) + v(t)\delta t + \frac{1}{3}a(t)\delta t + \frac{5}{6}a(t)\delta t - \frac{1}{6}a(t - \delta t)\delta t \tag{43}$$

The advantages and disadvantages of this algorithm have been summarized in Table 1.2.

1.1.2.6 Periodic Boundary Conditions

Small sample size means that, unless surface effects are of particular interest, periodic boundary conditions need to be used. Consider 1000 atoms arranged in a $10 \times 10 \times 10$ cube. Nearly half the atoms are on the outer faces, and these will have a large effect on the measured properties. Even for $10^6 = 1003$ atoms, the surface atoms amount to 6% of the total, which is still nontrivial. Surrounding the cube with replicas of itself takes care of this problem. Provided the potential range is not too long, we can adopt the

TABLE 1.2 Beeman's Algorithm Characteristic

Pros	Cons
• provides a more accurate expression for the velocities and better energy conservation.	• more complex expressions make the calculation more expensive.

TABLE 1.3 The Comparison Between Classical MD and Quantum MD

Classical MD		
Simulate/predict processes		**Characteristics**
1. polypeptide folding 2. biomolecular association 3. partitioning between solvents 4. membrane/micelle formation	thermodynamic equilibrium governed by weak (nonbonded) forces	• degrees of freedom: atomic (solute + solvent) • equations of motion: classical dynamics • governing theory: statistical mechanics
Quantum MD		
5. chemical reactions, enzyme catalysis 6. enzyme catalysis 7. photochemical reactions, electron transfer	chemical transformations governed by strong forces	• degrees of freedom: electronic, nuclear • equations of motion: quantum dynamics • governing theory: quantum statistical mechanics

minimum image convention that each atom interacts with the nearest atom or image in the periodic array. In the course of the simulation, if an atom leaves the basic simulation box, attention can be switched to the incoming image. This is shown in Fig. 1.7. However, it is important to bear in mind the imposed artificial periodicity when considering properties which are influenced by long-range correlations. Special attention must be paid to the case where the potential range is not short: for example for charged and dipolar systems [5, 76].

1.1.2.7 Neighbor List

Computing the nonbonded contribution to the interatomic forces in an MD simulation involves, in principle, a large number of pairwise calculations: we consider each atom I and loop over all other atoms j to calculate the

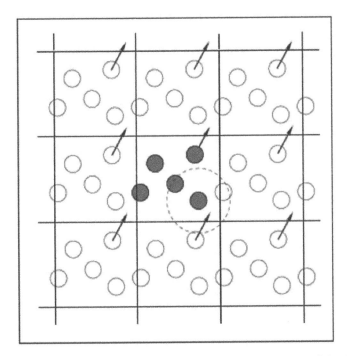

FIGURE 1.7 Periodic boundary conditions. As a particle moves out of the simulation box, an image particle moves in to replace it. In calculating particle interactions within the cutoff range, both real and image neighbors are included.

minimum image separations r_{ij}. Let us assume that the interaction potentials are of short range, $v(r_{ij}) = 0$ if $r_{ij} > r_{cut}$, the potential cutoff. In this case, the program skips the force calculation, avoiding expensive calculations, and considers the next candidate j. Nonetheless, the time to examine all pair separations is proportional to the number of distinct pairs, $\frac{1}{2}N(N-1)$ in an N-atom system, and for every pair one must compute at least r_{ij}^2; this still consumes a lot of time. Some economies result from the use of lists of nearby pairs of atoms. Verlet [7] suggested such a technique for improving the speed of a program. The potential cutoff sphere, of radius r_{cut}, around a particular atom is surrounded by a 'skin,' to give a larger sphere of radius r_{list}. At the first step in a simulation, a list is constructed of all the neighbors of each atom, for which the pair separation is within r_{list}. Over the next few MD time steps, only pairs appearing in the list are checked in the force routine. From time to time the list is reconstructed: it is important to do this before any unlisted pairs have crossed the safety zone and come within interaction range. It is possible to trigger the list reconstruction automatically, if a record is kept of the distance traveled by each atom since the last update. The choice of list cutoff distance r_{list} is a compromise: larger lists will need to be reconstructed less frequently, but will not give as much of a saving on cpu time as smaller lists. This choice can easily be made by experimentation.

For larger systems ($N \geq 1000$ or so, depending on the potential range) another technique becomes preferable. The cubic simulation box (extension to noncubic cases is possible) is divided into a regular lattice of $n_{cell} \times n_{cell} \times n_{cell}$ cells. These cells are chosen so that the side of the cell $l_{cell} = L/n_{cell}$ is greater than the potential cutoff distance rcut. If there is a separate list of atoms in each of those cells, then searching through the neighbors is a rapid process: it is only necessary to look at atoms in the same cell as the atom of interest, and in nearest neighbor cells. The cell structure may be set up and used by the method of linked lists[55, 43.] The first part of the method involves sorting all the atoms into their appropriate cells. This sorting is rapid, and may be performed every step. Then, within the force routine, pointers are used to scan through the contents of cells, and calculate pair forces. This approach is very efficient for large systems with short-range forces. A certain amount of unnecessary work is done because the search region is cubic, not (as for the Verlet list) spherical [5].

1.1.2.8 Time Dependence

A knowledge of time-dependent statistical mechanics is important in three general areas of simulation. Firstly, in recent years there have been significant advances in the understanding of molecular dynamics algorithms, which have arisen out of an appreciation of the formal operator approach to classical mechanics. Second, an understanding of equilibrium time correlation functions, their link with dynamical properties, and especially their connection with transport coefficients, is essential in making contact with experiment. Third, the last decade has seen a rapid development of the use of nonequilibrium molecular dynamics, with a better understanding of the formal aspects, particularly the link between the dynamical algorithm, dissipation, chaos, and fractal geometry [5].

1.1.2.9 Application and Achievements

Given the modeling capability of MD and the variety of techniques that have emerged, what kinds of problem can be studied? Certain applications can be eliminated owing to the classical nature of MD. There are also hardware imposed limitations on the amount of computation that can be performed over a given period of time – be it an hour or a month – thus restricting the number of molecules of a given complexity that can be handled, as well as storage limitations having similar consequences (to some extent, the passage of time helps alleviate hardware restrictions). The phenomena that can be explored must occur on length and time scales that are encompassed by the computation. Some classes of phenomena may require repeated runs based on different sets of initial conditions to sample adequately the kinds of behavior that can develop, adding to the computational demands. Small system size enhances the fluctuations and sets a limit on the measurement accuracy; finite-size effects – even the shape of the simulation region – can also influence certain results. Rare events present additional problems of observation and measurement. Liquids represent the state of matter most frequently studied by MD methods. This is due to historical reasons, since both solids and gases have well-developed theoretical foundations, but there is no general theory of liquids. For solids, theory begins by assuming that the atomic constituents undergo small oscillations about fixed lattice positions; for gases, independent atoms are assumed and interactions are introduced as weak perturbations. In the case of liquids, however, the interactions are as important as in the solid state, but there is no underlying ordered structure to begin with. The following list includes a somewhat random and far from complete assortment of ways in which MD simulation is used:

- Fundamental studies: equilibration, tests of molecular chaos, kinetic theory, diffusion, transport properties, size dependence, tests of models and potential functions.
- Phase transitions: first- and second-order, phase coexistence, order parameters, critical phenomena.
- Collective behavior: decay of space and time correlation functions, coupling of translational and rotational motion, vibration, spectroscopic measurements, orientation order, and dielectric properties.
- Complex fluids: structure and dynamics of glasses, molecular liquids, pure water and aqueous solutions, liquid crystals, ionic liquids, fluid interfaces, films and monolayers.
- Polymers: chains, rings and branched molecules, equilibrium conformation, relaxation and transport processes.
- Solids: defect formation and migration, fracture, grain boundaries, structural transformations, radiation damage, elastic and plastic mechanical properties, friction, shock waves, molecular crystals, epitaxial growth.
- Biomolecules: structure and dynamics of proteins, protein folding, micelles, membranes, docking of molecules.
- Fluid dynamics: laminar flow, boundary layers, rheology of non-Newtonian fluids, unstable flow. And there is much more.

The elements involved in an MD study, the way the problem is formulated, and the relation to the real world can be used to classify MD problems into various categories. Examples of this classification include whether the interactions are shorter long-ranged; whether the system is thermally and mechanically isolated or open to outside influence; whether, if in equilibrium, normal dynamical laws are used or the equations of motion are modified to produce a particular statistical mechanical ensemble; whether the constituent particles are simple structure less atoms or more complex molecules and, if the latter, whether the molecules are rigid or flexible;

whether simple interactions are represented by continuous potential functions or by step potentials; whether interactions involve just pairs of particles or multiparticle contributions as well; and so on [70].

Despite the successes, many challenges remain. Multiple phases introduce the issue of interfaces that often have a thickness comparable to the typical simulated region size. In homogeneities such as density or temperature gradients can be difficult to maintain in small systems, given the magnitude of the inherent fluctuations. Slow relaxation processes, such as those typical of the glassy state, diffusion that is hindered by structure as in polymer melts, and the very gradual appearance of spontaneously forming spatial organization, are all examples of problems involving temporal scales many orders of magnitude larger than those associated with the underlying molecular motion [70].

1.1.3   MONTE CARLO METHOD

In the MD method, the motion of molecules (particles) is simulated according to the equations of motion and therefore it is applicable to both thermodynamic equilibrium and nonequilibrium phenomena. In contrast, the MC method generates a series of microscopic states under a certain stochastic law, irrespective of the equations of motion of particles. Since the MC method does not use the equations of motion, it cannot include the concept of explicit time, and thus is only a simulation technique for phenomena in thermodynamic equilibrium. Hence, it is unsuitable for the MC method to deal with the dynamic properties of a system, which are dependent on time. In the following paragraphs, we explain important points of the concept of the MC method.

How do microscopic states arise for thermodynamic equilibrium in a practical situation? We discuss this problem by considering a two-particle attractive systemizing (Fig. 1.8). As shown in Fig. 1.8A, if the two particles overlap, then a repulsive force or a significant interaction energy arises. As shown in Fig. 1.8B, for the case of close proximity, the interaction energy becomes low and an attractive force acts on the particles. If the two particles are sufficiently distant, as shown in Fig. 1.8C, the interactive force is negligible and the interaction energy can be regarded as zero.

In actual phenomena, microscopic states which induce a significantly high energy, as shown in Fig. 1.8A, seldom appear, but microscopic states which give rise to a low-energy system, as shown in Fig. 1.8B, frequently arise. However, this does not mean that only microscopic states that induce a minimum energy system appear. Consider the fact that oxygen and nitrogen molecules do not gather in a limited area, but distribute uniformly in a room. It is seen from this discussion that, for thermodynamic equilibrium, microscopic states do not give rise to a minimum of the total system energy, but to a minimum free energy of a system. For example, in the case of a system specified by the number of particles N, temperature T, and volume of the system V, microscopic states arise such that the following Helmholtz free energy F becomes a minimum:

$$F = E - TS \qquad (44)$$

where E is the potential energy of the system, and S is the entropy. In the preceding example, the reason why oxygen or nitrogen molecules do not gather in a limited area can be explained by taking into account the entropy term on the right-hand side in Eq. (44). That is, the situation in which molecules do not gather together and form flocks but expand to fill a room gives rise to a large value of the entropy. Hence, according to the counterbalance relationship of the energy and the entropy, real microscopic states arise such that the free energy of a system is at minimum. [76]

Next, we consider how microscopic states arise stochastically. We here treat a system composed of N interacting spherical particles with temperature T and volume V of the system; these quantities are given values and assumed to be constant. If the position vector of an arbitrary particle i (i = 1, 2,..., N) is denoted by r_i, then the total interaction energy U of the system can be expressed as a function of the particle positions; that is, it can be

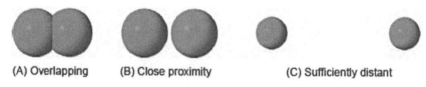

(A) Overlapping   (B) Close proximity   (C) Sufficiently distant

FIGURE 1.8   Typical energy situations for a two particle system.

expressed as $U = U(r_1, r_2,...,r_N)$. For the present system specified by given values of N, T, and V, the appearance of a microscopic state that the particle i (i = 1, 2,..., N) exits within the small rang of $r_i \sim (r_i + \Delta r_i)$ is governed by the probability density function $\rho(r_1, r_2,...,r_N)$. This can be expressed from statistical mechanics [57] as

$$\rho(r_1, r_2, ..., r_N) = \frac{exp\{-U(r_1,r_2,...r_N)/kT\}}{\int_V \cdots \int_V exp\{-U(r_1,r_2,...r_N)/kT\}dr_1 dr_2..dr_N} \tag{45}$$

If a series of microscopic states is generated with an occurrence according to this probability, a simulation may have physical meaning. However, this approach is impracticable, as it is extraordinarily difficult and almost impossible to evaluate analytically the definite integral of the denominator in Eq. (45). In fact, if we were able to evaluate this integral term analytically, we would not need a computer simulation because it would be possible to evaluate almost all physical quantities analytically. The "metropolis method" [58] overcomes this difficulty for MC simulations. In the metropolis method, the transition probability from microscopic states i to j, p_{ij}, is expressed as

$$p_{ij} = \begin{cases} 1 \ (for \ \rho_j/\rho_i \geq 1) \\ \frac{\rho_j}{\rho_i} \ (for \frac{\rho_j}{\rho_i} < 1) \end{cases} \tag{46}$$

where ρ_j and ρ_i are the probability density functions for microscopic states j and i appearing, respectively. The ratio ρ_j/ρ_i of is obtained from Eq. (45) as

$$\frac{\rho_j}{\rho_i} = exp\left\{-\frac{1}{kT}(U_j - U_I)\right\}$$
$$= exp\left[-\frac{1}{kT}\{U(r_1^j, r_2^j, ..., r_N^i) - U(r_1^i, r_2^i, .., r_N^i)\}\right] \tag{47}$$

In the above equations, U_i and U_j are the interaction energies of microscopic states i and j, respectively. The superscripts attached to the position vectors denote the same meanings concerning microscopic states. Equation (46) implies that, in the transition from microscopic states i to j, new microscopic state j is adopted if the system energy decreases, with the probability ρ_j/ρ_i [1] if the energy increases. As clearly demonstrated by Eq. (47), for ρ_j/ρ_i the denominator in Eq. (45) is not required in Eq. (47), because ρ_j is divided by ρ_i and the term is canceled through this operation. This is the main reason for the great success of the metropolis method for MC simulations. That a new microscopic state is adopted with the probability ρ_j/ρ_i, even in the case of the increase in the interaction energy, verifies the accomplishment of the minimum free-energy condition for the system. In other words, the adoption of microscopic states, yielding an increase in the system energy, corresponds to an increase in the entropy [76].

The above discussion is directly applicable to a system composed of nonspherical particles. The situation of nonspherical particles in thermodynamic equilibrium can be specified by the particle position of the mass center, $r_i (i = 1, 2,..., N)$, and the unit vector e_i $(i = 1, 2,..., N)$ denoting the particle direction. The transition probability from microscopic states i to j, p_{ij} can be written in similar form to Eq. (46). The exact expression of ρ_j/ρ_i becomes

$$\frac{\rho_j}{\rho_i} = exp\left\{-\frac{1}{kT}(U_j - U_i)\right\} =$$
$$exp\left[-\frac{1}{kT}\{U(r_1^j, r_2^j, r_N^j, e_1^j, e_2^j, ..., e_N^j) - U(r_1^j, r_2^j, r_N^j, e_1^j, e_2^j, ..., e_N^j)\}\right] \quad (48)$$

The main procedure for the MC simulation of a nonspherical particle system is as follows:
1. Specify the initial position and direction of all particles.
2. Regard this state as microscopic state i, and calculate the interaction energy U_i.
3. Choose an arbitrary particle in order or randomly and call this particle "particle α."
4. Make particle α move translationally using random numbers and calculate the interaction energy U_j for this new configuration.
5. Adopt this new microscopic state for the case of UjUi and go to Step 7.
6. Calculate ρ_j/ρ_i in Eq. (48) for the case of U_jU_i and take a random number R_1 from a uniform random number sequence distributed from zero to unity.
   - If $R_1\rho_j/\rho_i$, adopt this microscopic state j and go to Step 7.
   - If $R_1\rho_j/\rho_i$, reject this microscopic state, regard previous state *i* as new microscopic state j, and go to Step 7.

7. Change the direction of particle α using random numbers and calculate the interaction energy U_k for this new state.
8. If $U_k \leq U_j$, adopt this new microscopic state and repeat from Step 2.
9. If $U_k > U_j$, calculate ρ_k/ρ_j in Eq. (48) and take a random number R2 from the uniform random number sequence.
    - If $R_2 \leq \rho_k/\rho_j$, adopt this new microscopic state k and repeat from Step 2.
10. If $R_2 > \rho_k/\rho_j$, reject this new state, regard previous state j as new microscopic state k, and repeat from Step 2.

Although the treatment of the translational and rotational changes is carried out separately in the above algorithm, a simultaneous procedure is also possible in such a way that the position and direction of an arbitrary particle are simultaneously changed, and the new microscopic state is adopted according to the condition in Eq. (46). However, for a strongly interacting system, the separate treatment may be found to be more effective in many cases. We will now briefly explain how the translational move is made using random numbers during a simulation. If the position vector of an arbitrary particle α in microscopic state i is denoted by $r_\alpha = (x_\alpha, y_\alpha, z_\alpha)$, this particle is moved to a new position by the following equations using random numbers R_1, R_2, and R_3, taken from a random number sequence ranged from zero to unity:

$$\begin{cases} x'_\alpha = x_\alpha + R_1 \delta r_{max} \\ y'_\alpha = y_\alpha + R_2 \delta r_{max} \\ z'_\alpha = z_\alpha + R_3 \delta r_{max} \end{cases} \qquad (49)$$

These equations imply that the particle is moved to an arbitrary position, determined by random numbers, within a cube centered at the particle center with side length of $2\delta r_{max}$. A series of microscopic states is generated by moving the particles according to the above-mentioned procedure. Finally, we show the method of evaluating the average of a physical quantity in MC simulations. These averages, called "ensemble averages," are different from the time averages that are obtained from MD simulations. If a physical quantity A is a function of the microscopic states of a system, and A_n is the nth sampled value of this quantity in an MC simulation, then the ensemble average <A> can be evaluated from the equation

$$\langle A \rangle = \sum_{n=1}^{M} A_n / M \qquad (50)$$

where M is the total sampling number. In actual simulations, the sampling procedure is not conducted at each time step but at regular intervals. This may be more efficient because if the data have significant correlations they are less likely to be sampled by taking a longer interval for the sampling time. The ensemble averages obtained in this way may be compared with experimental data (see Appendix 2) [76].

1.1.4   LATTICE BOLTZMANN METHOD

Whether or not the lattice Boltzmann method is classified into the category of molecular simulation methods may depend on the researcher, but this method is expected to have a sufficient feasibility as a simulation technique for polymeric liquids and particle dispersions. In the lattice Boltzmann method [82], a fluid is assumed to be composed of virtual fluid particles, and such fluid particles move and collide with other fluid particles in a simulation region. A simulation area is regarded as a lattice system, and fluid particles move from site to site; that is, they do not move freely in a region. The most significant difference of this method in relation to the MD method is that the lattice Boltzmann method treats the particle distribution function of velocities rather than the positions and the velocities of the fluid particles. Figure 1.9 illustrates the lattice Boltzmann method for a two-dimensional system. Figure 1.9A shows that a simulation region is divided into a lattice system. Figure 9B is a magnification of a unit square lattice cell. Virtual fluid particles, which are regarded as groups or clusters of solvent molecules, are permitted to move only to their neighboring sites, not to other, more distant sites. That is, the fluid particles at site 0 are permitted to stay there or to move to sites 1, 2,..., 8 at the next time step. This implies that fluid particles for moving to sites 1, 2, 3, and 4 have the velocity $c = (\Delta x/\Delta t)$, and those for moving to sites 5, 6, 7, and 8 have

The velocity $\sqrt{2}c$, in which Δx is the lattice separation of the nearest two sites and Δt is the time interval for simulations. Since the movement speeds of fluid particles are known as $\sqrt{2}c$ or, macroscopic velocities of a

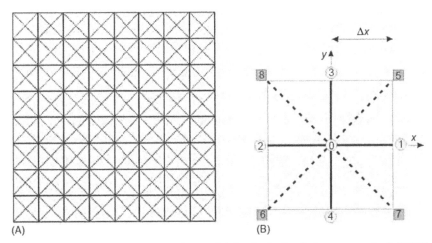

FIGURE 1.9   Two-dimensional lattice model for the lattice Boltzmann method (D2Q9 model).

fluid can be calculated by evaluating the number of particles moving to each neighboring lattice site. In the usual lattice Boltzmann method, we treat the particle distribution function, which is defined as a quantity such that the above-mentioned number is divided by the volume and multiplied by the mass occupied by each lattice site. This is the concept of the lattice Boltzmann method. The two-dimensional lattice model shown in Figure 1.9 is called the "D2Q9" model because fluid particles have nine possibilities of velocities, including the quiescent state (staying at the original site) [76].

The velocity vector for fluid particles moving to their neighboring site is usually denoted by cα and, for the case of the D2Q9 model, there are nine possibilities, such as, c_0, c_1, c_2,..., c_8. For example, the velocity of the movement in the left direction in Fig. 1.8B is denoted by c_2, and c_0 is zero vector for the quiescent state ($c_0 = 0$). We consider the particle distribution function $f_a(r,t)$ at the position r (at point 0 in Fig. 1.8B) at time t in the α-direction. Since $f_a(r,t)$ is equal to the number density of fluid particles moving in the α-direction, multiplied by the mass of a fluid particle, the summation of the particle distribution function concerning all the directions ($\alpha = 0, 1,..., 8$) leads to the macroscopic density $\rho(r,t)$:

$$\rho(r,t) = \sum_{\alpha=0}^{8} f_\alpha(r,t) \tag{51}$$

Similarly, the macroscopic velocity **u(r,t)** can be evaluated from the following relationship of the momentum per unit volume at the position :

$$\rho(r,t)u(r,t) = \sum_{\alpha=0}^{8} f_{\alpha}(r,t)C_{\alpha} \tag{52}$$

In Eqs. (51) and (52), the macroscopic density ρ(r,t) and velocity u(r,t) can be evaluated if the particle distribution function is known. Since fluid particles collide with the other fluid particles at each site, the rate of the number of particles moving to their neighboring sites changes. In the rarefied gas dynamics, the well-known Boltzmann equation is the basic equation specifying the velocity distribution function while taking into account the collision term due to the interactions of gaseous molecules; this collision term is a complicated integral expression. The Boltzmann equation is quite difficult to solve analytically, so an attempt has been made to simplify the collision term. One such simplified model is the Bhatnagar-Gross-Krook (BGK) collision model. It is well known that the BGK Boltzmann method gives rise to reasonably accurate solutions, although this collision model is expressed in quite simple form. We here show the lattice Boltzmann equation based on the BGK model. According to this model, the particle distribution function $f_a(r + c_a \Delta t, t + \Delta t)$ in the α-direction at the position $(r + c_a \Delta t)$ at time $(t + \Delta t)$ can be evaluated by the following equation: [76].

$$f_{\alpha}(r + c_{\alpha}\Delta t, t + \Delta t) = f_{\alpha}(r,t) + \frac{1}{\tau}\left\{f_{\alpha}^{(0)}(r,t) - f_{\alpha}(r,t)\right\} \tag{53}$$

This equation is sometimes expressed in separate expressions indicating explicitly the two different processes of collision and transformation:

$$f_{\alpha}(r + c_{\alpha}\Delta t, t + \Delta t) = \tilde{f}_{\alpha}(r,t)$$

$$\tilde{f}_{\alpha}(r,t) = f_{\alpha}(r,t) + \frac{1}{\tau}\left\{f_{\alpha}^{(0)}(r,t) - f_{\alpha}(r,t)\right\} \tag{54}$$

where τ is the relaxation time (dimensionless) and $f_{\alpha}^{(0)}$ is the equilibrium distribution, expressed for the D2Q9 model as

$$f_{\alpha}^{(0)} = \rho w_{\alpha}\left\{1 + 3\frac{c_{\alpha}.u}{c^2} - \frac{3u^2}{2c^2} + \frac{9}{2}.\frac{(c_{\alpha}.u)^2}{c^4}\right\}] \tag{55}$$

$$w_\alpha = \begin{cases} 4/9 \ for \ \alpha = 0 \\ 1/9 \ for \ \alpha = 1,2,3,4 \\ 1/36 \ for \ \alpha = 5.6.7.8 \end{cases} \quad |c_\alpha| = \begin{cases} 0 \ for \ \alpha = 0 \\ c \ for \ \alpha = 1,2,3,4 \\ \sqrt{2}c \ for \ \alpha = 5,6,7,8 \end{cases} \quad (56)$$

In these equations ρ is the local density at the position of interest, u is the fluid velocity (u = |U|), c = $\Delta x/\Delta t$, and w_α is the weighting constant. The important feature of the BGK model shown in Eq. (54) is that the particle distribution function in the α-direction is independent of the other directions. The particle distributions in the other directions indirectly influence $f_\alpha(r + c_\alpha \Delta t, t + \Delta t)$ through the fluid velocity u and the density ρ. The second expression in Eq. (55) implies that the particle distribution $f_\alpha(r,t)$ at the position r changes into $f_\alpha(r,t)$ after the collision at the site at time t, and the first expression implies that $f_\alpha(\mathbf{r}, t)$ becomes the distribution $f_\alpha(r + c_\alpha \Delta t, t + \Delta t)$ at $(r + c_\alpha \Delta t)$ after the time interval Δt [76].

The main procedure of the simulation is as follows:

1. Set appropriate fluid velocities and densities at each lattice site.
2. Calculate equilibrium particle densities $f_\alpha^{(0)}$ (α = 0, 1,..., 8) at each lattice site from Eq. (55) and regard these distributions as the initial distributions, $f_\alpha = f_\alpha^{(0)}$ (α = 0, 1,..., 8).
3. Calculate the collision terms $f_\alpha(\mathbf{r}, t)$ = (α = 0, 1,..., 8) at all sites from the second expression of Eq. (54).
4. Evaluate the distribution at the neighboring site in the α-direction $f_\alpha(r + c_\alpha \Delta t, t + \Delta t)$ from the first expression in Eq. (54).
5. Calculate the macroscopic velocities and densities from Eqs. (51) and (52), and repeat the procedures from Step 3.

In addition to the above-mentioned procedures, we need to handle the treatment at the boundaries of the simulation region. These procedures are relatively complex and are explained in detail in Chapter 8. For example, the periodic boundary condition, which is usually used in MD simulations, may be applicable. For the D3Q19 model shown in Fig. 1.9, which is applicable for three-dimensional simulations, the equilibrium distribution function is written in the same expression of Eq. (55), but the weighting constants are different from Eq. (56). The basic equations for $f_\alpha(r + c_\alpha \Delta t, t + \Delta t)$ are the same as Eq. (53) or (54), and the above-mentioned simulation procedure is also directly applicable to the D3Q19 model [76].

1.2   PART 2: INTRODUCTION ON METHODOLOGY

The term aerogel was first introduced by Kistler in 1932 to designate gels in which the liquid was replaced with a gas, without collapsing the gel solid network [50]. While wet gels were previously dried by evaporation, Kistler applied a new supercritical drying technique, according to which the liquid that impregnated the gels was evacuated after being transformed to a supercritical fluid. In practice, supercritical drying consisted in heating a gel in an autoclave, until the pressure and temperature exceeded the critical temperature T_c and pressure P_c of the liquid entrapped in the gel pores. This procedure prevented the formation of liquid–vapor meniscuses at the exit of the gel pores, responsible for a mechanical tension in the liquid and a pressure on the pore walls, which induced gel shrinkage. Besides, a supercritical fluid can be evacuated as gas, which in the end lets the "dry solid skeleton" of the initial wet material. The dry samples that were obtained had a very open porous texture, similar to the one they had in their wet stage. Overall, aerogels designate dry gels with a very high relative or specific pore volume, although the value of these characteristics depends on the nature of the solid and no official convention really exists. Typically, the relative pore volume is of the order of 90% in the most frequently studied silica aerogels [12].

1.2.1   SILICA AEROGEL AND ITS PROPERTIES

The first aerogels were synthesized from silica gels by replacing the liquid component with a gas. Silica aerogel, a highly porous material [50], is currently being produced using sol–gel processes. One such process is supercritical drying of tetramethoxysilane (TMOS), which is hydrolyzed to form silica and methanol [12, 34]. Several desirable properties of silica aerogels include very low densities (as low as 0.003 g cm^{-3}), surface area as large as 1000 m²/g, a refraction index only 5% greater than that of air, the lowest thermal conductivity among all solid materials, a speed of sound approximately three times smaller than that of air, and a dielectric constant only 10% greater than that of vacuum. [34]. These properties make silica aerogels very suitable for applications such as thermal and acoustic insulation in buildings and appliances, passive solar energy collection devices and dielectrics for integrated circuits [16], but probably one of the most notable uses was in Cherenkov radiators [4] as Cherenkov counters. Aerogels were also used to thermally insulate the 2003 Mars Exploration rovers, as well as to capture comet dust [30].

1.2.2   PREPARATION OF AEROGELS

Experimentally, aerogels and xerogels are made by sol–gel processing which is a core starting of an aerogel. An important boundary condition for the initial solution is the solubility of the reactants in the applied solvent. Therefore, it is important to study the miscibility diagram of the system under investigation (Fig. 1.10) or to perform corresponding tests to make

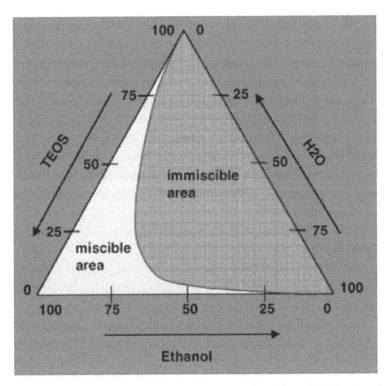

FIGURE 1.10   Miscibility diagram for the system tetraethoxysilane (TEOS), ethanol, and water.

sure that full solubility is assured [71]. Otherwise partial precipitation or inhomogeneous gels will result.

The sol–gel process generally comprises two major steps on the molecular level: (i) the creation or activation of reactive sites; and (ii) the condensation of monomers, oligomers, and clusters via cross-linking of the reactive sites. The kinetics of each of these two reactions is controlled by various parameters, for example, the concentration of the reactants, the solvent, the catalyst used, the catalyst concentration (pH value), the temperature, and the steric effects, etc. [14].

The relative kinetics of the two steps together with reverse reactions due to solubility effects determine the morphological characteristics of the resulting gel, for example, the size of the particles forming the solid backbone, the mass distribution of the backbone elements characterized, for example, by the so-called fractal dimension of the backbone [46], and the connectivity in the solid phase, which is seen in the mechanical properties and the electrical or thermal conductivity of the backbone structure at a given porosity. Depending on the governing reaction kinetics, the gel backbone can range from a network of colloids to interconnected, highly branched polymer-like clusters (Fig. 1.11). The rate of cross-linking on a molecular scale determines the macroscopically observable gel time that is the time where the viscosity of the initial solution increases by several orders of magnitude, changing its appearance from a low viscosity liquid, to a syrup-like substance, and finally to a gel body. Almost all metal oxides can be transformed into gels and corresponding aerogels. In aqueous solution, the reactants are the corresponding salts; alkoxide precursors are processed in organic solvents [71].

The processing starts with the hydrolysis of a silica precursor, typically tetramethyl ortho silane (TMOS, $Si(OCH3)4$) or tetraethyl orthosilane (TEOS, $Si(xOC2H5)4$). Typically, the solvents used in these cases are ethanol or methanol and a base or acid is added to control the reaction rates (Fig. 1.12).

In the alcoxysilane solution, the "activation" step is the partial or complete hydrolysis of the alcoxysilane, which is favored with increasing concentration of either an acid or a base in the initial solution (reaction with TMOS, Eq. (57)). The condensation rate (Eq. (58)) of the hydroxyl groups increases on both sides of the isoelectrical point (point of zero charge), but is suppressed at high Ph values [14].

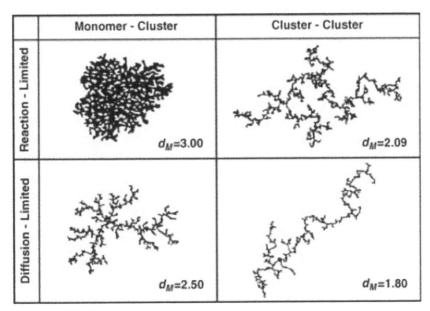

FIGURE 1.11   Morphology of the gel backbone in the case of different reaction kinetics [46]. The parameters dM are the fractal dimensions resulting from the combination of the different growth mechanisms, as indicated on the top and the left-hand side of the sketch.

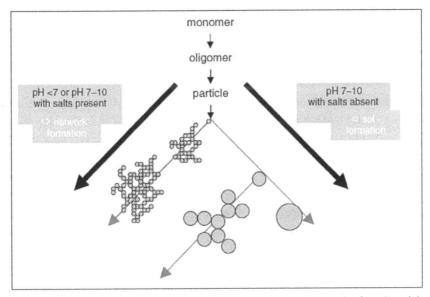

FIGURE 1.12   Impact of the pH and the presence of salts on the growth of a sol particle and the network formation upon polymerization of silica.

$$Si(OCH_3)_4 + XH_2O \; Si(OCH_3)_4 - X(OH)X + XCH_3OH \; \text{Hydrolysis} \quad (57)$$

$$Si(OCH_3)_3(OH) + Si(OCH_3)_3(OH) \; Si_2O(OCH_3)_6$$
$$+ H_2O \; \text{Condensation} \quad (58)$$

However, other precursors can be used, too. For example, Steven Kistler, who is considered to have discovered this type of materials [8], produced the first silica aerogel using sodium silicate (Na_2SiO_3) dissolved in a solution of hydrochloridric acid (HCl) and water [19].

$$Na_2SiO_3 + 2HCl \; [SiO_2.xH_2O] + 2NaCl \quad (59)$$

The pH of the solution can be controlled by adding acidic or basic additives; those will react with neither the silica precursor nor the water but will affect the rate at which the hydrolysis proceeds as well as the final structure of the porous solid. In other words, the two reaction steps hydrolysis and condensation on the pH enables a very sensitive control of the reaction, and thus resulting gel morphology. Besides the so-called "one-step" process, in which all reactants are mixed within a short period of time, the sensitivity of the kinetics to the pH can be used to partially decouple the hydrolysis and condensation step (two-step sol–gel process) [13, 51]. For example, the reactants in a concentrated starting solution can first be prehydrolyzed under acidic conditions, subsequently diluted to match the reactant concentration for a given target solid content of the gel to be and then subject to basic conditions to accelerate the rate of cross-linking of the hydrolyzed species and to form a gel [17, 18, 22, 71].

- Acid catalysis generally produces weakly cross-linked gels which easily compact under drying conditions, yielding low-porosity microporous (smaller than 2 nm) xerogel structures (Fig. 1.13a).
- Conditions of neutral to basic pH result in relatively mesoporous xerogels after drying, as rigid clusters a few nanometers across pack to form mesoporous. The clusters themselves may be microporous.
- Under some conditions, base-catalyzed and two-step acid-base catalyzed gels (initial polymerization under acidic conditions and further gelation under basic conditions) exhibit hierarchical structure and complex network topology (Fig. 1.13).

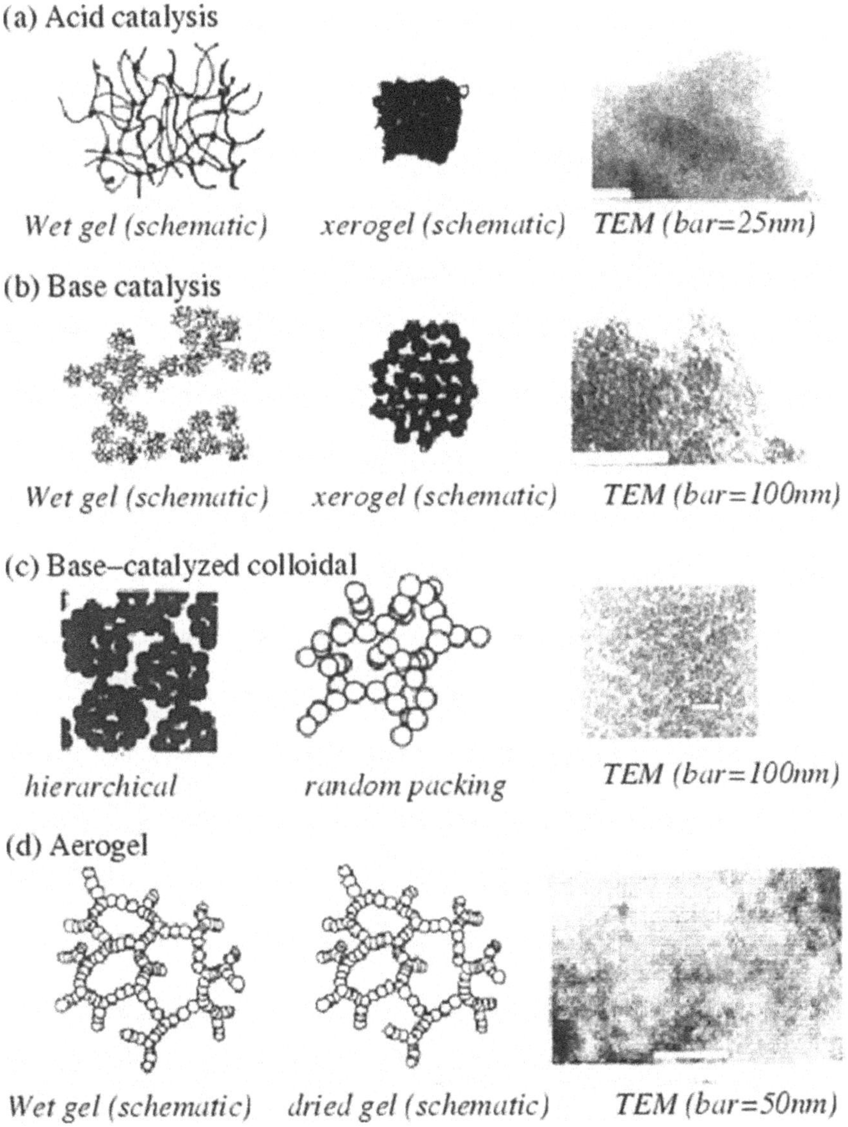

(a) Acid catalysis

Wet gel (schematic)   *xerogel (schematic)*   *TEM (bar=25nm)*

(b) Base catalysis

Wet gel (schematic)   *xerogel (schematic)*   *TEM (bar=100nm)*

(c) Base–catalyzed colloidal

hierarchical   *random packing*   *TEM (bar=100nm)*

(d) Aerogel

Wet gel (schematic)   *dried gel (schematic)*   *TEM (bar=50nm)*

FIGURE 1.13   Schematic wet and dry gel morphologies and representative transmission electron micrographs (2002, aerogel).

After the hydrolysis, polycondensation of silica occurs and the solution becomes a gel. A large change of the viscosity indicates the gelation of the solution. At this point a nanoporous solid skeleton of silica is formed inside the solution and fills almost all its volume. If the gel is dried at room temperature and atmospheric pressure, large shrinkage of the solid structure is caused along with large reduction of the porosity. The shrinkage is mainly due to the stresses exerted on the solid branches when the liquid evaporates. Materials obtained this way are called xerogels and have porosities as high as 50%, corresponding to a density of 1.1 g/cm³. To avoid excessive shrinkage, the pressure and the temperature of the gel are increased until the critical point of the liquid phase has been exceeded. Under those conditions there is no liquid/vapor interface and no surface tension. Therefore, no large stresses are exerted on the solid skeleton and the shrinkage is highly reduced. Solids processed at supercritical conditions reach porosities up to 99.8% and are called aerogels [1/9]. Figure 1.14 show the process steps to synthesizing an aerogel.

1.2.3   DRYING

The removal of the liquid from the interconnected pores of a gel is the second key process step to synthesizing an aerogel. The reason why it is so tricky is the combination of small pores and thus high capillary pressures up to 10 MPa (100 bar) during evaporation (Fig. 1.15) and a sparse, incompletely interconnected solid backbone that possesses only a low modulus of compression. These two facts typically result in a large permanent shrinkage upon drying unless specific measures are taken to prevent or at least suppress part of the effects. Possible means are:

- strengthening of the backbone to increase the stiffness of the back-bone.
- reduction of the surface tension of the pore liquid by either
- replacing the initially present pore liquid with a liquid of lower surface tension and/or reducing the contact angle (toward complete wetting) via a modification of the gel surface.
- eliminating the surface tension by taking the pore liquid above the critical point of the pore liquid prior to evaporation (supercritical drying) or sublimation of the pore liquid (cryogenic drying).

Alternatively, one can allow the gel to shrink during drying; if the inner surface of the gel is chemically inert and the drying is performed slowly

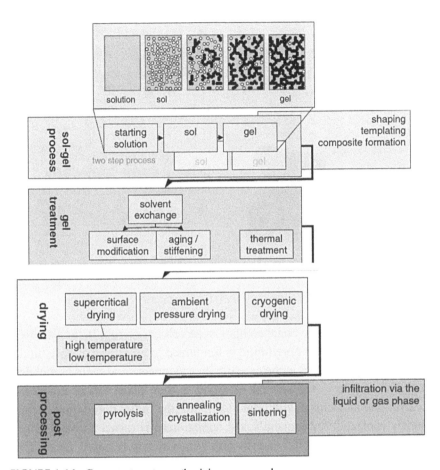

FIGURE 1.14   Process steps to synthesizing an aerogel.

enough to prevent large gradients in shrinkage across the sample and thus crack formation, the gel will at least partially reexpand when the capillary forces decrease in the second stage of the drying process (Fig. 1.10) [71].

1.2.4   SUPERCRITICAL DRYING

Supercritical drying is the extraction of a fluid in the supercritical state, thus avoiding the liquid–vapor coexistence regime and the associated capillary pressure. Figire 1.16 exemplarily shows the phase diagram of CO_2

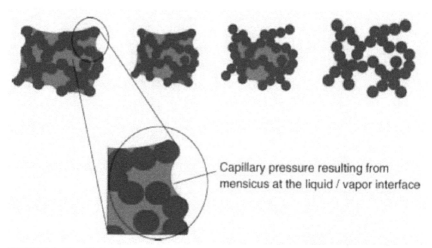

Capillary pressure resulting from
mensicus at the liquid / vapor interface

FIGURE 1.15 Shrinkage of the gel body due to capillary forces upon ambient pressure drying. As soon as the pores begin to empty, the pressure is released and the gel partially reexpands.

with a possible path to be taken upon supercritical drying: In the first step, the gel containing the pore liquid is immersed in an excess of pore liquid, for example, methanol in the case of silica alcogels, to prevent partial drying; the sample in the solvent reservoir is then placed in an autoclave, pressurized and rinsed with liquid CO_2 until the initial pore liquid is completely replaced. Subsequently, the temperature is raised above the critical point of CO_2 (318°C) and the pressure is slowly released via a valve. When fully depressurized, the autoclave with the sample is allowed to cool down to room temperature. This scheme is also valid if a fluid other than CO_2 is used. Depending on the value of the critical point of the respective fluid, one distinguishes between low and high temperature supercritical drying (e.g., critical point of ethanol: 2168°C). In the case of high temperature supercritical drying, additional aging effects are observed that are mainly controlled by the solubility of the solid phase (i.e., the backbone) at the given conditions [18, 90].

In theory, it is expected that upon supercritical drying no capillary pressure and thus no shrinkage of the gel body will be observed. In practice, linear shrinkages of the order up to 10–15% percent are detected [39] (Fig. 1.17). This effect shows, for example, upon CO_2 supercritical drying when

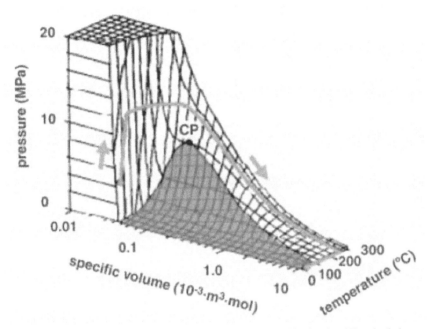

FIGURE 1.16 Phase diagram of CO_2; CP denotes the critical point. The shaded area marks the regime of coexistence of the liquid and the gas phase. The line and the arrows indicate the path taken upon supercritical drying.

the CO_2 pressure is released and seems to result from the change in surface tension of the gel backbone upon desorption of the CO_2 surface layer. The effect has been proven to be partially reversible [38]. A rapid supercritical drying process was developed by Poco and co-workers and further characterized by Gross and co-workers [36, 37]. Here, the autoclave vessel acts as the mold for the gel and supports the backbone when tensile strains arise during depressurization. This concept enables a solution to be processed to an aerogel within 1 h. This concept was further optimized by Scherer and co-workers [71, 77].

1.2.5 AMBIENT PRESSURE DRYING

Kawaguchi and co-workers [48] and Smith and co-workers [79] were among the first to look into the drying behavior of silica gels at ambient conditions. In particular Smith and co-workers [79] studied the effects

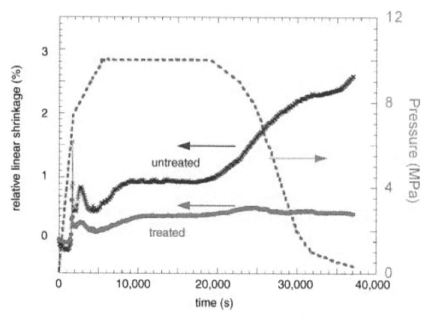

FIGURE 1.17 In situ measured relative linear shrinkage of an as prepared (untreated) and a heat treated silica alcogel (TMOS based) upon supercritical drying in CO_2.

of different pore liquids and surface methylation on the drying kinetics and shrinkage of the gels in great detail. Schwertfeger and Frank applied their findings and patented a process for the drying of low density silica aerogel granules at ambient pressure. Bhagat and co-workers [9] recently published a novel fast route for silica aerogel monoliths; hereby they use a co precursor method for the surface modification in silica hydrogels by adding trimethylchlorosilane [77] and hexamethyldisilazane (HMDS) to the starting solution of a diluted (water glass derived) ion exchanged silicic acid. By using this approach, they have been able to prepare silica aerogels with a diameter of ~10 mm and a density as low as 0.1 g/cm³ within ~30 h. Recently, Job and co-workers and Leonard and co-workers investigated in detail the option of applying convective drying at ambient pressure to convert monolithic organic aquagels into crack-free aerogels. The shrinkage of mesoporous aquagels upon ambient pressure drying can be significantly reduced by replacing the water in the pores by a liquid with a significantly lower surface tension, for example, acetone, isopropyl alcohol or ethanol [71].

The key to a fast crack-free drying is to avoid vapor pressure gradients across the macroscopic surface of the gel body that result in locally different rates of volume change and thus crack formation. Unexpectedly, at first sight, the fact that upon drying the first segment that is accompanied by significant shrinkage is not the most crucial phase of drying at ambient conditions. In this phase, the gel is still filled with pore liquid, and mass loss is exclusively due to liquid squeezed out of the gel by capillary forces originating from the evolution of menisci in pores at the outer surface of the sample only (Fig. 1.15). Unless the partial vapor pressure of the pore liquid is reduced faster than the liquid can be squeezed through the pores to the external surface of the gel, no macroscopic damage is observed. When the capillary forces are finally counterbalanced by the compressed gel body, the gel stops shrinking, the menisci invade the gel, and pores are progressively emptied. Consequently, the capillary forces decrease and the gel partially reexpands. In this phase, the drying rate is limited by transport in the gas phase and along the inner surface of the sample. The extent of reexpansion can, in theory, be 100%. In practice, however, the sample volume is only partially recovered due to the reaction of neighboring inner surfaces or creeping resulting in a plastic deformation of the gel backbone.

1.2.6   CRYOGENIC DRYING

To reduce the capillary pressure upon drying, the pore fluid can also be extracted via sublimation. For this purpose, the gel has to be taken below the freezing point of the pore liquid (note that with decreasing pore size the freezing point will be shifted toward lower temperatures). Besides the temperature, the cooling rate is an important factor in this process since a low cooling rate can result in a crystal growth that can severely modify or even destroy the gel body. The second important quantity is the volume change of the pore liquid upon freezing; in particular in the case of water its density anomaly results in a 10% expansion upon freezing that is not tolerated by the gel backbone and yields a disruption of the gel network. Therefore, prior to freezing, pore water is displaced by a liquid, for example, test-butanol that preferably also possesses a high vapor pressure and thus enables faster sublimation. With respect to the mesoporosity of the resulting aerogels, freeze drying seems to be superior to ambient pressure drying at least in the case of some organic gels; however, the mesopore volume detected is still lower for freeze-dried gels compared to their super critically dried counterparts [71].

1.2.7   PROPERTIES OF AEROGEL

Aerogel properties are usually correlated with density, rather than related to morphological features. One reason for this is the lack of suitable representation of aerogel morphology. They are characterized by a network of three-dimensionally (3D) interconnected particles that define a system of well-accessible meso or macropores (Fig. 1.18). Performance windows, supercapacitors, heat barriers, particle traps, ultrasound probes, and ion exchange media [8, 19, 23, 24].

Although the internal structure of aerogels and xerogels is the key factor for its excellent properties, such as large surface area and low thermal conductivity, their relatively poor mechanical response limits its final application. Hence, it is important to study and evaluate the mechanical properties of these materials.

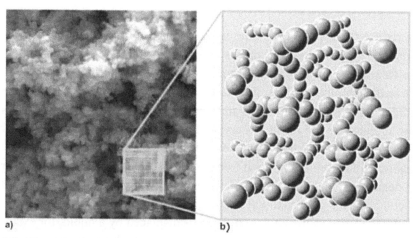

a)   b)

FIGURE 1.18   (a) The Scanning electron microscopy (SEM) image of an aerogel with a colloidal backbone morphology; (b) Corresponding sketch of the 3D interconnected skeleton.

1.2.7.1   Morphological Characteristics

The unique properties of aerogels are a consequence of the distribution and layout of the two phases, that is, the solid backbone and the pore network. Interestingly, the presence and properties of a void phase can significantly modify the properties of the overall material [84]. The impact of the total porosity on any macroscopic property can be predicted in a first step by using the Hashin–Shtrikman principle [84]. Here, the two extreme cases of a serial and a parallel connection of the two phases define the upper and lower bonds of the range of possible values for a physical property at a given porosity. In addition, the pore properties also play a crucial role with respect to the effective behavior of the porous medium. Aerogels per definition contain a network of pores that possess a connection to the outer surface of the gel (unless strong shrinkage takes place upon drying or sintering, resulting in the cut-off of pore connections). The main quantities characterizing the pore phase are (see Fig. 1.19): the overall porosity; the contribution of micro, meso, and macroporosity; the pore size distribution and the average pore size; the pore shape; the pore connectivity; and the physical and chemical properties of the solid–void interface. The highest porosities achieved for aerogels so far have even exceeded 99%.

Specific surface areas related to meson or macropores range between a few and several hundred meters squared per gram (m^2/g). In the case of aerogels with a microporous (pores <2 nm) backbone, for example, carbon, aerogels, total specific surface areas up to ~2000 m^2/g, can be provided. The average pore sizes (not including the subnanometer micropores) are between a few nanometers and several microns; here in particular silica aerogels seem to possess a narrow pore size distribution as suggested by the comparison of quantities with a different weighting of the pore size distribution like the chord length or pore size determined from small-angle X-ray scattering and sorption isotherms, respectively, and the hydraulic radius derived from gas permeation measurements. Several orders of magnitude in mean pores sizes can easily be covered even at medium porosity with organic aerogels and their carbon counterparts [66]. In porous solids, the solid phase also has an impact on the aerogel properties through its composition, its connectivity, and the size of the backbone segments (length, diameter). The backbone characteristics are reflected in

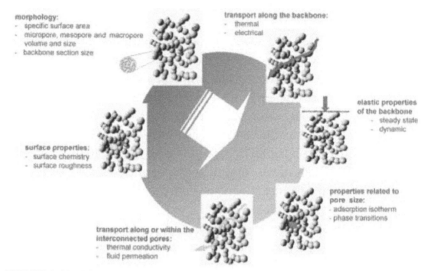

FIGURE 1.19   The morphological properties of the two phases of an aerogel, as well the characteristics of their interface (top left) determine deduced properties, for example, the transport properties in the solid and the pore phase, their mechanical behavior, and steady-state properties of the pore, for example, their adsorption characteristics.

macroscopic quantities, for example, the elastomechanical behavior and the electrical or thermal transport along the solid phase (Fig. 1.19).

1.2.7.1.1   Scaling Behavior

Typical for the change of the macroscopic properties with the density or porosity of an aerogel (at a given growth mechanism of the backbone) is their scaling behavior. For this reason, aerogels were popular model materials, in particular in the 1980s and 1990s, to theoretically investigate the predicted properties of percolating systems [2] and fractals. While scaling of properties with bulk density can be seen at least over a limited range for most aerogels, not all of them also exhibit fractal morphology. Porous materials with fractal morphology are characterized by a scale invariant structure, that is, they look alike under different magnifications. In practice, this behavior is only observed on a limited length scale with upper and lower bounds (Finite Size Scaling). Since a good deal of work has been performed on predicting the relationship of properties in the case of fractals, this concept will be used here to explain the intimate connection between different macroscopic properties. Similar results, however, are found for nonhierarchical, random heterogeneous media where the so-called critical exponents play a similar role as the exponents in the concept of fractal systems [84]. Important is the fact that different macroscopic properties are closely correlated for a given type of backbone and pore morphology connectivity. In some cases, the scaling concept is found to be only a coarse approximation for a small density range [31]; however, the concept is valuable for identifying trends and correlations even in these cases. Relevant quantities that describe aerogels with a fractal backbone, for example, silica aerogels, are the mass fractal dimension d_M that relates the total mass m within a sphere of radius R to the length scale R.

$$m(R) = R^{d_M} \tag{60}$$

the surface fractal dimension dS that reflects the surface roughness and thus the surface area S as a function of the scale length L

$$S(L) = L^{dS} \tag{61}$$

And the fraction exponent or spectral dimension d_0 of the material that relates the vibrational density of states $D(\omega)$ to the frequency of the vibration via

$$D(\omega) = \omega^{d'-1} \tag{62}$$

Equations (60) and (61) refer to the mass distribution and roughness. These properties can directly be determined from small-angle X-ray or neutron scattering data, since the exponents in the length scale regime (Eqs. (60) and (61)) transfer into a scaling of the scattering intensity with the scattering vector. Typically, the mass fractal dimensions d_M of aerogels range between 1.8 and 2.5. To compare and interpret scaling exponents, one has to keep in mind that these values are based on the assumption of a fixed underlying growth mechanism (e.g., diffusion-limited aggregation, cf. Fig. 1.11) that dominates over the full range of densities considered. If the mechanism changes with density, largely deviating exponents are to be expected.

Equation (62) contains information about the connectivity of the gel backbone segments. From a more general point of view, the knowledge of the density of vibrational states over the full frequency range from several hertz (Hz) to terahertz (THz) provides the full spectrum of localized and nonlocalized vibrations (phonons) possible in the aerogel under investigation. All "phonon" related properties, for example, the specific heat as a function of temperature, as well as acoustical properties and the heat transfer along the backbone, can be deduced from this quantity. Typically, the density of states of an aerogel consists of four characteristic regimes (Fig. 1.20):

1. The *Debye regime*, where the phonon wavelength is larger than the inhomogeneity's of the aerogel backbone.
2. The *fraction or localized modes* regime that corresponds to vibrational motions of the chains between the knots in the disordered interconnected backbone network.

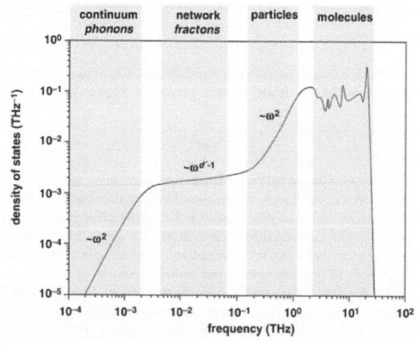

FIGURE 1.20 Typical vibrational density of states of an aerogel.

3. The *particle modes regime*, corresponding to the Eigen modes of the particles or struts forming the backbone.
4. The *molecular regime*, reflecting the motion of molecules attached to the backbone surface or in the bulk.

1.2.7.2 Macroscopic Properties: Mechanical Characteristics

In the case of fractals, the scaling exponents that describe the elastome-chanical behavior and the transport properties along the solid backbone can be calculated from the fractal dimension d_M and the fraction scaling exponent d, [87]. Scaling laws for the elastic modulus E of silica, as well as organic and carbon aerogels as a function of density ρ, were experimentally determined by Gross and co-workers:

$$E(\rho) = \rho^\alpha \text{ with } \alpha \text{ between } 2.0 \text{ and } 3.6 \qquad (63)$$

Theoretical assumptions yield for the scaling exponent a

$$\alpha = (5 - d_M)/(3 - d_M)(169) \qquad (64)$$

That is, with mass fractal dimensions d_M typical for aerogels between 1.8 and 2.5, values for a between 2.7 and 5 are expected; these values roughly match the experimentally determined data. Ma and co-workers investigated the relationship between the density and the mechanical stiffness of aerogels in a very sophisticated simulation of the mechanical behavior of gel backbones; hereby the solid phase of different aerogels, generated by diffusion-limited cluster–cluster aggregation (DLCA) algorithms, mimicked the experimentally observed morphological properties [54, 55]. The goal of this study was to understand the high, experimentally derived scaling exponent for Young's modulus with density (cf. Eq. (63)). One important result was the role of the dangling bonds in the network and the decrease in their relative mass contribution with increasing bulk density; in addition, the authors were able to simulate the local strain energy distribution in the gel network as a function of gel density (Fig. 1.22). Their data revealed that at low aerogel density, only a very small percentage of the network supports an externally imposed load. Gross and co-workers showed that determining the elastic properties for low density aerogels at ambient conditions via a dynamic measurement, for example, a sound velocity measurement, results in large deviations (towards larger sound velocities) from the scaling behavior deduced from higher densities. The reason for this effect is the finite compressibility of the gas in the pores of the aerogel that dominates the properties at low backbone stiffness [37, 38]; the authors showed that the scaling behavior is fully recovered when the sample is evacuated. Typical values for the sound velocity and Young's moduli at an aerogel density of ~0.3 g/cm^3 are 200–500 m/s and ~6–60 MPa, respectively (Fig. 1.21). The elastic properties of the backbone of aqua- and alcogels can be investigated with the beam-bending method to study, for example, the impact of additional treatments of gels prior to drying with the aim to increase the backbone stiffness [71].

The damping of sound waves in silica aerogels was studied by Zimmermann and Gross, as well as by Forest and Woignier; internal friction effects due to adsorption of molecules at the inner surface of the aerogels seem to control the damping behavior. The inverse quality factor Q^{-1} that combines the acoustic extinction coefficient with the wavelength of the phonons under investigation l was found to be 0.3–0.5 ×10^{-3} [71].

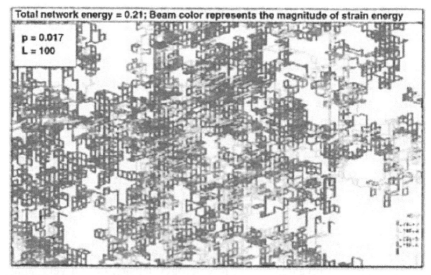

FIGURE 1.21 Distribution of the strain energy in a perfectly distributed, diffusion limited cluster aggregation network.

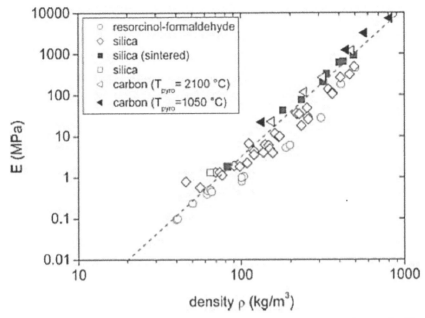

FIGURE 1.22   Young's modulus E versus density for different types of aerogels. The dashed line corresponds to a scaling exponent of 3.6.

$$Q^{-1} = \alpha . \lambda / \pi \text{ with } I(x) = I_0 . \exp(-\alpha . x) \tag{65}$$

Hereby I and I_0 are the acoustic intensities of the dumped and undumped wave, respectively.

1.2.7.3   Thermal and Electrical Conductivity

Scaling theory predicts, for the conductivity of the backbone, an increase with density with a scaling exponent that is always smaller than the one for Young's modulus. For the thermal and electrical conductivity of aerogels, λ and σ, respectively, experimentally derived scaling exponents of ~1.5–2.2 are reported at ambient conditions. The very detailed study by Bock reveals a scaling exponent for the electrical conductivity that drops from ~3.6 at 10 K to ~1.7 at temperatures 100 K. This result is due to the temperature dependence of the charge carrier mobility. Besides porosity, the thermal, as well as the electrical backbone conductivity, are controlled by the connectivity of the solid phase and its intrinsic conductivity properties. While in the case of thermal conductivity the transport occurs via phonons and, in case of an electrically conductive material, also via electrons, the electrical conductivity takes place via charge carriers only. Because at knots the aerogel backbone is chemically connected, the conductivity of the aerogel skeleton is often higher than in a system of powder particles with the same solid content and the same chemical composition. The fact that the scaling exponent is well above unity reflects the increasing amount of dangling mass with increasing porosity in aerogels [31]. The effective conductivities of aerogels can be strongly enhanced when annealing results in a phase change or molecular rearrangement of the backbone material, dangling ends are reconnected to the continuous backbone or necks at the joints of adjacent backbone particles are smeared out. Other options to affect the backbone conductivity are additives on the molecular (doping) or mesocopic (particles, fibers) scale. Pure carbon aerogels and xerogels typically show electrical conductivities at room temperatures between ~1 S/cm (for densities of ~100 kg/m^3) and 50 S/cm (for densities of ~1000 kg/m^3) [71].

The different methods that are typically applied to characterize the morphological and other basic properties of aerogels are summarized in Fig. 1.23. Note that some of them carry the tag "restrictions," denoting that special care has to be taken to make sure that the aerogel under investigation is neither temporarily nor permanently modified upon analysis. The restrictions for applying electron microscopy refer to the fact that the backbone units may be partially sintered due to local heating effects. Mercury porosimetry is an established technique for characterizing open porous materials with pore sizes ranging from a few nanometers to ~300 m. It makes use of the fact that mercury does not wet most substances. To force the mercury into pores, an external pressure is needed that is inversely proportional to the size of the pores to be infiltrated; to force mercury into pores of ~50 nm, a pressure of ~20 MPa (200 bar) is necessary. Especially the low density aerogels, however, are compliant and are therefore largely compressed before the mercury actually enters the pore network.

A similar problem can occur upon nitrogen sorption analysis, a technique that enables the characteristics of accessible micro and mesopores

Technique	Length scale
Electron microscopy (SEM) (restrictions!)	Millimeter
light scattering	
Mercury porosimetry (restrictions!)	
small angle X-ray or neutron scattering	
N2-Sorption (restrictions!)	
Electron microscopy (TEM) (restrictions!)	Angstrom
X-ray diffraction	

Further methods:	
Beam bending:	Elastic properties, fluid permeation of the aqua or alcogel
Sound velocity:	Elastic properties
Gas permeability:	Gas permeation, average pore size
Thermal conductivity:	Connectivity of the gel backbone, average pore size
Electrical conductivity:	Connectivity of the gel backbone

FIGURE 1.23 Summary of methods typically applied to characterize the morphology and other basic properties of aerogels.

to be determined. Upon nitrogen sorption, the liquid–vapor interface of nitrogen condensed in the mesopores of the sample under investigation forms a meniscus; the related capillary forces, upon both adsorption and desorption, can significantly compress the gel backbone; this results in an elastic or plastic deformation of the upon analysis and even worse a sorption isotherm that is highly affected by the compression effect rather than reflecting the pore size distribution of the uncompressed backbone as expected. One positive side effect is that, in the case of compliant aerogels, the shape of the sorption hysteresis reveals the modulus of compression and allows for a zero-order correction of the extracted pore size.

Light, X-ray, and neutron scattering are nondestructive methods that are sensitive to length scales of microns to angstroms. These methods are complementary, for example, to nitrogen sorption, since they also detect density fluctuations and closed porosity in aerogels (e.g., micropores in carbon aerogels) and even allow the morphological changes upon gelation or the morphology of a wet gel to be analyzed.

Another very valuable method to investigate the properties of the aqua- or alcogel is the beam-bending technique. Here, a rod-like shaped gel is bent in a three-point bending set-up and the force needed to keep the sample bent is measured as a function of time. Since the pores of the gel are liquid filled, the pressure gradient within the sample caused by the bending of the gel rod initiates a redistribution of the pore liquid so that the pore liquid is finally free of force. The velocity of this process is reflected in the decrease in the external force measured. Accordingly, the effective pore size can be derived from these data, if the sample geometry is well defined and the viscosity of the pore liquid and the deflection of the rod are known. The residual equilibrium force and the deflection of the gel rod finally provide the modulus of the gel backbone.

The mechanical properties of the aerogel can be quantified by measuring the time needed for an ultrasonic pulse to travel a defined path through the sample. Using the density ρ of the aerogel, Young's modulus E can be calculated from the sound velocity v via $E = v.\rho^2$. The sound velocity reflects the properties of the gel backbone and in particular its connectivity at a given porosity. It has to be noted, however, that at low sound velocities, the impact of the gas in the pores is no longer negligible and will result in a density-independent sound velocity.

The backbone connectivity can also be investigated via the thermal or the electrical conductivity of the backbone. Both sound velocity and electrical conductivity measurements (of conductive aerogels) are fast methods of extracting additional information beyond the macroscopic density. The gas pressure dependence of the thermal conductivity at pressures 0.1 MPa (1 bar) can be exploited to extract average pore sizes for pores 100 nm [71].

1.2.7.4 Properties of Aerogel

Aerogels are a promising material for a host of applications [8, 19] due to their thermal, optical and mechanical properties [8]. They are among the best thermal insulating solid materials known [6, 9, 17]. It is important to link aerogel properties to their complex internal microstructure, and to understand how such properties can be optimized for a given application [8].

After Monsanto closed down the aerogel plant in the 1970s, small companies like Airglass (Sweden) took over the production of aerogels on a small scale, mainly serving Cerenkov detector applications and research projects on window glazing's. Nanopore Inc. is another company that has been continuously active in the field of aerogels over the past 20 years. Two "breakthroughs" in the 1990s triggered serious interest in aerogels from further companies: the discovery of carbon aerogels by Pekala and Kong and the patent for a new process to synthesize granular silica aerogels in 1997 by Schwertfeger and Frank. Though the potential applications of aerogels are numerous (Table 1.4) and cover almost all technical fields, it has only been in recent years that commercialization has become very active. The current worldwide promotion of nanotechnology is supporting this trend. Examples of commercially available products containing aerogel are listed in Table 1.5.

Thermal insulation and shock absorber applications significantly benefited from the development of silica aerogels with respect to space applications, for example, the thermal insulation of the Pathfinder on its mission to Mars or capturing micrometeorites in the STARDUST mission (Fig. 1.24). Still of scientific and commercial interest are double glazing systems filled with transparent silica aerogel tiles.

Current aerogel activities focus on particular catalyst supports based on carbon or other types of aerogels, molds for metal casting, preparation of aerogel particles and aerogel-comprising composites for electrodes or other functional materials. Increasingly, aerogel films and coating are synthesized and investigated to optimize them as dielectric or sensor components or to provide additional functionalities in layered systems. Recently, aerogels are also discussed in the context of hydrogen storage either as the matrix for the actual storage material or as the storage component itself.

TABLE 1.4 Summery of Aerogel Application

Thermal insulations	Day lighting components	Mechanical properties and Sensors
• Aspen Aerogels, Inc.: Technical insulation.	• Cabot Corporation together with Wasco Products, Inc., and Kalwall	• Dunlop: Tennis racquet with aerogel composite
• Corpo Nove and Aspen Aerogels, Inc.: Jacket for severe weather conditions.	• Mechanical properties	• General Dynamics
	• Dunlop: Tennis racquet with aerogel composite	
• Nanopore Inc.: Shipping containers.	• Sensors	
• Cabot Corporation together with Corus: HPHT pipe in pipe systems.	• General Dynamics	
• Toasty Feet: Shoe insoles.		
Catalyst support:		
• Aerogel Composite, LLC: Carbon aerogel-based electrocatalyst for fuel cells	• Taasi aerogels: Air filters.	• CDT Systems, Inc.
	• CDT Systems: Water purification.	• Cooper Power Systems: Supercapacitors.

1.2.8 CHARACTERIZATION OF AEROGELS USING MOLECULAR DYNAMICS SIMULATION

Aerogels are known to have mass and surface distributions consistent with fractal behavior over certain length scales. At sufficiently small (atomic) length scales their structure is determined by chemical considerations and is therefore not fractal, and at sufficiently large length scales they are homogeneous [31]. The evolution of this structure and the corresponding effects of relaxation, chemical kinetics, and atomic coordination have been of particular interest in modeling and simulation studies. Computer simulations and experiments are widely used for the studies of the structure and properties of aerogels and xerogels.

TABLE 1.5   Commercially Available Products

Permeation or adsorption	Filter
	Gas separation
	Waste Water Treatment
	Chromatography
Mechanical or Acoustical	Shock Absorber
	Acoustic Impedance Matching
Optical	IR-reflector or –absorber
	Light Management
Electrical	Porous Electrodes
	Dielectric Layers
Carrier, Support or Matrix	For catalysts
	Drug Release
	Bio/ Medical Components
	Sensors
	Explosives
Thermal	Insulation Management

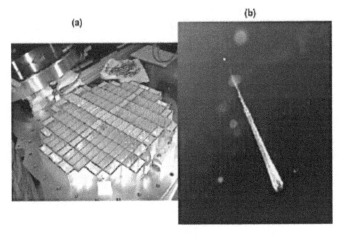

FIGURE 1.24   (a) Aerogel dust collector under construction. (b) Particle captured in aerogel.

1.2.8.1   Modeling of Aerogel Through Different Techniques

In a computational model, the material structure is known exactly and completely, and so structure/property relationships can be determined and understood directly [1]. the interatomic potential is the key component of any MD simulation. The model specifies the actual "fundamental" objects simulated, be they electrons, atoms, molecules, or sol particles, and their interactions, expressed through some kind of potential energy function [1]. It represents the most important interactions among atoms, that is, bonding interactions, interactions with nonbonding neighbors, and the extension of those interactions. If the interatomic potential does not accurately describe the interatomic interactions, the simulation results will not be representative of the actual material [73].

Having prepared a model structure, simulation studies have generally focused on structural characterization and the relationship between aerogel structure and mechanical properties. Global measures including fractal dimensions, surface areas, and pore size distributions can be directly calculated from the model structure and compared with experimental data. Microscopic measurements such as the distributions of bond lengths and bond angles and the number of bridging oxygen's bound to each silicon atom are also measured. Finally mechanical properties, including moduli, shrinkage upon drying, and the vibrational density of states can be determined and correlated with the gel structure [8]. Other topics that have been the focus of modeling studies include the prediction and analysis of gelation kinetics and aerogel mechanical properties, for which a considerable amount of experimental data is available [61–63].

The most challenging part of such a computational study is obtaining the model structure itself; Techniques applied in the case of aerogels include both "mimetic" simulations, in which the experimental preparation of an aerogel is imitated using dynamical simulations, and reconstructions, in which available experimental data is used to generate a statistically representative structure.

Simulating the formation and properties of aerogels, like all other applications of molecular simulation, involves choosing both a model and a simulation technique. The model specifies the actual "fundamental" objects simulated, be they electrons, atoms, molecules, or sol particles, and their interactions, expressed through some kind of potential energy function. Models discussed below include both atomistic descriptions, in which each atom is treated individually, and coarse-grained descriptions, which treat larger objects. At the atomic scale, there are two classes of potential used. In quantum-mechanical potentials, calculation of the energy of a configuration of atoms is accomplished by determining the electronic wave function (or density, in the case of density functional theory [24] and associated energy [25, 26]. In empirical potentials, also known as force fields, the energy is built up as a sum over different types of interactions: core-repulsions, bond stretches and bends, torsions, coulombic interactions, hydrogen bonds, dispersion forces, etc., each of which is described using a relatively simple function that has been parameterized against either quantum-mechanical results or experimental data [27–33]. [h].

Once a model has been specified, different calculations may be performed. Of particular relevance here are dynamical calculations, which generate a trajectory according to specified equations of motion. In the simplest case, Newton's equations are used. The trajectory thus generated conserves total energy and, under equilibrium conditions, samples the microcanonical ensemble. The equations of motion can be modified to enforce constant-temperature and/or constant-pressure conditions. These are collectively referred to as molecular dynamics simulations. Stochastic dynamics, based on the Langevin equation (or derived results), are used in order to avoid the simulation of solvent molecules [27, 33]. In these techniques, friction terms and random impulses are used to model the interaction of solute molecules with the solvent; in dilute solutions, this can reduce the number of objects simulated by several orders of magnitude. In the particular case of sol–gel processing in aqueous media, the use of such an "implicit" solvent is problematic because water itself is both a product of and catalyst for the siloxane condensation reaction. Other, "nonsimulation" operations include energy minimization and transition-state location, and are primarily used in quantum-mechanical studies of chemical reactions. [1].

Generally, the model structures (Fig. 1.25) have been described as below:

1.   Atomistic: atom is treated individually.
   a)   Quantum-mechanical potential: calculation of the energy of a configuration of atoms is accomplished by determining the

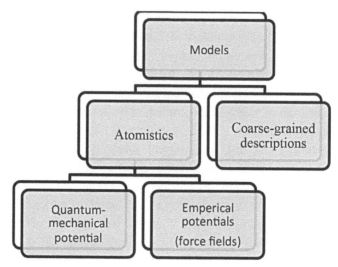

FIGURE 1.25 Modeling structure.

electronic wave function (or density, in the case of density functional theory and associated energy.

b) the energy is built up as a sum over different types of interactions: core-repulsions, bond stretches and bends, torsions, columbic interactions, hydrogen bonds, dispersion forces, etc., each of which is described using a relatively simple function that has been parameterized against either quantum-mechanical results or experimental data.

2. Coarse-grained descriptions: treat larger objects.

Atomistic modeling with empirical potentials is much less expensive and also scales better with increasing system size. Such calculations are therefore capable of accessing larger systems, typically as many as 105 atoms, and occasionally much larger. Likewise, they can be extended to longer times; simulations of more than a microsecond have been accomplished using commonly available computers, although most force-field-based molecular dynamics simulations to date are shorter than 10 ns or so. However, in sol-gel processing due to the making and breaking of the chemical bonds biochemical simulation are unable to describe such processes. Hence, many empirical modeling required the development and validation of new potentials [1].

To access length scales larger than tens of nanometers and time scales longer than nanoseconds, simulators turn to coarse-grained models. In such simulations the "primary" particle simulated is no longer a single atom or small molecule, but a greatly simplified representation of an assembly of many atoms or molecules. These objects interact with each other through effective potentials obtained from either theoretical consideration, matching to higher-resolution atomistic simulations, or fitting against experimental data. Such models can be extended to much larger length and time scales, for several reasons. Reduction in the number of degrees of freedom in the simulation greatly reduces the computational effort required for energy and force calculations. Furthermore, because these particles are heavy (compared with atoms or small molecules) and interact through somewhat softer potentials, the equations of motion can be integrated with much larger time steps. Although many hybrids of these basic approaches are also known, the physical and chemical processes underlying aerogel preparation and properties are thought to be relatively friendly to the separation of scales and methods just described, and studies to date have exploited this. The aqueous hydrolysis/condensation chemistry underlying silicate precursor oligomerization has been studied using quantum mechanical methods. Formation of larger oligomers and complete gelation of dense aerogels has been studied with atomistic simulations of empirical force-field models. Since such simulations cannot access the time or length scales associated with the gelation of low-concentration precursors, a variety of coarse-grained models have been proposed for these processes. In this section we review these studies and discuss both the methods and models used and the conclusions reached.

An alternative approach to modeling complex porous materials is reconstruction, in which experimental data are inverted to generate a computational realization of the material structure. There is insufficient data in any experimental characterization short of high resolution three-dimensional tomography to exactly reconstruct an amorphous material. Indeed, most experiments provide either a one-dimensional dataset, such as a structure factor or adsorption isotherm, or a two-dimensional dataset, such as an electron microscope image. Stochastic reconstruction generates representative structures that fit such limited input data but are otherwise random and isotropic. There are in general many possible model structures that will fit the input data equally well, which is known as the "problem of uniqueness." Although we do not review these techniques or their applications in detail, we do note that they have been applied to sol–gel materials in a number of recent studies. Quintanilla et al., applied the method of Gaussian random fields to generate stochastic reconstructions of TEOS-derived aerogels. Eschricht et al. used evolutionary optimization to reconstruct the commercially available "GelSil 200" material from a combination of small-angle neutron scattering data (SANS) and a pore-size distribution derived from gas adsorption data. Steriotis et al. used SANS data to guide a "process-based" atomic-scale reconstruction of a silica xerogel [1].

Finally, an entirely different approach to simulating gelation is the "Dynamic Monte Carlo" (DMC) method, in which chemical reactions are modeled by stochastic integration of phenomenological kinetic rate laws. This has been used successfully to understand the onset of gel formation, first-shell substitution effects, and the influence of cyclization in silicon alkoxide systems. However, this approach has not so far been extended to include the instantaneous positions and diffusion of each oligomer, which would be necessary in order for the calculation to generate an actual model of an aerogel that could be used in subsequent simulations [1].

These different computational and simulation techniques have their own advantages and disadvantages. For instance, although fully quantum-mechanical calculations can be very accurate, they can be applied only to a few molecules, and when used in dynamical calculations they can access only picosecond time-scales. As the system size increases the cost rises steeply. Hence, they impose high computational cost of calculating the total energy of an assembly of atoms using either wave function-based or density-functional techniques. This restricts such models only to small systems [1].

1.2.8.2   Simulation of a Silica Aerogel Through Sol-Gel Processing

Mostly simulation of an aerogel follows this procedure:
1. Introduction of the MD simulation procedures used to prepare the aerogel and xerogel samples. Once a model has been specified, different calculations may be performed. Of particular relevance here are dynamical calculations, which generate a trajectory according

to specified equations of motion. In the simplest case, Newton's equations are used.

2. Description of methods for obtaining structural characterization and the relationship between aerogel structure and the fractal dimension and the thermal and mechanical properties of the samples.

3. Discussion about the characteristics of the samples including: fractal dimensions, surface areas, pore size distributions, thermal properties, elastic modulus, and strength in relation to density and comparing the results with experimental data. Moreover, microscopic measurements such as the distributions of bond lengths and bond angles and the number of bridging oxygen's bound to each silicon atom are also measured. Finally mechanical properties, including moduli, shrinkage upon drying, and the vibrational density of states can be determined and correlated with the gel structure [1, 73].

Simulating the preparation of xerogels and aerogels involves separate treatment of gelation, aging, drying, and for nonporous materials, consolidation. Consolidation, or heating at high temperatures, is used to generate densified, nonporous materials for optics and other applications. The aqueous hydrolysis/condensation chemistry underlying silicate precursor oligomerization and reaction mechanisms and energetics has been studied using quantum mechanical methods. Formation of larger oligomers and complete gelation of dense aerogels has been studied with atomistic simulations of empirical force-field models. Since such simulations cannot access the time or length scales associated with the gelation of low-concentration precursors, a variety of coarse-grained models have been proposed for these processes [1].

1.2.8.2.1 Gelation

As we explained before, in the gelatin step, alkoxide gel precursors in aqueous solution are hydrolyzed and polymerize through alcohol or water producing condensations:

$$\equiv Si - OR + H_2O \leftrightarrow \equiv Si - OH + ROH$$
$$\equiv Si - OR + OH - Si \equiv \leftrightarrow \equiv Si - O - Si \equiv + ROH$$
$$\equiv Si - OH + OH - Si \equiv \leftrightarrow \equiv Si - O - Si \equiv + H_2O$$

The gel morphology is influenced by temperature, the concentrations of each species (attention focuses on r, the water/alkoxide molar ratio, typically between 1 and 50), and especially acidity. An entirely different approach to simulating gelation is the "Dynamic Monte Carlo" (DMC) method, in which chemical reactions are modeled by stochastic integration of phenomenological kinetic rate laws [15].

1.2.8.2.2   Aging

Gel aging is an extension of the gelation step in which the gel network is reinforced through further polymerization, possibly at different temperature and solvent conditions. Simulating aging requires the use of an approach which can access long time scales. The "activation-relaxation technique" (ART) is being implemented for this purpose.

1.2.8.2.3   Drying

The gel drying process consists of removal of water from the gel system, with simultaneous collapse of the gel structure, under conditions of constant temperature, pressure, and humidity. In the coarse-grained model the equation of state is trivially calculable, and drying is easily modeled by choosing the solvent chemical potential to favor the vapor phase and allowing the particle positions and cell volume to slowly relax under the influence of solvent capillary forces. At the molecular scale, we can model this process using an extension of the "Gibbs Ensemble Monte Carlo" technique for binary mixtures, where the mixture consists of water and atmosphere.

1.2.8.2.4   Consolidation

Xerogels are higher in free energy than conventional amorphous silica (glass) and crystalline silica, as they have a substantial internal surface area and associated surface tension. Simulations of consolidation will use molecular models. Both isothermal conditions and constant heating rates can be accessed with standard molecular dynamics simulations and ART as above.

1.2.8.2.5 Aerogels

Aerogel systems do not collapse (much) under drying conditions and supercritical gel drying will be easier to simulate than the subcritical process. High-porosity aerogels are only accessible via the meso-scale model. The initial stages of gelation, when the average cluster size is very small, are best modeled with a purely atomistic approach. Hierarchically structured gels and low-density gels cannot be directly treated with molecular models; a meso-scale approach must be used in this case. Relatively dense gels can be modeled with either atomistic simulations or coarse-grained simulations [25, 27, 91].

1.2.8.3 Interatomic Potential

Atomistic modeling and simulation have been used both in studies of the hydrolysis/condensation chemistry underlying silica sol–gel processing, and in dynamical simulations of the formation of sol particles and even gels. Studies of reaction mechanisms and energetics have made extensive use of quantum mechanical methods, while the larger length scales and time scales of oligomerization and gelation have been handled using empirical potentials, as reviewed in this section.

Vashishta et al. [88, 89] developed an interatomic potential for amorphous silica. This potential involved terms representing the interaction between two atoms (two-body component); which, accounts for the potential energy due to the distance between them. The potential also includes the potential energy due to the change of orientation and bonding angle of triplets of atoms (three-body component). The two body and three-body components of the potential are given by

$$V_{ij}^{(2)} = \frac{H_{ij}}{r_{ij}^{\eta_{ij}}} + \frac{z_i z_j}{r_{ij}} \exp\left(\frac{-r_{ij}}{r_{1s}}\right) - \frac{P_{ij}}{r_{ij}^A} \exp\left(\frac{-r_{ij}}{r_{4s}}\right) \qquad (66)$$

$$V_{jik}^{(3)} = B_{jik} f\left(r_{ij} r_{ik}\right) \left(\left(\cos\theta_{jik} - \cos\theta_{\widetilde{jik}}\right)\right)^2 \qquad (67)$$

In the previous equations r_{ij} is the distance between the atoms i and j. The first term of Eq. (66) represents the steric repulsion due to the atomic size.

The parameters H_{ij} and η_{ij} are the strength and exponent of steric repulsion. The second term corresponds to the Coulomb interactions between the atoms and accounts for the electric charge transfer, where Z^i is the effective charge of the i_{th} ion. The third term includes the charge–dipole interactions. It takes into account the electric polarizability of the atoms through the variable P_{ij}, which is given by

$$P_{ij} = \frac{1}{2}\left(\alpha_i Z_j^2 + \alpha_j Z_i^2\right) \qquad (68)$$

where α_i is the electric polarizability of the ith ion. The parameters $r_{1\,s}$ and $r_{4\,s}$ are cut-off values for the interactions.

In Eq. (67), B_{jik} is the strength of the three-body interaction; θ_{jik} is the angle between the vector position of the atoms, that is, r_{ij} and r_{ik}. The function f represents the effect of bond stretching and the component containing $(\cos\theta_{jik})$ takes into account the bending of the bonds, and $\tilde{\theta}_{jik}$ is a reference angle for the respective interaction. Although, there are six possible three-body interactions in the system, this potential only considers the most dominant ones, which are related to (Si–O–Si) and (O–Si–O) angles. More information about the potential, including the values of the parameters, is available in Refs. [88, 89].

The first molecular dynamics studies of oligomerization and the silica sol–gel process were carried out by Garofalini and Melman [20]. A solution consisting of water, silicic acid monomers, and silicic acid dimers was simulated. The potential consisted of a modified Born–Mayer–Huggins (BMH) model for the Si–Si, Si–O, and O–O interactions in nonwater molecules, and a modified Rahman–Stillinger–Lemberg (RSL2) potential [31] for O–O, O–H, and H–H interactions in water molecules.

In 1990, Feuston and Garofalini revised the model in a subsequent study [26]. The modified BMH potential was applied to interactions between all atoms, with H–X interactions supplemented by terms from the RSL2 potential [80].A subsequent study by Garofalini and Martin extended this to larger systems (216 silicic acid monomers) at densities between 1.4 and 1.6 g/cm³ [87].

Martin and Garofalini have also further studied the details of the condensation reaction in this model [56], observing that when there are multiple bridging oxygen in the intermediate, there is a tendency to break a silanol bond in creating a leaving group, rather than lose a nonbridging oxygen. This is consistent with the experimental finding that acid-catalyzed conditions favor the formation of linear clusters [14].

The systems described in all these studies were quite small ($N \leq 216$) and of relatively high density (1.3 g/cm³ and above). The temperature and pressure conditions used are comparable with an autoclave process used to produce aerogels. These are not typical laboratory conditions for the preparation of sol–gel derived materials. [14, 56].

Rao and Gelb used the Feuston and Garofalini model in simulations closer to these conditions [28]. In this work, 729 silicic acid monomers were placed in water at liquid density with r between 0 and 26, with most work performed at r 11. Simulation times were as long as 12 ns. High temperatures (1,500–2,500 K) were again used to promote reaction, and the system volume held fixed. A series of snapshots from one of these simulations is shown in Fig. 1.26. These studies confirmed the action of water as a catalyst, and further illustrated the processes involved in the conversion of monomer precursors to "sol" particles. At short times, monomers quickly dimerized, and further growth occurring via monomer addition. At longer times (after several nanoseconds), condensation between larger oligomers was also observed. The model developed by Feuston and Garofalini, while successfully encompassing much of the important chemistry involved in sol–gel processing, could nonetheless be improved in many ways.

Bhattacharya and Kieffer introduced a more realistic all-atom reactive "charge-transfer" potential for simulating silica sol–gel processing [10]. In this approach, a certain amount of charge is transferred between bonded atoms; this allows for more realistic electrostatic interactions between molecules, and correctly charged ions upon dissociation.

MD simulations are used in Kieffer and Angell and Nakano et al. [49, 64, and 65] to generate surrogate (computer models) of silica aerogels. Kieffer and Angell [19] propose to use MD to model silica aerogel and xerogel starting from a sample of silica glass that is gradually expanded. The expansion causes breaking of the Si–O bonds and leads to the formation of a fractal structure characterized by a fractal dimension that changes linearly with density. Silica was modeled using the Born–Mayer empirical potential, and up to 1,500 atoms were included in the simulations. The quantity of greatest interest However the linearity is not sustained for samples with densities higher than 1 g/cm³.

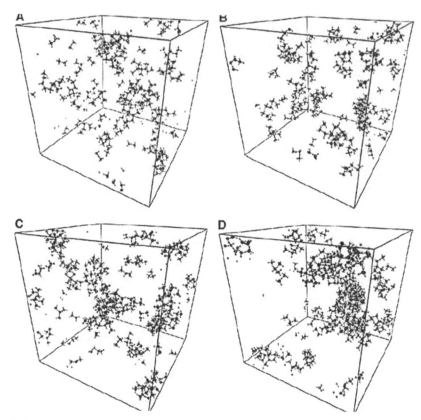

FIGURE 1.26 Snapshots from molecular dynamics simulations of oligomerization in aqueous silicic acid with r=11. Snapshots are taken from the simulation at times. (A) 1.0, (B) 2.0, (C) 3.0, and (D) 4.8 ns. The evolution of small oligomers into compact sol particles is clearly visible.

Very similar simulation protocols were used in subsequent work by Nakano et al. [24]. In these studies, however, simulation cells containing over 40,000 atoms were used, such that much larger length scales could be probed. An example of these large-system results is shown in Fig. 1.27.

Comparing the results of studies, fractal dimensions found in the (low-temperature) expansion simulations of Nakano et al. [24] and Bhattacharya and Kieffer [10,49] are not in perfect agreement, again suggesting that either or both of the potential and the simulation protocol can also affect the final network morphology.

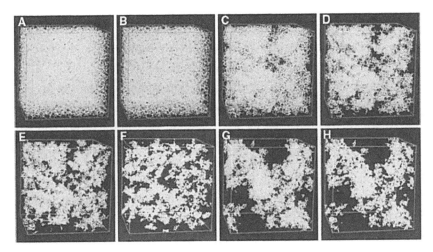

FIGURE 1.27 Snapshots of molecular dynamics of porous SiO2 glasses (A) 2.2, (B) 1.6, (C) 0.8, (D) 0.4, (E) 0.2, (F) 0.1 g/cm3 prepared at 300 K, (G) 0.2 (H) 0.1 g/cm^3 prepared at 1,000 (K) Yellow lines represent Si–O bonds.

Since the atoms are initially set at distances larger than their equilibrium distances and have high kinetic energy, they can diffuse to farther regions in the simulation volume and could potentially form any kind of structure. That is an advantage of this procedure over the gradual expansion, where the atoms can only move short distances and are bounded to remain in the same region [73].

Muralidharan et al. [61] summarizes the elastic modulus and the strength that can be obtained using four traditional potentials to model dense SiO_2 systems, namely Soules potential (S-potential), Born–Meyer–Huggins potential (BMH), Feuston–Garofalini potential (FG) and van Beest–Kramer-van Santeen potential (BKS). The minimum deviation between the simulated value and the experimental one is 39% obtained using BKS-potential.

Kieffer and Angell [18] use an interaction potential of the Born–Mayer form, which accounts for two body interactions, to model a system containing between 300 and 1500 particles. Nakano et al. [64] use a potential including two-body and three body interaction components to model porous silica with densities varying from 2.2 g/cm^3 to 0.2 g/cm^3 in a system composed of 41,472 particles. Similar to Kieffer and Angell [19], in studies of Nakano et al. [64] the system was gradually expanded until the desired density is reached and the dependence of the fractal dimension, internal surface area, the pore to volume ratio, pore size distribution, correlation length, and mean particle size on the density of the sample were investigated.

Campbell et al. [19], using the same potential as Nakano et al. [64], generate porous samples by placing spherical clusters of dense silica glass in a large volume and sintering the system at constant pressure and temperature. With that approach they generate samples of densities varying from 1.67 g/cm^3 to 2.2 g/cm^3, a range of density corresponding to xerogels, and study the changes on the short-range and intermediate range order of the structure with density. They also study the effect of densification on the elastic modulus of the samples, finding a power law relation between modulus and density with an exponent of 3.5 ± 0.2.

Murillo et al. used the same potential as Nakano et al. [64], however, the preparation of the samples was through different approach. The procedure to generate the porous samples analyzed in this study follows the steps listed below:

1. Placing the atoms at the crystalline sites of β-cristobalite with a lattice constant corresponding to the desired density.
2. Heating up of the system to 3000 K.
3. Cooling down of the system allowing relaxation at several temperatures.

The volume of a dense crystalline sample (β-cristobalite) is expanded in one step to the desired density. Then the temperature is increased to give the atoms enough energy to diffuse through the system, and finally it is cooled down in a stepwise process. Keeping the system at high temperature, along with the cooling scheme, eliminates the effect of the initial positions of the atoms. This process resembles a diffusion limited aggregation. The goal of this process is to create a uniform distribution of atoms across the volume. The interatomic potential was the same as the studies of Vashishta et al. This potential can accurately reproduced the structural parameters for dense silica glass. Moreover, comparing the experimental results of neutron scattering with computer simulations, this potential produces a smallest error of only 4.4%, using the factor R_x [49]. This potential also calculated the elasticity of silica glass and the results were in good agreement with experimental value reported by Muralidharan et al. [62].

Murillo et al. used an in-house Fortran program for the investigation of structural and mechanical properties of silicon aerogel and xerogel. In this code the Velocity–Verlet algorithm is used to solve the equations of motion of the particles, using a time step of 0.5 fs. The selection of a small time step ensures that the atoms do not fly off, which is important in this case because large free surfaces are generated inside the samples. Langevin dynamics is implemented and all the particles in the system are used to control its temperature accordingly [73].

1.2.8.4 Simulation Results

The results of simulation in Murillo et al. show that, increasing the temperature of the system at initial stages of the preparation process lead to an increase in the kinetic energies of the atoms. By reducing the temperature of the system, however the atoms are still able to move in the system, their numbers increases per group. At low temperature the atoms find equilibrium positions and the structure is locked. Finally the simulated samples have branched structures formed by groups of atoms interconnected by small bridges. Figure 1.28 represent the formation of clusters interconnected by small chains [73].

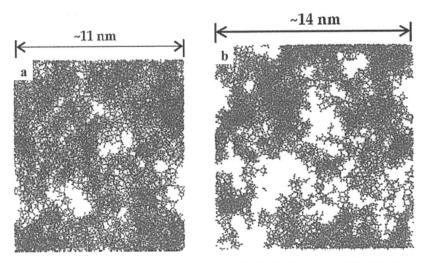

FIGURE 1.28   Simulated samples: (a) density of 0.58 g/cm³, porosity of 74%, b) density of 0.28 g/cm³, porosity of 87%.

Yeo et al. [96] simulations are performed on the LAMMPS [67] software to attain a closer fit of the thermal conductivity. The interaction potential used is the Tersoff potential [83], reparameterized to model interactions between silicon and oxygen [83]. Comparing the BKS potential and Tersoff potential, it is found that the Tersoff potential is more suitable for thermal conductivity studies, while, the BKS potential augmented with a "24–6" Lennard Jones potential can prevent uncontrollable dynamics at very high temperatures [89]. After producing aerogels in different density, RNEMD was used to determine the thermal conductivity at each densities [96].

Coarse-grained computer models of silica aerogels (and other gels) may be roughly divided into two categories, depending on the simulation algorithm used and type of interactions included. In hard-sphere aggregation models, which have been extensively studied, aggregates are formed out of simple particles according to one of several procedures. Nonbonded particles interact as "billiard balls," without soft attractive or repulsive forces, while bonded particles are held rigidly together at the point of contact. In flexible models, equations of motion similar to those used in atomistic simulations are applied to objects with rather more complex interactions, including soft nonbonded interactions and deformable and/ or breakable bonds [1].

1.2.8.4.1 Calculation of the Pair Distribution Function (PDF)

After the generation of the samples using MD, the fractal dimension can be obtained based on its relation to the pair distribution function (PDF) of the samples. The PDF represents the probability of finding a pair of atoms separated by a distance r in the structure, relative to the probability expected for a randomly distributed structure having the same density [*]. The PDF is calculated as:

$$g_{\alpha\beta}(r_1 r_2) = \frac{V^2}{N_\alpha N_\beta} \langle \sum_{i\in\{\alpha\}}^{N_\alpha} \sum_{i\in\{\beta\}}^{N_\beta} \delta(r_1 - r_i)\delta(r_2 - r_j) \rangle$$

$$= \rho_\alpha^{-1}\rho_\beta^{-1} \langle \sum_{i\in\{\alpha\}}^{N_\alpha} \sum_{i\in\{\beta\}}^{N_\beta} \delta(r_1 - r_i)\delta(r_2 - r_j) \rangle \tag{69}$$

where V is the volume of the system, N_α and N_β are the numbers of particles of the entities of type α and β, respectively, ρ_α and ρ_β are the corresponding densities of α and β subsystems, and the symbol $\langle\rangle$ means ensemble average. Therefore, several configurations of the system should be used to compute the PDF. The term inside the $\langle\rangle$ symbol indicates that when finding the distribution of distances between the atoms of type α and β one must locate each atom of type α and obtain the distance from it to each one of the atoms of type β. Figure 1.29 is an example of the PDF for one of the systems created. The decay of the PDF for large values of the distance between the atoms is shown in Fig. 1.29. That decay can be used to estimate the fractal dimension, d_f, of porous systems [49]

$$d_f = 3 + \frac{dLog(g(r))}{dLog(r)} \tag{70}$$

Fractal dimension of a structure can also be obtained through a simulated scattering experiment [19, 23]. To simulate a scattering experiment one needs to calculate the scattering intensity, I, corresponding to different wavelengths of radiation shined to the sample, represented by their wave number, q. An expression to calculate I, when the positions of all the particles in the system are known, is given by Refs. [23, 26]:

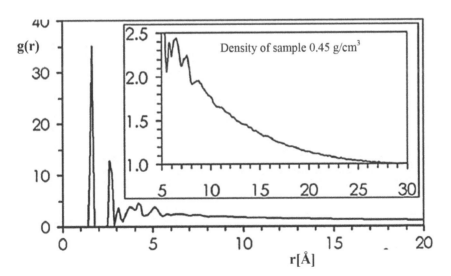

FIGURE 1.29   Pair distribution function.

$$\frac{I(q)}{I_0} = \Sigma_{ij} \frac{\sin(qr_{ij})}{qr_{ij}} \tag{71}$$

where I is the scattering intensity, I_0 is a reference value of intensity, q is the wave number and r_{ij} is the distance between two atoms i and j. For fractal structures, the scattering intensity can be related to the wave number by a power law, where the exponent is the negative of the fractal dimension of the structure, that is, $I \propto q^{-df}$ [62, 67]. An example of the I(q) is plotted in the Fig. 1.30. For large wave numbers the scattering intensity corresponds to the individual particles and remains constant; while for small wave numbers, clusters of particles are responsible for the scattering. The limiting value of wave number in the scattering is determined by the size of the sample modeled [73].

Murillo et al. [1] found that in the short-range order, features under 5 Å, the computational samples show characteristics similar to those of silica samples and the bonding distance (Si–O atoms), the nearest neighbors distances O to O atoms and Si to Si atoms are in good agreement with experimental results (Table 1.6). At higher ranges between 9 and 25 Å, approximately, the simulated samples exhibit a fractal behavior, which agrees with the fractal range found in simulations done by Nakano et al. [24], 5–25 Å. In aerogel samples studied by Vacher et al. the fractal range of the samples was identified extending from features as small as 4 Å to

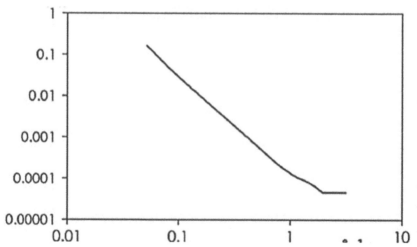

FIGURE 1.30 Log–Log plot of the scattering intensity vs. wavenumber for a sample of density of 0.45 g/cm³.

TABLE 1.6   The Results of Pair Distribution Function of the Samples

| Bonding | Ref | Distance (Å) | | |
		Simulation	Ref	Experimental
Si–O atoms	[1]	1.63 ±0.03	[2]	1.61±0.05
O to O atoms		2.65±0.03		2.632±0.089
Si to Si atoms		3.08±0.03		2.632±0.089

TABLE 1.7   Fractal Range Found Through Simulation and Reported by Different Literatures

References	Fractal Ranges (Å)
Murillo et al. [1]	Under 5
	9 and 25
Nakano et al. [9]	5–25
Vacher et al. [5]	4–200
Courtens and Vacher [20]	10 to 1000

features larger than 200 Å, depending on the density of the sample. Courtens and Vacher [*] report that the fractal range of silica aerogels extends from 10–1000 Å, approximately (Table 1.7).

Murillo et al. [73] report that the fractal dimension increases with the density toward a limiting value of 3, corresponding to bulk silica. Their results are shown in Fig. 1.31. For larger densities the scattering experiments lead to fractal dimensions slightly smaller than those obtained using the decay of the PDF. For lighter samples the opposite occurs [73].

Yeo et al. [96], simulated silica aerogel on a cubic system of 52,728 atoms with densities in ranges of 0.3 to 1 g/cm^3. Determination of fractal dimension was determined in accordance to the following studies [94, 95]. In this method the total radial distributions for each density are calculated, and power-law decays are superimposed on the peak structures to determine the fractal dimensions. These results were in accordance with previous theoretical studies by Murillo et al. [73].

1.2.8.4.2   Thermal Properties

Experimental studies have shown that, thermal conductivity and transport mechanisms are scaled with density via a power law:

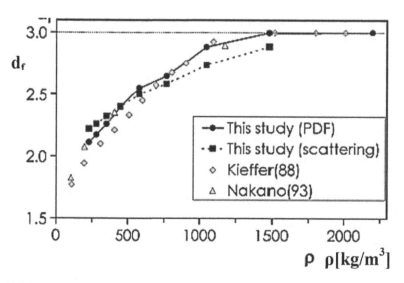

FIGURE 1.31   Fractal dimension of silica aerogels samples with different densities.

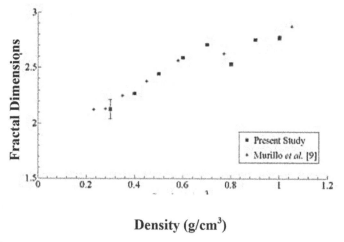

Density (g/cm³)

FIGURE 1.32   Decreasing fractal dimensions as density decreased.

$$\lambda_s = C\rho^\alpha \tag{72}$$

where α was approximately 1.6 for densities between 0.3 to 1.0 g/cm³ [24].

Ng et al. [42] employed negative pressure rupturing with the van Beest, Kramer and van Santen (BKS) potential [64, 65], and determined their thermal conductivities. It was found that the power-law fit of the data corresponds to experimental bulk sintered aerogel. Yeo et al. [96] studied the solid thermal conductivity of silica aerogels using reverse nonequilibrium MD (RNEMD) simulations, following the method Murillo et al. [73].

In Yeo et al. [96] studies amorphous silica samples was generated by quenching β-cristobalite, from 5000 K to 300 K. For validation of MD methods the thermal conductivities of increasing length of amorphous silica was compared with experimental results. They concluded that the Tersoff potential can give a much better estimation of bulk thermal properties than BKS potential. This has been shown in Fig. 1.33, that the Tersoff potential shows an almost linear dependence with increasing length scales while BKS potential overshoots the thermal conductivity of bulk amorphous silica [40] In order to further analysis, each potential (BKS and Tersoff) were examined with vibrational density of states (vDOS) which are obtained through the discrete Fourier transform of the velocity autocorrelation function (VACF). In Tersoff potential the peaks were clearly close to those obtained from experimental and theoretical results. However, the BKS potential showed no apparent peaks. It was concluded that the reparameterized Tersoff

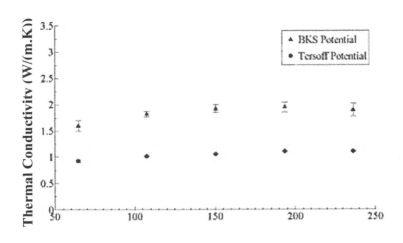

Length (Angstrom)

FIGURE 1.33 Amorphous silica of various lengths and their thermal conductivities.

potential is a far superior alternative in the thermal characterization of bulk amorphous [96].

The solid thermal conductivity of silica aerogels can be determined using reverse nonequilibrium MD (RNEMD) simulations. Yeo et al., [96] generated porous samples according to the methods Murillo et al. used [29]. Thermal conductivities were found to increase nonlinearly as the density decreased. The power-law exponent, α, of experimental bulk aerogel (1.6) in the density range of 0.3 to 1.0 g/cm³was in accordance with α value obtained from RNEMD model (1.61). [24]. The advantage of this model is the accessibility of very low densities (lower than 0.1 g/cm³) without any adverse phenomena. However, the significant difference between thermal conductivities of simulated and experimental aerogels makes a limitation for this model. This mostly could be due to inability to attain micropores [96].

Experimentally, using small angle neutron scattering (SANS) and nuclear magnetic resonance (NMR), it has been observed that aerogels and xerogels are fractal structures characterized by a fractal dimension varying from 3 for very low porosity (almost dense glass) to 1.8 for very light specimens. The fractal dimension is slightly affected by the pH of the solution prior to gelation [23, 73, 87, 93].

Aerogel samples with densities smaller than 0.43 g/cm³ was prepared through SANS under acidic and neutral conditions have a fractal dimension around 2.4±0.03 [87]. Devreux et al. [22, 23] prepared aerogels with densities around 0.17 g/cm³ prepared under acidic conditions and a fractal dimension of 2.2. Woignier et al. [93, 94], have studied preparation of aerogel under acidic, basic and neutral phi They found that under basic conditions aerogels have fractal dimensions close to 1.8± 0.1, while under acidic or neutral conditions the samples had a fractal dimension around 2.2±0.1 and 2.4±0.1, respectively.

1.2.8.4.3   Mechanical Characterization

A tension test should be simulated by the stretching the sample along one direction at strain rates larger than the one used in laboratory tests. Because at higher strain rate, the speed of the atoms, and consequently the temperature of the sample, tend to increase. Moreover, the temperature is uncontrolled when the stretching speed is near the speed of sound in the material [63]. Murillo et al. used Langevin dynamics [8] to control the speed of all the atoms in the system. Langevin dynamics is based on the following equation,

$$m_i \ddot{r_i} = F_i - m_i \gamma_i \dot{r_i} + R_i \tag{73}$$

where m_i is the mass of the atom, are the position, velocity and acceleration of the particle, is the force exerted on the particle by all the other particles in the system, is a stochastic force applied to the particle, and γ_i is a damping coefficient [49].

Murillo et al. [73] showed that the relation between elastic modulus and density of samples can be described by power law. The obtained exponent for this relation is 3.11 ± 0.21. Figure 1.34 shows that a power law relation between strength and density exist. The main differences in predicted strength through simulation and experimental is due to the using other potentials for silica (Table 1.8). The calculated exponent for that relation was 2.53 ± 0.15 that compares well with previously published values. Yeo et al. investigate the Young's modulus with the strain rate 0.0005 ps^{-1} for

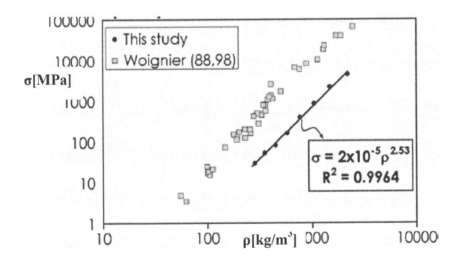

FIGURE 1.34 Strength vs. density for silica aerogels. Woignier [22, 24].

200 ps, the tension tests are carried out on samples of different densities. The exponent of the relationship between Young's modulus and density of Yeo et al. [96] studies was lower than the one that Murillo et al. has been mentioned. This is caused from different interactive potential. The comparison of reported values is summarized in Table 1.9.

The procedure Murillo et al. [73] used to generate the porous samples does not correspond to the events happening during the experimental gelation of a solution; nevertheless it produces samples having geometrical features that are similar to those of real aerogel and xerogel material.

1.3   SUMMARY

A wide variety of modeling techniques have been developed over the years, and those relevant for work at the molecular level include, in addition to MD, classical Monte Carlo, quantum based techniques and MC methods, and MD combined with discrete approaches such as lattice–Boltzmann method. MD simulation is a technique for computing the equilibrium and properties of a system and helps to better understanding the properties of

TABLE 1.8   Exponent for the Power Law Relation Between Elastic Modulus and Density

References	Exponent	Density [g/cm³]	Methods
Campbell et al. [20]	3.5±0.2	1.67–2.2	Computer simulation
Yeo et al. [A]	2.4313	0.3–1	Computer simulation
Murillo et al. [2]	3.11±0.21	0.23–2.2	Computer simulation
Groß et al. [10]	3.49±0.07	0.14–2.7	Exp.
Groß et al. [10]	2.97±0.05	0.08–1.2	Exp. sintered aerogels
Woignier et al. [8]	3.8±0.2	0.1–0.4 (approx. values)	Exp. pH neutral
Woignier et al. [7]	3.7±0.2	0.055–0.5 (approx. values)	Exp. pH neutral, acidic and basic
Woignier et al. [7]	3.2±0.2	0.42–2.2 (approx. values)	Exp. partially densified samples

TABLE 1.9   Potentials lead to values strength and elastic modulus of silica [61]

Potential	Strength (GPa)	Elastic Modulus (GPa)
BMH-potential	30 and 65	220
S-potential	24–35	220
BKS-potential	12 and 22	12 and 22
FG-potential	12 and 21	125

assemblies of molecules in terms of their structure and the microscopic interactions between them.

Recent work at all levels of description, from quantum mechanical to empirical atomistic modeling to coarse-grained modeling, has confirmed the broad picture developed by previous studies but also illustrated the many challenges remaining in developing a complete description and understanding of silica aerogels. MD simulation has proved effective for obtaining useful structural information on sol–gel process for silica systems. The latest first-principles calculations clearly indicate that the roles of water and counter ions in the underlying chemistry are more important and more complex than previously thought. Atomistic simulations have illustrated that structure at small scales is dramatically affected by degree of hydration and that the small clusters produced by short-time oligomerization have a complex distribution of size, shape, and structure. Recent attempt at more quantitative coarse-graining indicates that quite large system sizes are required to obtain reliable results for mechanical properties of even reasonably dense aerogels.

Although atomistic empirical (or even quantum-mechanical) simulations of increasingly large scale and length would seem to be a route forward, there is no practical way that these can access the actual time- and length-scales required. Near-complete gelation was observed in many atomistic simulations, but this was only possible due to the tiny simulation cells and high densities used; with more reasonable system sizes, gelation would not have occurred on a "simulable" time-scale. In addition, important parameters, including the pH and pKas, of many of the proposed atomistic models are unknown, which makes correlation with experiment difficult. The geometrical features of the modeled samples are characterized by the fractal dimension, determined from the pair distribution function of the samples as well as from simulated scattering experiments and are found to be in good agreement with computational and experimental data. Furthermore, the mechanical properties, namely the elastic modulus and strength, of the porous samples are found to scale with density following a power law, which is expected for fractal structures.

Results showed that direct expansion of crystalline samples of β-cristobalite to reach densities between 2.2 g/cm³ and 0.23 g/cm³, along with thermal processing, leads to fractal structures which allow to investigate the properties of silica aerogels and xerogels.MD and the reparameterized Tersoff potential were used to model the porous structures of silica aerogels demonstrate that this potential is suitable for modeling thermal properties in amorphous silica. Using a quenching and expanding process, different densities of aerogel samples could be generated. Through RNEMD, their thermal conductivity is determined and the power-law fit of our data corresponds well with experimental studies.

It appears clear that further quantitative improvements in computational aerogel models are most likely to result from a multiscale approach, incorporating all three levels of descriptions studied here. Only with such combinations will modelers be able to reach the necessary system sizes and simulation times while still connected to an atomistic description of the underlying chemistry. Empirical models could be further improved by parametrization against quantum-mechanical simulation results. Large-scale atomistic simulations could be used to generate chemically realistic distributions of sol particles, even if gelation cannot be directly observed. Finally, atomistic models of all types could be used as reference systems for developing better coarse-graining techniques. Multi-scale techniques are of considerable interest in the wider simulation community at this time, and developments in other areas may be expected to help further the goal of realistic simulation and modeling of aerogels.

APPENDIX 1: VELOCITY –VERLET ALGORITHMS

In this appendix, the case studies which integrated the equations of motion by means of Verlet algorithms have been mentioned. As mentioned in Section 1.2 studies of the structure and properties of aerogels and xerogels are done via experiments and computer simulations. Generally, after introducing the simulation procedure for the preparation of the aerogel and xerogel samples, methods for obtaining the physical properties of the samples are described. After which, the characteristics of the samples (fractal dimension, elastic modulus and strength, thermal properties, etc.) are discussed. Considerably, MD simulation using the velocity Verlet method is a powerful computational tool that can be used to simulate many physical phenomena. The sol–gel process is an important method for producing various commercial silica gels such as fibers, coating films, adsorbents, and catalyst supports. Typically, alkoxysiloxanes are used as the silica source. The polymerization processes of silica gel have been described in details in Section 1.2. Such polymerization process consists of the following three functional group reactions. Each reaction has a rate constants as, k_h, k_{cw}, and k_{ca} which are the rate constants of hydrolysis and water- and alcohol-producing condensations, respectively.

The optical and physical properties of product gels strongly depend on the conditions of reaction such as initial composition, temperature and pH through the three-dimensional network structures obtained. However, the details of the structural development in gel formation process are not yet well understood. For instances, Vasher et al. [87] have found through SANS that the fractal dimension is slightly affected by the pH of the solution prior to gelation. The aerogel samples prepared under acidic and neutral conditions have a fractal dimension around 2.4±0.03. A fractal dimension of 2.2 was found by Devreux et al. [23] for aerogels with densities around 0.17 g/cm^3 prepared under acidic conditions. Confirmation of those results is provided by Woignier et al. [93,94], who studied aerogels prepared under acidic, basic and neutral pH finding that under basic conditions aerogels have a fractal dimension close to 1.8± 0.1, while under acidic or neutral conditions the samples had a fractal dimension around 2.2±0.1 and 2.4±0.1, respectively.

Yamahara et al. investigated the microscopic features of structural development using MD simulation for two characteristic systems: silicic acid, $Si(OH)_4$, and monomethoxy silicic acid, $Si(OMe.)(OH)_3$. Hereafter, the former is called THS (tetrahydroxysilane) and the latter MMS. The two systems were regarded as differing from each other in the extent of hydrolysis. The simulation results were compared with the H NMR data reported by Kay and Assink, in terms of the Q_n species distribution as a function of SiOSi:Si ratio. Moreover, maximum cluster size, cluster type, ring size distribution, and network structure were examined to make clear the characteristic of polymerization process. The effect of the remaining methoxy groups on the structural development was also found from a detailed comparison of the simulation results for both systems.

The interatomic potential employed for this study contains a two-body term and a three-body term. The former term has a short-range repulsion and a screened Coulomb for all atom pairs, and a specific term for all pairs involving hydrogen atoms. The latter term confines each bond angle to the equilibrium value. The potential is given below:

$$U_{ij}(r_{ij}) = A_{ij} exp(-r_{ij}/\rho_{ij}) + Z_i Z_j e^2/r_{ij} erfc(r_{ij}/\beta_{ij}) \tag{74}$$

$$U_{jik}(r_{ij}, r_{ik}, \theta_{jik}) = \lambda_{jik} \exp\left[\frac{\gamma_{ij}}{(r_{ij}-r_{ij}^0)} + \frac{\gamma_{ik}}{(r_{ik}-r_{ik}^0)}\right] \times (\cos\theta_{jik} - \cos\theta_{jik}^0)^2 \tag{75}$$

$$if \ r_{ij} < r_{ij}^0 \ or \ r_{ik} < r_{ik}^0$$

$$U_{jik}(r_{ij}, r_{ik}, \theta_{jik}) = 0, \qquad if \ r_{ij} > r_{ij}^0 \ or \ r_{ik} > r_{ik}^0 \tag{76}$$

$$U_H = a_{ij}\{1 + \exp(b_{ij}(r_{ij} - c_{ij}))\}, for \ Si - H, O - H \ and \ H - H$$

where Z_i and Z_j are the formal ionic charges on atoms i and j, e the elementary electric charge, r_{ij} the interatomic distance, $\theta\xi_{ik}$ the angle between bonds ij and ik, and A, ρ, β, a, b, c, r^0 and θ^0 are j,k the empirical parameters. In addition, for the system of MMS an extra term was included in the potential to account for the existence of the stable Me–O bond, which is not broken in the sol–gel process. This research team paid much attention to the effect of the remaining methoxy groups on the structural development and the gel structure. For this purpose, a harmonic term with a large force constant was introduced:

$$U_{Me} = 1/2 K(r_{ij} - r_0)^2 , for \ Me - O \tag{77}$$

where K is the harmonic force constant, r_0 the equilibrium length. Methyl groups were treated as united atoms. It is not always necessary that

various potential parameters associated with methoxy groups are determined exactly. In fact, several MD simulations using different parameter
sets associated with ethoxy groups, where the polymerization reaction
occurred, gave common characteristic features as described later. The
charge of a methyl group attached to nonbridging oxygen strongly influenced the progress of polymerization. MD simulations were carried out
using the MD program MXDORTO developed by Kawamura. The equations of motion were integrated by means of a Verlet's algorithm. The time
step used in MD runs was 0.4 fs. The systems composed of 216 molecules
were simulated for THS and MMS. Densities were varied from 1.40 to
1.60 g cm^{-3} for THS and from 1.83 to 2.20 g cm^{-3} for MMS. This results
in a periodic cubic cell with a length of 27.6 on a side. After the molecules were distributed randomly at 500 K, polymerization runs of $4 10^5$
time steps were performed at 2500 K. Temperature was controlled by scaling particle velocities. The high density and high temperature conditions
were adopted to accelerate polymerization and reduce computation time
drastically.

Results of this project showed that the calculated Q_n species distribution as a function of SiOSi:Si ratio was in good agreement with that
observed by H NMR. Microscopic features of the structural development
were extracted from a series of the simulation results. There was a slow
cluster growth during the early stage, a rapid cluster growth followed
this due to cluster–cluster aggregation, and the network structure became
denser with time through the repeated bond-formation and bond-breaking.
The present simulation performed gave a reasonable picture from a microscopic point of view. Having some problems to be examined about set
temperature, system size and dilution effect, MD simulation has proved
effective for obtaining useful structural information on sol–gel process for
silica systems.

In another study, Murillo et al. [62] investigated the structure and properties of aerogels and xerogels (fractal dimension, mechanical properties)
via the potential used by Nakano et al. [64]. The simulations done to generate and characterize the samples of silica aerogel and xerogel considered for this study were performed using an in-house Fortran program,
which has been used successfully to investigate structural and mechanical properties of silicon nitrate and silicon carbide systems. In this code

the Velocity–Verlet algorithm is used to solve the equations of motion of the particles, using a time step of 0.5 fs. Langevin dynamics is implemented and all the particles in the system are used to control its temperature accordingly.

Murillo et al. [62] made a development in preparing the samples. In which, the volume of a dense crystalline sample (β-cristobalite) is expanded in one step to the desired density. Then the temperature is increased to give the atoms enough energy to diffuse through the system, and finally it is cooled down in a stepwise process. This process resembles a diffusion limited aggregation. Since the atoms are initially set at distances larger than their equilibrium distances and have high kinetic energy, they can diffuse to farther regions in the simulation volume and could potentially form any kind of structure. That is an advantage of this procedure over the gradual expansion, where the atoms can only move short distances and are bounded to remain in the same region. For this study samples with densities varying from 0.23 g/cm³ to 2.2 g/cm³ (dense glass) are produced. The fractal dimensions for those samples are calculated based on the decay of the pair distribution functions (PDF) and simulated scattering experiments. The results obtained in this study, geometrical and mechanical properties of porous silica are in the ranges found experimentally.

The interatomic potential used for this study is the one developed by Vashishta et al. [89] for amorphous silica. It involves terms representing the interaction between two atoms (two-body component); which, disregarding if the atoms bond or not, accounts for the potential energy due to the distance between them. The potential also includes the potential energy due to the change of orientation and bonding angle of triplets of atoms (three-body component). The two-body and three-body components of the potential are given by

$$V_{ij}^{(2)} = \frac{H_{ij}}{r_{ij}^{\eta_{ij}}} + \frac{Z_i Z_j}{r_{ij}} exp\left(\frac{-r_{ij}}{r_{1s}}\right) - \frac{P_{ij}}{r_{ij}^A} exp\left(\frac{-r_{ij}}{r_{4s}}\right) \tag{78}$$

$$V_{jik}^{(3)} = B_{jik} f\left(r_{ij} r_{ik}\right) \left(\left(cos\theta_{jik} - cos\tilde{\theta}_{jik}\right)\right)^2 \tag{79}$$

In the previous equations r_{ij} is the distance between the atoms i and j. The first term of Eq. (78) represents the steric repulsion due to the atomic size. The parameters H_{ij} and η_{ij} are the strength and exponent of steric repulsion. The second term corresponds to the Coulomb interactions between the atoms and accounts for the electric charge transfer, where Zi is the effective charge of the i_{th} ion. The third term includes the charge–dipole interactions. It takes into account the electric polarizability of the atoms through the variable P_{ij}, which is given by

$$P_{ij} = \frac{1}{2}\left(\alpha_i Z_j^2 + \alpha_j Z_i^2\right) \tag{80}$$

where α_i is the electric polarizability of the i_{th} ion. The parameters $r_{1\,s}$ and $r_{4\,s}$ are cut-off values for the interactions.

In Eq. (79), B_{jik} is the strength of the three-body interaction; θ_{jik} is the angle between the vector position of the atoms, that is, r_{ij} and r_{ik}. The function f represents the effect of bond stretching and the component containing $(\cos \theta_{jik})$ takes into account the bending of the bonds, and $\tilde{\theta}_{jik}$ is a reference angle for the respective interaction. Although, there are six possible three-body interactions in the system, this potential only considers the most dominant ones, which are related to (Si–O–Si) and (O–Si–O) angles.

APPENDIX 2: MONTE-CARLO

Modeling the competition between drying and curing processes in polymerizing films is of great importance to many existing and developing materials synthesis processes. These processes involve multiple length and time scales ranging from molecular to macroscopic, and are challenging to fully model in situations where the polymerization is nonideal, such as sol-gel silica thin film formation. A comprehensive model of sol-gel silica film formation should link macroscopic flow and drying (controlled by process parameters) to film microstructure (which dictates the properties of the films).

Li et al. studied multiscale model in which dynamic Monte Carlo (DMC) polymerization simulations are coupled to a continuum model of drying. Unlike statistical methods, DMC simulations track the entire molecular structure distribution to allow the calculation not only of molecular weight but also of cycle ranks and topological indices related to molecular size and shape. The entire DMC simulation (containing 106 monomers) is treated as a particle of sol whose position and composition are tracked in the continuum mass transport model of drying. The validity of the multiscale model is verified by the good agreement of the conversion evolution of DMC and continuum simulations for ideal polycondensation and first shell substitution effect (FSSE) cases.

This model allows cyclic and cage-like siloxanes to form; it is better able to predict the silica gelation conversion than other reported kinetic models. By studying the competition between molecular growth and cyclization, and the competition between mass transfer (drying) and reaction (gelation) on the drying process of the sol-gel silica film, they observe that cyclization delays gelation, shrink the molecular size, increase the likelihood of skin formation, and leads to a molecular structure gradient inside the film. It also has been observed that compared with a model with only 3-membered rings, the molecular structure is more complicated and the structure gradients in the films are larger with 4-membered rings. This simulation would allow better prediction of the formation of structure gradients in sol-gel derived ceramics and other nonideal multifunctional polycondensation products, and that this will help in developing procedures to reduce coating defects.

Krakoviack et al. present a theoretical study of the phase diagram and the structure of a fluid adsorbed in high-porosity aerogels by means of an integral-equation approach combined with the replica formalism. In this study, in order to obtain another assessment of approximation for the present situation of an "out-of-equilibrium" matrix the Henry's constant K_H of the reference system was computed by direct Monte-Carlo integration for various realizations of the model aerogels at the three working densities. Then, the theoretical value by using the (scaled-particle-theory) charging process expression was computed (remembering that $K_H = \exp[-\beta\mu^R_{ex} (\rho_1 = 0)]$)

$$K_H = \exp(-\rho_0 \int_0^\sigma G_{01} (r; r; \rho_1 = 0)) \mid \qquad (81)$$

where $G_{01}(r; r; \rho_1 = 0)$ is the contact value at zero adsorbate density of the fluid-matrix pair distribution function when the corresponding hard-core diameter equals r.

To simulate a realistic gel environment, an aerogel structure factor was used which were obtained from an off-lattice diffusion-limited cluster-cluster aggregation process. The predictions of the theory are in qualitative agreement with the experimental results, showing a substantial narrowing of the gas-liquid coexistence curve (compared to that of the bulk fluid), associated with weak changes in the critical density and temperature. The influence of the aerogel structure (nontrivial short-range correlations due to connectedness, long-range fractal behavior of the silica strands) is shown to be important at low fluid densities.

In Esquivias et al. study gel structure was depicted as a hierarchy at several levels by means of models using the Monte-Carlo technique, on the basis of random close packing (RCP). premises. The pore volume distributions were calculated from the largest sphere radius inscribed within the interstices. These distributions were compared to the pore volume distributions of classic models and to pore volume distributions of a series of aerogels measured by means of the Brunauer–Emmett–Teller (BET) and H_g intrusion methods. Data on the pore volumes associated with micro, meso-, or macropores were obtained.

NOMENCLATURE

a	Acceleration of particle
A	Atom-type dependent constants
B	Strength of the three-body interaction
b_0	Ideal bond length
C_{ik}	Constants
D	Effective dielectric function
d_f	Fractal dimension of porous systems
E	Potential energy
E_{kin}	Kinetic energy
F	Force
H	Hamiltonian
H	Strength
H_{ij}	Strength
k_B	Boltzmann constant, 1.3806×10^{-23} J/K
K_b	Force constant for bound atoms
K_θ	Force constant for bond angels theta
m	Mass of particle
M	Total sampling number

N	Number of atoms
P	Momentum
q	Atomic charges
r	Distance between two atoms
S	Entropy
t	Time
T	Temperature
V	Potential
v	Velocity
x	Direction
Z_i	Formal ionic charges
Δt	Time interval
	Potential energy

Greek Symbols

$\emptyset_{LJ}$	Lennard-Jones potential
$\emptyset_{ijkl}$	Torsion angles
α	Electric polarizability
ε	Energy
ε_0	Permittivity of free space
η	Exponent of steric repulsion
η_{ij}	Exponent of steric repulsion
θ	Angle between the vector position of the atoms
θ_{ijk}	Band angel
ρ	Density
σ	Length parameters
τ	Relaxation time

KEYWORDS

- **aerogels**
- **basic concepts**
- **fractals**
- **mechanical and thermal properties**
- **molecular dynamics**

- **nanostructured materials**
- **porosity**
- **research methodology**
- **simulations**
- **thermal insulate**

REFERENCES

1. Aegerter, M. A., Leventis, N., and Koebel, MM. (2011). Advances in Sol-gel Derived Materials and Technologies, *Springer*, 565–580.
2. Aharony, A. and Stauffer, D., (2003). Introduction to percolation theory, *Taylor & Francis*.
3. Alder, B. and Wainwright, T., (1957). "Phase transition for a hard sphere system." *The Journal of chemical physics* 27(5): 1208.
4. Alemán, J., Chadwick, A., He, J., Hess, M., Horie, K., Jones, R.G., *Kratochvíl, P.*, Meisel, I., Mita, I., Moad, G., Penczek, S., and Stepto, R.F.T., (2007). "Definitions of terms relating to the structure and processing of sols, gels, networks, and inorganic-organic hybrid materials (IUPAC Recommendations 2007)." *Pure and Applied Chemistry* 79(10): 1801–1829.
5. Allen, M. and Tildesley, D., (1989) "Computer simulation of liquids," *Clarendon Press, Oxford, paperback edition.*
6. Allinger, N. L., Chen, K., and Lii, H-J., (1996). "An improved force field (MM4) for saturated hydrocarbons." *Journal of Computational Chemistry* 17(5-6): 642–668.
7. Allinger, N. L., Yuh, Y. H., Lii, H-J. (1989). "Molecular mechanics. The MM3 force field for hydrocarbons. 1." *Journal of the American Chemical Society* 111(23): 8551–8566.
8. Berendsen, H. J., Postma, J. P. M., van Gunsteren, W. F., DiNola, A., and Haak, J. R., (1984). "Molecular dynamics with coupling to an external bath." *The Journal of chemical physics* 81(8): 3684–3690.
9. Bhagat, S. D., Kim, Y.-H., Moon, M.J., Ahn, YS, and Yeo, J.G., (2007). "Rapid synthesis of water-glass based aerogels by in situ surface modification of the hydrogels." *Applied Surface Science* 253(6): 3231–3236.
10. Bhattacharya, S. and Kieffer, J., (2005). "Fractal dimensions of silica gels generated using reactive molecular dynamics simulations." *The Journal of chemical physics* 122(9): 094715.
11. Bhattacharya, S. and Kieffer, J., (2008). "Molecular dynamics simulation study of growth regimes during polycondensation of silicic acid: from silica nanoparticles to porous gels." *The Journal of Physical Chemistry C* 112(6): 1764–1771.
12. Brinker, C., Drotning, W. Scherer, G.W., (1984). A comparison between the densification kinetics of colloidal and polymeric silica gels. *MRS Proceedings, Cambridge University Press* 32.

13. Brinker, C.J, Keefer K., Schaefer, D.W., and Ashley, C.S, (1982). "Sol-gel transition in simple silicates." *Journal of Non-Crystalline Solids* 48(1): 47–64.
14. Brinker, C. J. and Scherer G. W., (1990). "Sol–gel science:the physics and chemistry of sol-gel processing. Gulf Professional Publishing" *Gulf Professional Publishing*: 841–870.
15. Brooks, B. R., Bruccoleri, R. E., Robert, E. Bruccoleri, Olafson, B. D., States, D. J., Swaminathan, S., and Karplus, M., (1983). "CHARMM: A program for macromolecular energy, minimization, and dynamics calculations." *Journal of Computational Chemistry* 4(2): 187–217.
16. Buisson, P. and Pierre, A. C., (2006). "Immobilization in quartz fiber felt reinforced silica aerogel improves the activity of Candida rugosa lipase in organic solvents." *Journal of Molecular Catalysis B: Enzymatic* 39(1): 77–82.
17. Cai, W. (2005). "Handout 1. An Overview of Molecular Simulation." Stanford University: 1–11.
18. Calas, S. and Sempere, R. (1998). "Textural properties of densified aerogels." *Journal of Non-Crystalline Solids* 225: 215–219.
19. Campbell, T., Kalia, R. K., Nakano, A., Vashishta, P., Ogata, s., and Rodgers, S., (1999). "Structural correlations and mechanical behavior in nanophase silica glasses." *Physical review letters* 82(20): 4018.
20. Carofalini, S. and Melman, H., (1986). Applications of molecular dynamics simulations to sol-gel processing. MRS Proceedings, *Cambridge Univ Press*.
21. Cotter, C. J. and Reich, S., (2004)."Time stepping algorithms for classical molecular dynamics." *Computational Nanotechnology*. American Scientific Publishers: 2–18.
22. d'Arjuzon, R. J., Frith, W., and Melrose, J.R., (2003). "Brownian dynamics simulations of aging colloidal gels." *Physical Review* E 67(6): 061404.
23. Devreux, F., Boilot, J., Chaput, F., and Lecomte, A., (1990). "NMR determination of the fractal dimension in silica aerogels." *Physical review letters* 65(5): 614.
24. Emmerling, A. and Fricke, J., (1997). "Scaling properties and structure of aerogels." *Journal of Sol-Gel Science and Technology* 8(1–3): 781–788.
25. Feuston, B. and Garofalini, S., (1988). "Empirical three-body potential for vitreous silica." *The Journal of chemical physics* 89(9): 5818–5824.
26. Feuston, B. and Garofalini, S., (1990). "Oligomerization in silica sols." *Journal of Physical Chemistry* 94(13): 5351–5356.
27. Feuston, B. and Garofalini, S. H., (1990). "Water-induced relaxation of the vitreous silica surface." Journal of applied physics 68(9): 4830–4836.
28. Field, M. J., (1999). "A practical introduction to the simulation of molecular systems," *Cambridge University Press*.
29. Frenkel, D. and Smit, B., (2001). "Understanding molecular simulation: from algorithms to applications," *Academic press* 1: 7–80.
30. Freundlich, H. and Hatfield, H. S., (1926). "Colloid and capillary chemistry."Methuen, London, 576–578.
31. Fricke, J. and Emmerling, A. (1998). "Aerogels—recent progress in production techniques and novel applications." *Journal of Sol-Gel Science and Technology* 13(1–3): 299–303.

32. Garcia, E., Glaser, M. A., Clark, N., and Walba, D., (1999). "HFF: a force field for liquid crystal molecules." *Journal of Molecular Structure: THEOCHEM* 464(1): 39–48.

33. Garofalini, S. H. and Martin, G., (1994). "Molecular simulations of the polymerization of silicic acid molecules and network formation." *The Journal of Physical Chemistry* 98(4): 1311–1316.

34. Gash, A. E., Tillotson, T. M., and Satcher, J. H., (2001). "New sol–gel synthetic route to transition and main-group metal oxide aerogels using inorganic salt precursors." *Journal of Non-Crystalline Solids* 285(1): 22–28.

35. GEAR, C. (1966). "The numerical integration of ordinary differential equations of various order (Ordinary differential equation." *Mathematics of Computation* 21.98: 146–156.

36. Gross, J., Coronado, P. R., (1998). "Elastic properties of silica aerogels from a new rapid supercritical extraction process." *Journal of Non-Crystalline Solids* 225: 282–286.

37. Gross, J. and Fricke, J., (1992). "Ultrasonic velocity measurements in silica, carbon and organic aerogels." *Journal of Non-Crystalline Solids* 145: 217–222.

38. Gross, J. and Scherer, G. W., (2003). "Dynamic pressurization: novel method for measuring fluid permeability." *Journal of Non-Crystalline Solids* 325(1): 34–47.

39. Haereid, S., Dahle, M., and Einarsrud, M.-A., (1995). "Preparation and properties of monolithic silica xerogels from TEOS-based alcogels aged in silane solutions." *Journal of Non-Crystalline Solids* 186: 96–103.

40. Haile, J. M. (1992)."Molecular dynamics simulation: elementary methods," *John Wiley & Sons*, Inc: 1–512.

41. Hairer, E., Lubich, C., Lubich, C., and Wanner, G., (2003). "Geometric numerical integration illustrated by the Störmer–Verlet method." *Acta Numerica* 12: 399–450.

42. Himmel, B., Bürger, H., Holzhuter, G., and Olbertz, A., (1995). "Structural characterization of SiO_2 aerogels." *Journal of Non-Crystalline Solids* 185(1): 56–66.

43. Jabbarzadeh, A. and Tanner, R. I, (2006). "Molecular dynamics simulation and its application to nano-rheology." *Rheology Reviews* 2006: 165.

44. Jones, J. E. (1924). "On the determination of molecular fields. I. From the variation of the viscosity of a gas with temperature." *Proceedings of the Royal Society of London. Series A* 106(738): 441–462.

45. Jorgensen, W. L., Maxwell, D. S., and Tirado-Rivas, J., (1996). "Development and testing of the OPLS all-atom force field on conformational energetics and properties of organic liquids." *Journal of the American Chemical Society* 118(45): 11225–11236.

46. Jullien, R., Meakin, P., Pavlovith, A., (1992). "Three-dimensional model for particle-size segregation by shaking." *Physical review letters* 69(4): 640.

47. Kallala, M., Jullien, R.,and Cabane, B. (1992). "Crossover from gelation to precipitation." *Journal de Physique II* 2(1): 7–25.

48. Kawaguchi, T., J. Iura, Taneda, N., and Hishikura, H., (1986). "Structural changes of monolithic silica gel during the gel-to-glass transition." *Journal of Non-Crystalline Solids* 82(1): 50–56.

49. Kieffer, J. and Angell, C. A., (1988). "Generation of fractal structures by negative pressure rupturing of SiO_2 glass." *Journal of Non-Crystalline Solids* 106(1): 336–342.
50. Kistler, S. S. (1931). "Coherent Expanded Aerogels and Jellies." *Nature* 127: 741.
51. Kocon, L., Despetis, F., and Phalippou. J., (1998). "Ultralow density silica aerogels by alcohol supercritical drying." *Journal of Non-Crystalline Solids* 225: 96–100.
52. Leimkuhler, B. and Reich, S., (2004). "Geometric numerical methods for Hamiltonian mechanics." Cambridge Monographs on Applied and Computational Mathematics, *Cambridge University Press*.
53. Lii, J. H. and Allinger, N. L. (1989). "Molecular mechanics. The MM3 force field for hydrocarbons. 3. The van der Waals' potentials and crystal data for aliphatic and aromatic hydrocarbons." *Journal of the American Chemical Society* 111(23): 8576–8582.
54. Ma, H.-S., J.-H. Prévost, Jullien, R., Scherer, G.W., (2001). "Computer simulation of mechanical structure–property relationship of aerogels." *Journal of Non-Crystalline Solids* 285(1): 216–221.
55. Ma, H.-S., Roberts, A. P., Prevost, J.H., Jullien, R., Scherer, G.W., (2000). "Mechanical structure–property relationship of aerogels." *Journal of Non-Crystalline Solids* 277(2): 127–141.
56. Martin, G. E. and Garofalini S. H., (1994). "Sol-gel polymerization: analysis of molecular mechanisms and the effect of hydrogen." *Journal of Non-Crystalline Solids* 171(1): 68–79.
57. McQuarrie, D. "Statistical mechanics, 1976." *Happer and Row*, New York.
58. Metropolis, N., Rosenbluth, A. W., Rosenbluth, M. N., and Teller, A. H., (2004). "Equation of state calculations by fast computing machines." *The Journal of chemical physics* 21(6): 1087–1092.
59. Mikes, J., and Dusek, K., (1982). "Simulation of polymer network formation by the Monte Carlo method." *Macromolecules* 15(1): 93–99.
60. Munetoh, S., Motooka, T., Moriguchib, K., and Shintani, A., (2007). "Interatomic potential for Si–O systems using Tersoff parameterization." *Computational Materials Science* 39(2): 334–339.
61. Muralidharan, K., Simmons, J., Deymier, P. A., Runge, K., (2005). "Molecular dynamics studies of brittle fracture in vitreous silica: review and recent progress." *Journal of Non-Crystalline Solids* 351(18): 1532–1542.
62. Murillo, J. S. R., Barbero, E. J., (2008). Towards Toughening and Understanding of Silica Aerogels: MD Simulations. *ASME 2008* International Mechanical Engineering Congress and Exposition, American Society of Mechanical Engineers 679–648.
63. Mylvaganam, K., and Zhang L., (2004). "Important issues in a molecular dynamics simulation for characterizing the mechanical properties of carbon nanotubes." *Carbon* 42(10): 2025–2032.
64. Nakano, A., Bi, Rajiv, K., Vashishta, P., (1993). "Structural correlations in porous silica: Molecular dynamics simulation on a parallel computer." *Physical review letters* 71(1): 85.
65. Nakayama, T., Yakubo, K., Orbach, RL., (1994). "Dynamical properties of fractal networks: Scaling, numerical simulations, and physical realizations." *Reviews of modern physics* 66(2): 381.

66. Petričević, R., Reichenauer, G., Bock, V., Emmerling, A., Fricke, J., (1998). "Structure of carbon aerogels near the gelation limit of the resorcinol–formaldehyde precursor." *Journal of Non-Crystalline Solids* 225: 41–45.

67. Plimpton, S. (1995). "Fast parallel algorithms for short-range molecular dynamics." *Journal of Computational Physics* 117(1): 1–19.

68. Pütz, M., K. Kremer, et al. (2000). "What is the entanglement length in a polymer melt?" EPL (Europhysics Letters) 49(6): 735.

69. Rao, N. Z. and Gelb, L. D., (2004). "Molecular dynamics simulations of the polymerization of aqueous silicic acid and analysis of the effects of concentration on silica polymorph distributions, growth mechanisms, and reaction kinetics." *The Journal of Physical Chemistry B* 108(33): 12418–12428.

70. Rapaport, D. C. (2004). "The art of molecular dynamics simulation, *Cambridge university press:*" 1–80.

71. Reichenauer, G. "Aerogels." Kirk-Othmer Encyclopedia of Chemical Technology. DOI: 10.1002/0471238961.

72. Reith, D., Pütz, M., Muller-Plathe, F., (2003). "Deriving effective mesoscale potentials from atomistic simulations." *Journal of Computational Chemistry* 24(13): 1624–1636.

73. Rivas Murillo, J. S., Bachlechner M. E., Campo, F. A., Barbero, J., (2010). "Structure and mechanical properties of silica aerogels and xerogels modeled by molecular dynamics simulation." *Journal of Non-Crystalline Solids* 356(25): 1325–1331.

74. Ryckaert, J.-P. and Bellemans A., (1975). "Molecular dynamics of liquid n-butane near its boiling point." *Chemical Physics Letters* 30(1): 123–125.

75. Saliger, R., Bock, v., Petricevic, R., Tillotson, T., Geis, S., Fricke, J., (1997). "Carbon aerogels from dilute catalysis of resorcinol with formaldehyde." *Journal of Non-Crystalline Solids* 221(2): 144–150.

76. Satoh, A. (2010). Introduction to practice of molecular simulation: molecular dynamics, Monte Carlo, Brownian dynamics, Lattice Boltzmann and dissipative particle dynamics, *Elsevier* 1–172.

77. Scherer, G. W. (1998). "Characterization of aerogels." *Advances in colloid and interface science* 76: 321–339.

78. Scherer, G. W., J. Gross, Hrubesh, L., Coronado, P.R., (2002). "Optimization of the rapid supercritical extraction process for aerogels." *Journal of Non-Crystalline Solids* 311(3): 259–272.

79. Smith, D. M., Stein, D., Anderson, J. M., Ackerman, W., (1995). "Preparation of low-density xerogels at ambient pressure." *Journal of Non-Crystalline Solids* 186: 104–112.

80. Stillinger, F. and Rahman, A., (2008). "Revised central force potentials for water." The *Journal of chemical physics* 68(2): 666–670.

81. Stone, A., (2013). The theory of intermolecular forces, *Oxford University Press,* Second Edition, 352.

82. Succi, S. (2001). The lattice Boltzmann equation: for fluid dynamics and beyond, *Oxford university press.*

83. Tersoff, J., (1989). "Modeling solid-state chemistry: Interatomic potentials for multicomponent systems." *Physical Review B* 39(8): 5566–5568.

84. Torquato, S., (2002). Random heterogeneous materials: *microstructure and macroscopic properties,* Vol. 16. Springer.

85. Tsige, M., Curro J. G., Grest, G.S., (2003). "Molecular dynamics simulations and integral equation theory of alkane chains: comparison of explicit and united atom models." *Macromolecules* 36(6): 2158–2164.

86. Vacher, R., Courtens, E., Coddens, G., Heidemann, A., Tsujimi, Y., Pelous, J., (1990). "Crossovers in the density of states of fractal silica aerogels." *Physical review letters* 65(8): 1008.

87. Vacher, R., Woignier, T., Pelous, J., and Courtens, E., (1988). "Structure and self-similarity of silica aerogels." *Physical Review B* 37: 6500–6503.

88. Vashishta, P., Kalia, R. K., Li W., Nakano A., Omeltchenko A., (1996). "Molecular dynamics methods and large-scale simulations of amorphous materials." *Amorphous Insulators and Semiconductors* 33: 151–213.

89. Vashishta, P., Kalia, R. K., Rajiv, K., Rino, J. P., Ebbsjo, I., (1990). "Interaction potential for SiO_2: a molecular-dynamics study of structural correlations." *Physical Review B* 41(17): 12197.

90. Wang, P., Emmerling, A., Tappert, W., Spormann, O., Fricke, J., and Haubold, H.-G. (1991). "High-temperature and low-temperature supercritical drying of aerogels-structural investigations with SAXS." *Journal of Applied Crystallography* 24(5): 777–780.

91. Webb III, E. B. and Garofalini, S. H., (1998). "Relaxation of silica glass surfaces before and after stress modification in a wet and dry atmosphere: molecular dynamics simulations." *Journal of Non-Crystalline Solids* 226(1): 47–57.

92. Weiner, S. J., Kollman, P. A., Case, D. A., Singh, U. C., Ghio, C., Alagona, G., Profeta, S., Weiner, Jr. P., (1984). "A new force field for molecular mechanical simulation of nucleic acids and proteins." *Journal of the American Chemical Society* 106 (3): 765–784.

93. Woignier, T., Phalippou, J., Phalippou, J., Dussossoy, J. L., Jacquet-Francillon, J., (1990). "Different kinds of fractal structures in silica aerogels." *Journal of Non-Crystalline Solids* 121(1): 198–201.

94. Woignier, T., Reynes, J., Hafidi Alaoui, A., Beurroies, I., Phalippou, J., (1998). "Different kinds of structure in aerogels: relationships with the mechanical properties." *Journal of Non-Crystalline Solids* 241(1): 45–52.

95. Wright, A. C. (1993). "The comparison of molecular dynamics simulations with diffraction experiments." *Journal of Non-Crystalline Solids* 159(3): 264–268.

96. Yeo, J. and Lei, J., (2013). Characterization of Mechanical and Physical Properties of Silica Aerogels Using Molecular Dynamics Simulation. *13th International Conference on Fracture June 16–21, 2013, Beijing, China.*

CHAPTER 2

QUANTUM-CHEMICAL CALCULATION OF PENTACONTACENE AND HECTACONTACENE BY MNDO IN APPROXIMATION OF THE LINEAR MOLECULAR MODEL OF GRAPHENE

V. A. BABKIN,[1] A. V. IGNATOV,[1] V. V. TRIFONOV,[1]
V. YU. DMITRIEV,[1] E. S. TITOVA,[2] A. I. RAKHIMOV,[2]
N. A. RAKHIMOVA,[2] and G. E. ZAIKOV[3]

[1]Volgograd State Architect-Build University Sebryakov Department

[2]Volgograd State Technical University

[3]Institute of Biochemical Physics, Russian Academy of Sciences

CONTENTS

ABSTRACT

Quantum-chemical calculation of molecules pentacontacene and hectacontacene was done by method MNDO. Optimized by all parameters geometric and electronic structures of these compounds were received. Each of these molecular models has a universal factor of acidity equal to 33 (pKa = 33). They all pertain to class of very weak H-acids (pKa>14).

2.1 AIMS AND BACKGROUNDS

The aim of this chapter is a quantum-chemical calculation of molecules pentacontacene and hectacontacene by MNDO method in approximation of the linear molecular model of graphene, which was discovered by Novoselov and Game in 2004 [1].

2.2 METHODICAL PART

The calculation was done with optimization of all parameters by standard gradient method MNDO building in PC GAMESS [2]. The calculation was executed in approach the insulated molecule in gas phase. Program Mac-MolPlt was used for visual presentation of the model of the molecule [3].

2.3 THE RESULTS

Optimized geometric and electronic structures, general and electronic energies of molecules pentacontacene and hectacontacene were received by method MNDO and are shown on Figs. 2.1 and 2.2 and in Tables 2.1–2.3. The universal factor of acidity was calculated by formula: pKa = 42.11– 147.18 $\times$ $q_{max}H^+$ [4] (where, $q_{max}H^+$ – a maximum positive charge on atom of the hydrogen R = 0.97, R– a coefficient of correlations, q_{max}^{H+} = +0.06) pKa = 33. This formula was successfully used in the following articles [5–7]:

Quantum-chemical calculation of molecules pentacontacene and hectacontacene by method MNDO was executed for the first time. Optimized geometric and electronic structures of these compounds were received.

Acid power of molecules pentacontacene and hectacontacene was theoretically evaluated (pKa = 33). These compounds pertain to class of very
weak H-acids (pKa>14).

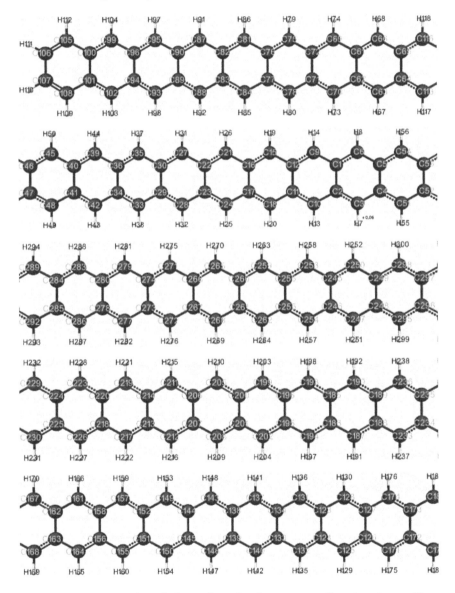

FIGURE 2.1   Geometric and electronic molecular structure of pentacontacene (E_0 =
−2,629,630 kDg/mol, $E_{эл}$ = −36,709,538 kDg/mol).

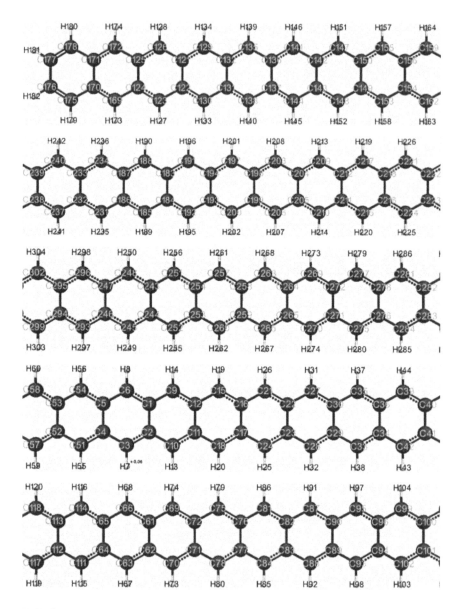

FIGURE 2.2   Geometric and electronic molecular structure of hectacontacene (E_0 = −5,228,911 kDg/mol, $E_{эл}$ = −85,264,384 kDg/mol).

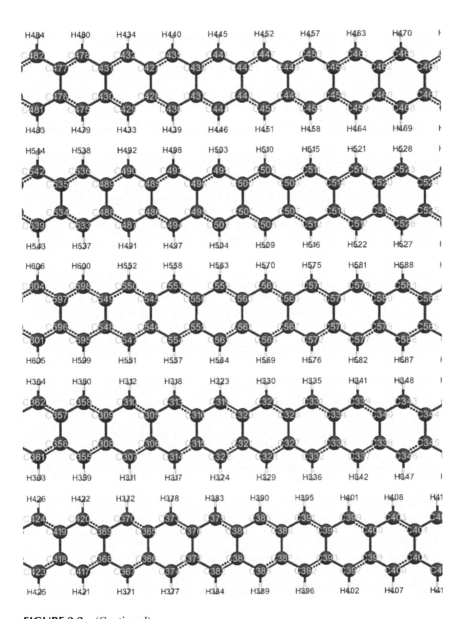

FIGURE 2.2   (Continued).

TABLE 2.1   Optimized Bond Lengths, Valence Corners and Charges on Atoms of the Molecule of Pentacontacene

Bond lengths	R,A	Valent corners	Grad	Atom	Charge (by Milliken)
C(2)-C(1)	1.47	C(1)-C(2)-C(3)	119	C(1)	−0.04
C(3)-C(2)	1.38	C(10)-C(2)-C(3)	123	C(2)	−0.04
C(4)-C(3)	1.46	C(5)-C(4)-C(3)	118	C(3)	−0.01
C(4)-C(5)	1.47	C(2)-C(3)-C(4)	123	C(4)	−0.04
C(5)-C(6)	1.46	C(6)-C(5)-C(4)	118	C(5)	−0.04
C(6)-C(1)	1.38	C(54)-C(5)-C(4)	120	C(6)	−0.01
H(7)-C(3)	1.09	C(1)-C(6)-C(5)	123	H(7)	0.06
H(8)-C(6)	1.09	C(2)-C(1)-C(6)	120	H(8)	0.06
C(9)-C(1)	1.46	C(2)-C(3)-H(7)	120	C(9)	−0.01
C(10)-C(2)	1.46	C(1)-C(6)-H(8)	120	C(10)	−0.01
C(11)-C(10)	1.38	C(2)-C(1)-C(9)	118	C(11)	−0.04
C(11)-C(12)	1.47	C(1)-C(2)-C(10)	118	C(12)	−0.04
C(11)-C(18)	1.46	C(12)-C(11)-C(10)	119	H(13)	0.06
C(12)-C(9)	1.38	C(18)-C(11)-C(10)	123	H(14)	0.06
H(13)-C(10)	1.09	C(2)-C(10)-C(11)	123	C(15)	−0.01
H(14)-C(9)	1.09	C(9)-C(12)-C(11)	120	C(16)	−0.04
C(15)-C(12)	1.46	C(17)-C(18)-C(11)	123	C(17)	−0.04
C(16)-C(15)	1.38	C(15)-C(12)-C(11)	118	C(18)	−0.01
C(16)-C(17)	1.47	C(1)-C(9)-C(12)	123	H(19)	0.06
C(17)-C(24)	1.46	C(18)-C(11)-C(12)	118	H(20)	0.06
C(18)-C(17)	1.38	C(2)-C(10)-H(13)	117	C(21)	−0.01
H(19)-C(15)	1.09	C(1)-C(9)-H(14)	117	C(22)	−0.04
H(20)-C(18)	1.09	C(9)-C(12)-C(15)	123	C(23)	−0.04
C(21)-C(16)	1.46	C(17)-C(16)-C(15)	120	C(24)	−0.01
C(22)-C(21)	1.38	C(12)-C(15)-C(16)	123	H(25)	0.06
C(23)-C(22)	1.47	C(24)-C(17)-C(16)	118	H(26)	0.06
C(24)-C(23)	1.38	C(18)-C(17)-C(16)	119	C(27)	−0.01
H(25)-C(24)	1.09	C(23)-C(24)-C(17)	123	C(28)	−0.01
H(26)-C(21)	1.09	C(24)-C(17)-C(18)	123	C(29)	−0.04
C(27)-C(22)	1.46	C(12)-C(15)-H(19)	117	C(30)	−0.04
C(28)-C(23)	1.46	C(17)-C(18)-H(20)	120	H(31)	0.06
C(29)-C(28)	1.38	C(15)-C(16)-C(21)	123	H(32)	0.06

TABLE 2.1   (Continued).

Bond lengths	R,A	Valent corners	Grad	Atom	Charge (by Milliken)
C(29)-C(30)	1.47	C(17)-C(16)-C(21)	118	C(33)	−0.01
C(30)-C(27)	1.38	C(16)-C(21)-C(22)	123	C(34)	−0.04
H(31)-C(27)	1.09	C(21)-C(22)-C(23)	120	C(35)	−0.01
H(32)-C(28)	1.09	C(27)-C(22)-C(23)	118	C(36)	−0.04
C(33)-C(29)	1.46	C(22)-C(23)-C(24)	119	H(37)	0.06
C(34)-C(33)	1.38	C(28)-C(23)-C(24)	123	H(38)	0.06
C(34)-C(36)	1.47	C(23)-C(24)-H(25)	120	C(39)	−0.01
C(34)-C(42)	1.46	C(16)-C(21)-H(26)	117	C(40)	−0.04
C(35)-C(30)	1.46	C(21)-C(22)-C(27)	123	C(41)	−0.04
C(36)-C(35)	1.38	C(22)-C(23)-C(28)	118	C(42)	−0.01
H(37)-C(35)	1.09	C(30)-C(29)-C(28)	119	H(43)	0.06
H(38)-C(33)	1.09	C(23)-C(28)-C(29)	123	H(44)	0.06
C(39)-C(36)	1.46	C(27)-C(30)-C(29)	120	C(45)	−0.01
C(40)-C(39)	1.38	C(35)-C(30)-C(29)	118	C(46)	−0.04
C(40)-C(41)	1.47	C(22)-C(27)-C(30)	123	C(47)	−0.04
C(41)-C(48)	1.46	C(22)-C(27)-H(31)	117	C(48)	−0.01
C(42)-C(41)	1.38	C(23)-C(28)-H(32)	117	H(49)	0.06
H(43)-C(42)	1.09	C(28)-C(29)-C(33)	123	H(50)	0.06
H(44)-C(39)	1.09	C(30)-C(29)-C(33)	118	C(51)	−0.01
C(45)-C(40)	1.46	C(36)-C(34)-C(33)	119	C(52)	−0.04
C(46)-C(45)	1.38	C(42)-C(34)-C(33)	123	C(53)	−0.04
C(46)-C(47)	1.47	C(29)-C(33)-C(34)	123	C(54)	−0.01
C(46)-C(120)	1.46	C(35)-C(36)-C(34)	120	H(55)	0.06
C(47)-C(119)	1.46	C(41)-C(42)-C(34)	123	H(56)	0.06
C(48)-C(47)	1.38	C(39)-C(36)-C(34)	118	C(57)	−0.01
H(49)-C(48)	1.09	C(27)-C(30)-C(35)	123	C(58)	−0.01
H(50)-C(45)	1.09	C(30)-C(35)-C(36)	123	H(59)	0.06
C(51)-C(4)	1.38	C(42)-C(34)-C(36)	118	H(60)	0.06
C(52)-C(51)	1.46	C(30)-C(35)-H(37)	117	C(61)	−0.04
C(52)-C(53)	1.47	C(29)-C(33)-H(38)	117	C(62)	−0.04
C(53)-C(54)	1.46	C(35)-C(36)-C(39)	123	C(63)	−0.01
C(54)-C(5)	1.38	C(41)-C(40)-C(39)	120	C(64)	−0.04

TABLE 2.1　(Continued).

Bond lengths	R,A	Valent corners	Grad	Atom	Charge (by Milliken)
H(55)-C(51)	1.09	C(36)-C(39)-C(40)	123	C(65)	−0.04
H(56)-C(54)	1.09	C(48)-C(41)-C(40)	118	C(66)	−0.01
C(57)-C(52)	1.38	C(42)-C(41)-C(40)	119	H(67)	0.06
C(58)-C(53)	1.38	C(47)-C(48)-C(41)	123	H(68)	0.06
C(58)-C(290)	1.46	C(48)-C(41)-C(42)	123	C(69)	−0.01
H(59)-C(57)	1.09	C(41)-C(42)-H(43)	120	C(70)	−0.01
H(60)-C(58)	1.09	C(36)-C(39)-H(44)	117	C(71)	−0.04
C(61)-C(66)	1.38	C(39)-C(40)-C(45)	123	C(72)	−0.04
C(62)-C(61)	1.47	C(41)-C(40)-C(45)	118	H(73)	0.06
C(63)-C(62)	1.38	C(47)-C(46)-C(45)	120	H(74)	0.06
C(64)-C(63)	1.46	C(120)-C(46)-C(45)	123	C(75)	−0.01
C(65)-C(64)	1.47	C(40)-C(45)-C(46)	123	C(76)	−0.04
C(66)-C(65)	1.46	C(119)-C(47)-C(46)	118	C(77)	−0.04
H(67)-C(63)	1.09	C(115)-C(120)-C(46)	123	C(78)	−0.01
H(68)-C(66)	1.09	C(48)-C(47)-C(46)	119	H(79)	0.06
C(69)-C(61)	1.46	C(114)-C(119)-C(47)	123	H(80)	0.06
C(70)-C(62)	1.46	C(120)-C(46)-C(47)	118	C(81)	−0.01
C(71)-C(70)	1.38	C(119)-C(47)-C(48)	123	C(82)	−0.04
C(71)-C(72)	1.47	C(47)-C(48)-H(49)	120	C(83)	−0.04
C(71)-C(78)	1.46	C(40)-C(45)-H(50)	117	C(84)	−0.01
C(72)-C(69)	1.38	C(3)-C(4)-C(51)	123	H(85)	0.06
H(73)-C(70)	1.09	C(5)-C(4)-C(51)	119	H(86)	0.06
H(74)-C(69)	1.09	C(53)-C(52)-C(51)	118	C(87)	−0.01
C(75)-C(72)	1.46	C(4)-C(51)-C(52)	123	C(88)	−0.01
C(76)-C(75)	1.38	C(54)-C(53)-C(52)	118	C(89)	−0.04
C(76)-C(77)	1.47	C(58)-C(53)-C(52)	120	C(90)	−0.04
C(77)-C(84)	1.46	C(5)-C(54)-C(53)	123	H(91)	0.06
C(78)-C(77)	1.38	C(290)-C(58)-C(53)	123	H(92)	0.06
H(79)-C(75)	1.09	C(6)-C(5)-C(54)	123	C(93)	−0.01
H(80)-C(78)	1.09	C(4)-C(51)-H(55)	120	C(94)	−0.04
C(81)-C(76)	1.46	C(5)-C(54)-H(56)	120	C(95)	−0.01
C(82)-C(81)	1.38	C(51)-C(52)-C(57)	123	C(96)	−0.04
C(83)-C(82)	1.47	C(53)-C(52)-C(57)	119	H(97)	0.06

TABLE 2.1   (Continued).

Bond lengths	R,A	Valent corners	Grad	Atom	Charge (by Milliken)
C(84)-C(83)	1.38	C(54)-C(53)-C(58)	123	H(98)	0.06
H(85)-C(84)	1.09	C(289)-C(290)-C(58)	123	C(99)	−0.02
H(86)-C(81)	1.09	C(291)-C(290)-C(58)	118	C(100)	−0.04
C(87)-C(82)	1.46	C(52)-C(57)-H(59)	120	C(101)	−0.04
C(88)-C(83)	1.46	C(53)-C(58)-H(60)	120	C(102)	−0.02
C(89)-C(88)	1.38	C(290)-C(58)-H(60)	117	H(103)	0.05
C(89)-C(90)	1.47	C(65)-C(66)-C(61)	123	H(104)	0.05
C(90)-C(87)	1.38	C(66)-C(61)-C(62)	119	C(105)	−0.04
H(91)-C(87)	1.09	C(69)-C(61)-C(62)	118	C(106)	−0.06
H(92)-C(88)	1.09	C(61)-C(62)-C(63)	119	C(107)	−0.06
C(93)-C(89)	1.46	C(70)-C(62)-C(63)	123	C(108)	−0.04
C(94)-C(93)	1.38	C(62)-C(63)-C(64)	123	H(109)	0.06
C(94)-C(96)	1.48	C(63)-C(64)-C(65)	118	H(110)	0.06
C(94)-C(102)	1.46	C(113)-C(64)-C(65)	119	H(111)	0.06
C(95)-C(90)	1.46	C(64)-C(65)-C(66)	118	H(112)	0.06
C(96)-C(95)	1.38	C(116)-C(65)-C(66)	123	C(113)	−0.01
H(97)-C(95)	1.09	C(62)-C(63)-H(67)	120	C(114)	−0.04
H(98)-C(93)	1.09	C(65)-C(66)-H(68)	117	C(115)	−0.04
C(99)-C(96)	1.46	C(66)-C(61)-C(69)	123	C(116)	−0.01
C(100)-C(99)	1.37	C(61)-C(62)-C(70)	118	H(117)	0.06
C(100)-C(101)	1.48	C(72)-C(71)-C(70)	119	H(118)	0.06
C(101)-C(108)	1.47	C(78)-C(71)-C(70)	123	C(119)	−0.01
C(102)-C(101)	1.37	C(62)-C(70)-C(71)	123	C(120)	−0.01
H(103)-C(102)	1.09	C(69)-C(72)-C(71)	119	H(121)	0.06
H(104)-C(99)	1.09	C(77)-C(78)-C(71)	123	H(122)	0.06
C(105)-C(100)	1.47	C(75)-C(72)-C(71)	118	C(123)	−0.04
C(106)-C(105)	1.36	C(61)-C(69)-C(72)	123	C(124)	−0.04
C(107)-C(106)	1.45	C(78)-C(71)-C(72)	118	C(125)	−0.01
C(108)-C(107)	1.36	C(62)-C(70)-H(73)	117	C(126)	−0.04
H(109)-C(108)	1.09	C(61)-C(69)-H(74)	117	C(127)	−0.04
H(110)-C(107)	1.09	C(69)-C(72)-C(75)	123	C(128)	−0.01
H(111)-C(106)	1.09	C(77)-C(76)-C(75)	119	H(129)	0.06
H(112)-C(105)	1.09	C(72)-C(75)-C(76)	123	H(130)	0.06

TABLE 2.1   (Continued).

Bond lengths	R,A	Valent corners	Grad	Atom	Charge (by Milliken)
C(113)-C(64)	1.38	C(84)-C(77)-C(76)	118	C(131)	−0.01
C(114)-C(113)	1.46	C(78)-C(77)-C(76)	120	C(132)	−0.01
C(114)-C(115)	1.47	C(83)-C(84)-C(77)	123	C(133)	−0.04
C(115)-C(116)	1.46	C(84)-C(77)-C(78)	123	C(134)	−0.04
C(116)-C(65)	1.38	C(72)-C(75)-H(79)	117	H(135)	0.06
H(117)-C(113)	1.09	C(77)-C(78)-H(80)	120	H(136)	0.06
H(118)-C(116)	1.09	C(75)-C(76)-C(81)	123	C(137)	−0.01
C(119)-C(114)	1.38	C(77)-C(76)-C(81)	118	C(138)	−0.04
C(120)-C(115)	1.38	C(76)-C(81)-C(82)	123	C(139)	−0.04
H(121)-C(119)	1.09	C(81)-C(82)-C(83)	119	C(140)	−0.01
H(122)-C(120)	1.09	C(87)-C(82)-C(83)	118	H(141)	0.06
C(123)-C(128)	1.46	C(82)-C(83)-C(84)	120	H(142)	0.06
C(124)-C(123)	1.47	C(88)-C(83)-C(84)	123	C(143)	−0.01
C(125)-C(124)	1.46	C(83)-C(84)-H(85)	120	C(144)	−0.04
C(126)-C(125)	1.38	C(76)-C(81)-H(86)	117	C(145)	−0.04
C(126)-C(127)	1.48	C(81)-C(82)-C(87)	123	C(146)	−0.01
C(127)-C(174)	1.46	C(82)-C(83)-C(88)	118	H(147)	0.06
C(128)-C(127)	1.38	C(90)-C(89)-C(88)	120	H(148)	0.06
H(129)-C(125)	1.09	C(83)-C(88)-C(89)	123	C(149)	−0.01
H(130)-C(128)	1.09	C(87)-C(90)-C(89)	119	C(150)	−0.01
C(131)-C(123)	1.38	C(95)-C(90)-C(89)	118	C(151)	−0.04
C(132)-C(124)	1.38	C(82)-C(87)-C(90)	123	C(152)	−0.04
C(133)-C(132)	1.46	C(82)-C(87)-H(91)	117	H(153)	0.06
C(133)-C(134)	1.47	C(83)-C(88)-H(92)	117	H(154)	0.06
C(134)-C(131)	1.46	C(88)-C(89)-C(93)	123	C(155)	−0.01
H(135)-C(132)	1.09	C(90)-C(89)-C(93)	118	C(156)	−0.04
H(136)-C(131)	1.09	C(96)-C(94)-C(93)	120	C(157)	−0.01
C(137)-C(134)	1.38	C(102)-C(94)-C(93)	123	C(158)	−0.04
C(138)-C(137)	1.46	C(89)-C(93)-C(94)	123	H(159)	0.06
C(138)-C(139)	1.47	C(95)-C(96)-C(94)	120	H(160)	0.06
C(139)-C(140)	1.46	C(101)-C(102)-C(94)	123	C(161)	−0.01
C(140)-C(133)	1.38	C(99)-C(96)-C(94)	118	C(162)	−0.04
H(141)-C(137)	1.09	C(87)-C(90)-C(95)	123	C(163)	−0.04

TABLE 2.1   (Continued).

Bond lengths	R,A	Valent corners	Grad	Atom	Charge (by Milliken)
H(142)-C(140)	1.09	C(90)-C(95)-C(96)	123	C(164)	−0.01
C(143)-C(138)	1.38	C(102)-C(94)-C(96)	118	H(165)	0.06
C(144)-C(143)	1.46	C(90)-C(95)-H(97)	117	H(166)	0.06
C(144)-C(145)	1.47	C(89)-C(93)-H(98)	117	C(167)	−0.01
C(145)-C(146)	1.46	C(95)-C(96)-C(99)	123	C(168)	−0.01
C(146)-C(139)	1.38	C(101)-C(100)-C(99)	120	H(169)	0.06
H(147)-C(146)	1.09	C(96)-C(99)-C(100)	123	H(170)	0.06
H(148)-C(143)	1.09	C(108)-C(101)-C(100)	118	C(171)	−0.02
C(149)-C(144)	1.38	C(102)-C(101)-C(100)	120	C(172)	−0.04
C(150)-C(145)	1.38	C(107)-C(108)-C(101)	122	C(173)	−0.04
C(151)-C(150)	1.46	C(108)-C(101)-C(102)	123	C(174)	−0.02
C(151)-C(152)	1.47	C(101)-C(102)-H(103)	120	H(175)	0.05
C(152)-C(149)	1.46	C(96)-C(99)-H(104)	117	H(176)	0.05
H(153)-C(149)	1.09	C(99)-C(100)-C(105)	123	C(177)	−0.04
H(154)-C(150)	1.09	C(101)-C(100)-C(105)	118	C(178)	−0.06
C(155)-C(151)	1.38	C(100)-C(105)-C(106)	122	C(179)	−0.06
C(156)-C(155)	1.46	C(105)-C(106)-C(107)	121	C(180)	−0.04
C(156)-C(158)	1.47	C(106)-C(107)-C(108)	121	H(181)	0.06
C(157)-C(152)	1.38	C(107)-C(108)-H(109)	120	H(182)	0.06
C(158)-C(157)	1.46	C(106)-C(107)-H(110)	118	H(183)	0.06
H(159)-C(157)	1.09	C(105)-C(106)-H(111)	121	H(184)	0.06
H(160)-C(155)	1.09	C(100)-C(105)-H(112)	118	C(185)	−0.04
C(161)-C(158)	1.38	C(63)-C(64)-C(113)	123	C(186)	−0.04
C(162)-C(161)	1.46	C(115)-C(114)-C(113)	118	C(187)	−0.01
C(162)-C(163)	1.47	C(64)-C(113)-C(114)	123	C(188)	−0.04
C(163)-C(164)	1.46	C(116)-C(115)-C(114)	118	C(189)	−0.04
C(164)-C(156)	1.38	C(120)-C(115)-C(114)	119	C(190)	−0.01
H(165)-C(164)	1.09	C(65)-C(116)-C(115)	123	H(191)	0.06
H(166)-C(161)	1.09	C(64)-C(65)-C(116)	119	H(192)	0.06
C(167)-C(162)	1.38	C(64)-C(113)-H(117)	120	C(193)	−0.01
C(168)-C(163)	1.38	C(65)-C(116)-H(118)	120	C(194)	−0.01
C(168)-C(240)	1.46	C(113)-C(114)-C(119)	123	C(195)	−0.04
H(169)-C(168)	1.09	C(115)-C(114)-C(119)	119	C(196)	−0.04

TABLE 2.1 (Continued).

Bond lengths	R,A	Valent corners	Grad	Atom	Charge (by Milliken)
H(170)-C(167)	1.09	C(116)-C(115)-C(120)	123	H(197)	0.06
C(171)-C(126)	1.46	C(114)-C(119)-H(121)	120	H(198)	0.06
C(172)-C(171)	1.37	C(115)-C(120)-H(122)	120	C(199)	−0.01
C(172)-C(173)	1.48	C(127)-C(128)-C(123)	123	C(200)	−0.04
C(173)-C(180)	1.47	C(128)-C(123)-C(124)	118	C(201)	−0.04
C(174)-C(173)	1.37	C(131)-C(123)-C(124)	120	C(202)	−0.01
H(175)-C(171)	1.09	C(123)-C(124)-C(125)	118	H(203)	0.06
H(176)-C(174)	1.09	C(132)-C(124)-C(125)	123	H(204)	0.06
C(177)-C(172)	1.47	C(127)-C(126)-C(125)	120	C(205)	−0.01
C(178)-C(177)	1.36	C(124)-C(125)-C(126)	123	C(206)	−0.04
C(179)-C(178)	1.45	C(174)-C(127)-C(126)	118	C(207)	−0.04
C(180)-C(179)	1.36	C(128)-C(127)-C(126)	120	C(208)	−0.01
H(181)-C(177)	1.09	C(173)-C(174)-C(127)	123	H(209)	0.06
H(182)-C(180)	1.09	C(174)-C(127)-C(128)	123	H(210)	0.06
H(183)-C(179)	1.09	C(124)-C(125)-H(129)	117	C(211)	−0.01
H(184)-C(178)	1.09	C(127)-C(128)-H(130)	120	C(212)	−0.01
C(185)-C(190)	1.46	C(128)-C(123)-C(131)	123	C(213)	−0.04
C(186)-C(185)	1.47	C(123)-C(124)-C(132)	120	C(214)	−0.04
C(187)-C(186)	1.46	C(134)-C(133)-C(132)	118	H(215)	0.06
C(188)-C(187)	1.38	C(124)-C(132)-C(133)	123	H(216)	0.06
C(188)-C(189)	1.47	C(131)-C(134)-C(133)	118	C(217)	−0.01
C(189)-C(236)	1.46	C(137)-C(134)-C(133)	119	C(218)	−0.04
C(190)-C(189)	1.38	C(123)-C(131)-C(134)	123	C(219)	−0.01
H(191)-C(187)	1.09	C(124)-C(132)-H(135)	120	C(220)	−0.04
H(192)-C(190)	1.09	C(123)-C(131)-H(136)	120	H(221)	0.06
C(193)-C(185)	1.38	C(131)-C(134)-C(137)	123	H(222)	0.06
C(194)-C(186)	1.38	C(139)-C(138)-C(137)	118	C(223)	−0.01
C(195)-C(194)	1.46	C(134)-C(137)-C(138)	123	C(224)	−0.04
C(195)-C(196)	1.47	C(140)-C(139)-C(138)	118	C(225)	−0.04
C(196)-C(193)	1.46	C(146)-C(139)-C(138)	119	C(226)	−0.01
H(197)-C(194)	1.09	C(133)-C(140)-C(139)	123	H(227)	0.06
H(198)-C(193)	1.09	C(132)-C(133)-C(140)	123	H(228)	0.06
C(199)-C(196)	1.38	C(134)-C(133)-C(140)	119	C(229)	−0.01

TABLE 2.1   (Continued).

Bond lengths	R,A	Valent corners	Grad	Atom	Charge (by Milliken)
C(200)-C(199)	1.46	C(134)-C(137)-H(141)	120	C(230)	−0.01
C(200)-C(201)	1.47	C(133)-C(140)-H(142)	120	H(231)	0.06
C(201)-C(202)	1.46	C(137)-C(138)-C(143)	123	H(232)	0.06
C(202)-C(195)	1.38	C(139)-C(138)-C(143)	119	C(233)	−0.01
H(203)-C(199)	1.09	C(145)-C(144)-C(143)	118	C(234)	−0.04
H(204)-C(202)	1.09	C(138)-C(143)-C(144)	123	C(235)	−0.04
C(205)-C(200)	1.38	C(146)-C(145)-C(144)	118	C(236)	−0.01
C(206)-C(205)	1.46	C(150)-C(145)-C(144)	119	H(237)	0.06
C(206)-C(207)	1.47	C(139)-C(146)-C(145)	123	H(238)	0.06
C(207)-C(208)	1.46	C(140)-C(139)-C(146)	123	C(239)	−0.01
C(208)-C(201)	1.38	C(139)-C(146)-H(147)	120	C(240)	−0.04
H(209)-C(208)	1.09	C(138)-C(143)-H(148)	120	C(241)	−0.04
H(210)-C(205)	1.09	C(143)-C(144)-C(149)	123	C(242)	−0.01
C(211)-C(206)	1.38	C(145)-C(144)-C(149)	119	H(243)	0.06
C(212)-C(207)	1.38	C(146)-C(145)-C(150)	123	H(244)	0.06
C(213)-C(212)	1.46	C(152)-C(151)-C(150)	118	C(245)	−0.04
C(213)-C(214)	1.47	C(145)-C(150)-C(151)	123	C(246)	−0.04
C(214)-C(211)	1.46	C(149)-C(152)-C(151)	118	C(247)	−0.02
H(215)-C(211)	1.09	C(157)-C(152)-C(151)	119	C(248)	−0.03
H(216)-C(212)	1.09	C(144)-C(149)-C(152)	123	C(249)	−0.03
C(217)-C(213)	1.38	C(144)-C(149)-H(153)	120	C(250)	−0.02
C(218)-C(217)	1.46	C(145)-C(150)-H(154)	120	H(251)	0.06
C(218)-C(220)	1.47	C(150)-C(151)-C(155)	123	H(252)	0.06
C(219)-C(214)	1.38	C(152)-C(151)-C(155)	119	C(253)	−0.01
C(220)-C(219)	1.46	C(158)-C(156)-C(155)	118	C(254)	−0.01
H(221)-C(219)	1.09	C(151)-C(155)-C(156)	123	C(255)	−0.04
H(222)-C(217)	1.09	C(157)-C(158)-C(156)	118	C(256)	−0.04
C(223)-C(220)	1.38	C(161)-C(158)-C(156)	120	H(257)	0.06
C(224)-C(223)	1.45	C(149)-C(152)-C(157)	123	H(258)	0.06
C(224)-C(225)	1.47	C(152)-C(157)-C(158)	123	C(259)	−0.01
C(225)-C(226)	1.45	C(152)-C(157)-H(159)	120	C(260)	−0.04
C(226)-C(218)	1.38	C(151)-C(155)-H(160)	120	C(261)	−0.04
H(227)-C(226)	1.09	C(157)-C(158)-C(161)	123	C(262)	−0.01

TABLE 2.1   (Continued).

Bond lengths	R,A	Valent corners	Grad	Atom	Charge (by Milliken)
H(228)-C(223)	1.09	C(163)-C(162)-C(161)	118	H(263)	0.06
C(229)-C(224)	1.39	C(158)-C(161)-C(162)	123	H(264)	0.06
C(230)-C(225)	1.39	C(164)-C(163)-C(162)	118	C(265)	−0.01
C(230)-C(302)	1.45	C(168)-C(163)-C(162)	119	C(266)	−0.04
H(231)-C(230)	1.09	C(156)-C(164)-C(163)	123	C(267)	−0.04
H(232)-C(229)	1.09	C(240)-C(168)-C(163)	123	C(268)	−0.01
C(233)-C(188)	1.46	C(155)-C(156)-C(164)	123	H(269)	0.06
C(234)-C(233)	1.38	C(158)-C(156)-C(164)	119	H(270)	0.06
C(234)-C(235)	1.47	C(156)-C(164)-H(165)	120	C(271)	−0.01
C(235)-C(242)	1.46	C(158)-C(161)-H(166)	120	C(272)	−0.01
C(236)-C(235)	1.38	C(161)-C(162)-C(167)	123	C(273)	−0.04
H(237)-C(233)	1.09	C(163)-C(162)-C(167)	120	C(274)	−0.04
H(238)-C(236)	1.09	C(164)-C(163)-C(168)	123	H(275)	0.06
C(239)-C(234)	1.46	C(239)-C(240)-C(168)	123	H(276)	0.06
C(240)-C(239)	1.38	C(241)-C(240)-C(168)	118	C(277)	−0.01
C(240)-C(241)	1.47	C(163)-C(168)-H(169)	120	C(278)	−0.04
C(241)-C(167)	1.46	C(240)-C(168)-H(169)	117	C(279)	−0.01
C(242)-C(241)	1.38	C(162)-C(167)-H(170)	120	C(280)	−0.04
H(243)-C(239)	1.09	C(125)-C(126)-C(171)	123	H(281)	0.06
H(244)-C(242)	1.09	C(127)-C(126)-C(171)	118	H(282)	0.06
C(245)-C(250)	1.4	C(173)-C(172)-C(171)	120	C(283)	−0.01
C(246)-C(245)	1.46	C(126)-C(171)-C(172)	123	C(284)	−0.04
C(247)-C(246)	1.4	C(180)-C(173)-C(172)	118	C(285)	−0.04
C(248)-C(247)	1.43	C(174)-C(173)-C(172)	120	C(286)	−0.01
C(249)-C(248)	1.45	C(179)-C(180)-C(173)	122	H(287)	0.06
C(250)-C(249)	1.43	C(180)-C(173)-C(174)	123	H(288)	0.06
H(251)-C(247)	1.09	C(126)-C(171)-H(175)	117	C(289)	−0.01
H(252)-C(250)	1.09	C(173)-C(174)-H(176)	120	C(290)	−0.04
C(253)-C(245)	1.45	C(171)-C(172)-C(177)	123	C(291)	−0.04
C(254)-C(246)	1.45	C(173)-C(172)-C(177)	118	C(292)	−0.01
C(255)-C(254)	1.39	C(172)-C(177)-C(178)	122	H(293)	0.06
C(255)-C(256)	1.47	C(177)-C(178)-C(179)	121	H(294)	0.06
C(255)-C(262)	1.45	C(178)-C(179)-C(180)	121	C(295)	−0.02

TABLE 2.1   (Continued).

Bond lengths	R,A	Valent corners	Grad	Atom	Charge (by Milliken)
C(256)-C(253)	1.39	C(172)-C(177)-H(181)	118	C(296)	−0.03
H(257)-C(254)	1.09	C(179)-C(180)-H(182)	120	C(297)	−0.03
H(258)-C(253)	1.09	C(178)-C(179)-H(183)	118	C(298)	−0.02
C(259)-C(256)	1.45	C(177)-C(178)-H(184)	121	H(299)	0.06
C(260)-C(259)	1.38	C(189)-C(190)-C(185)	123	H(300)	0.06
C(260)-C(261)	1.47	C(190)-C(185)-C(186)	118	C(301)	−0.02
C(261)-C(268)	1.46	C(193)-C(185)-C(186)	120	C(302)	−0.04
C(262)-C(261)	1.38	C(185)-C(186)-C(187)	118	C(303)	−0.04
H(263)-C(259)	1.09	C(194)-C(186)-C(187)	123	C(304)	−0.02
H(264)-C(262)	1.09	C(189)-C(188)-C(187)	119	H(305)	0.06
C(265)-C(260)	1.46	C(186)-C(187)-C(188)	123	H(306)	0.06
C(266)-C(265)	1.38	C(236)-C(189)-C(188)	118		
C(267)-C(266)	1.47	C(190)-C(189)-C(188)	120		
C(268)-C(267)	1.38	C(235)-C(236)-C(189)	123		
H(269)-C(268)	1.09	C(236)-C(189)-C(190)	123		
H(270)-C(265)	1.09	C(186)-C(187)-H(191)	117		
C(271)-C(266)	1.46	C(189)-C(190)-H(192)	120		
C(272)-C(267)	1.46	C(190)-C(185)-C(193)	123		
C(273)-C(272)	1.38	C(185)-C(186)-C(194)	119		
C(273)-C(274)	1.47	C(196)-C(195)-C(194)	118		
C(274)-C(271)	1.38	C(186)-C(194)-C(195)	123		
H(275)-C(271)	1.09	C(193)-C(196)-C(195)	118		
H(276)-C(272)	1.09	C(199)-C(196)-C(195)	120		
C(277)-C(273)	1.46	C(185)-C(193)-C(196)	123		
C(278)-C(277)	1.38	C(186)-C(194)-H(197)	120		
C(278)-C(280)	1.47	C(185)-C(193)-H(198)	120		
C(278)-C(286)	1.46	C(193)-C(196)-C(199)	123		
C(279)-C(274)	1.46	C(201)-C(200)-C(199)	118		
C(280)-C(279)	1.38	C(196)-C(199)-C(200)	123		
H(281)-C(279)	1.09	C(202)-C(201)-C(200)	118		
H(282)-C(277)	1.09	C(208)-C(201)-C(200)	119		
C(283)-C(280)	1.46	C(195)-C(202)-C(201)	123		
C(284)-C(283)	1.38	C(194)-C(195)-C(202)	123		

TABLE 2.1 (Continued).

Bond lengths	R,A	Valent corners	Grad	Atom	Charge (by Milliken)
C(284)-C(285)	1.47	C(196)-C(195)-C(202)	119		
C(285)-C(292)	1.46	C(196)-C(199)-H(203)	120		
C(286)-C(285)	1.38	C(195)-C(202)-H(204)	120		
H(287)-C(286)	1.09	C(199)-C(200)-C(205)	123		
H(288)-C(283)	1.09	C(201)-C(200)-C(205)	119		
C(289)-C(284)	1.46	C(207)-C(206)-C(205)	118		
C(290)-C(289)	1.38	C(200)-C(205)-C(206)	123		
C(290)-C(291)	1.47	C(208)-C(207)-C(206)	118		
C(291)-C(57)	1.46	C(212)-C(207)-C(206)	119		
C(292)-C(291)	1.38	C(201)-C(208)-C(207)	123		
H(293)-C(292)	1.09	C(202)-C(201)-C(208)	123		
H(294)-C(289)	1.09	C(201)-C(208)-H(209)	120		
C(295)-C(248)	1.41	C(200)-C(205)-H(210)	120		
C(296)-C(295)	1.42	C(205)-C(206)-C(211)	123		
C(296)-C(297)	1.45	C(207)-C(206)-C(211)	119		
C(297)-C(298)	1.42	C(208)-C(207)-C(212)	123		
C(297)-C(304)	1.43	C(214)-C(213)-C(212)	118		
C(298)-C(249)	1.41	C(207)-C(212)-C(213)	123		
H(299)-C(295)	1.09	C(211)-C(214)-C(213)	118		
H(300)-C(298)	1.09	C(219)-C(214)-C(213)	119		
C(301)-C(296)	1.43	C(206)-C(211)-C(214)	123		
C(302)-C(301)	1.4	C(206)-C(211)-H(215)	120		
C(302)-C(303)	1.46	C(207)-C(212)-H(216)	120		
C(303)-C(229)	1.45	C(212)-C(213)-C(217)	123		
C(304)-C(303)	1.4	C(214)-C(213)-C(217)	119		
H(305)-C(301)	1.09	C(220)-C(218)-C(217)	118		
H(306)-C(304)	1.09	C(213)-C(217)-C(218)	123		
		C(219)-C(220)-C(218)	118		
		C(223)-C(220)-C(218)	119		
		C(211)-C(214)-C(219)	123		
		C(214)-C(219)-C(220)	123		
		C(214)-C(219)-H(221)	120		
		C(213)-C(217)-H(222)	120		

TABLE 2.1   (Continued).

Bond lengths	R,A	Valent corners	Grad	Atom	Charge (by Milliken)
		C(219)-C(220)-C(223)	123		
		C(225)-C(224)-C(223)	118		
		C(220)-C(223)-C(224)	123		
		C(226)-C(225)-C(224)	118		
		C(230)-C(225)-C(224)	119		
		C(218)-C(226)-C(225)	123		
		C(302)-C(230)-C(225)	122		
		C(217)-C(218)-C(226)	123		
		C(220)-C(218)-C(226)	119		
		C(218)-C(226)-H(227)	120		
		C(220)-C(223)-H(228)	120		
		C(223)-C(224)-C(229)	123		
		C(225)-C(224)-C(229)	119		
		C(226)-C(225)-C(230)	123		
		C(301)-C(302)-C(230)	123		
		C(303)-C(302)-C(230)	118		
		C(225)-C(230)-H(231)	120		
		C(302)-C(230)-H(231)	118		
		C(224)-C(229)-H(232)	120		
		C(187)-C(188)-C(233)	123		
		C(189)-C(188)-C(233)	118		
		C(235)-C(234)-C(233)	119		
		C(188)-C(233)-C(234)	123		
		C(242)-C(235)-C(234)	118		
		C(236)-C(235)-C(234)	120		
		C(241)-C(242)-C(235)	123		
		C(242)-C(235)-C(236)	123		
		C(188)-C(233)-H(237)	117		
		C(235)-C(236)-H(238)	120		
		C(233)-C(234)-C(239)	123		
		C(235)-C(234)-C(239)	118		
		C(241)-C(240)-C(239)	119		
		C(234)-C(239)-C(240)	123		

TABLE 2.1 (Continued).

Bond lengths	R,A	Valent corners	Grad	Atom	Charge (by Milliken)
		C(167)-C(241)-C(240)	118		
		C(242)-C(241)-C(240)	120		
		C(162)-C(167)-C(241)	123		
		C(167)-C(241)-C(242)	123		
		C(234)-C(239)-H(243)	117		
		C(241)-C(242)-H(244)	120		
		C(249)-C(250)-C(245)	122		
		C(250)-C(245)-C(246)	119		
		C(253)-C(245)-C(246)	118		
		C(245)-C(246)-C(247)	119		
		C(254)-C(246)-C(247)	123		
		C(246)-C(247)-C(248)	122		
		C(247)-C(248)-C(249)	119		
		C(295)-C(248)-C(249)	119		
		C(248)-C(249)-C(250)	119		
		C(298)-C(249)-C(250)	123		
		C(246)-C(247)-H(251)	120		
		C(249)-C(250)-H(252)	118		
		C(250)-C(245)-C(253)	123		
		C(245)-C(246)-C(254)	118		
		C(256)-C(255)-C(254)	119		
		C(262)-C(255)-C(254)	123		
		C(246)-C(254)-C(255)	122		
		C(253)-C(256)-C(255)	119		
		C(261)-C(262)-C(255)	123		
		C(259)-C(256)-C(255)	118		
		C(245)-C(253)-C(256)	122		
		C(262)-C(255)-C(256)	118		
		C(246)-C(254)-H(257)	118		
		C(245)-C(253)-H(258)	118		
		C(253)-C(256)-C(259)	123		
		C(261)-C(260)-C(259)	119		
		C(256)-C(259)-C(260)	123		

TABLE 2.1   (Continued).

Bond lengths	R,A	Valent corners	Grad	Atom	Charge (by Milliken)
		C(268)-C(261)-C(260)	118		
		C(262)-C(261)-C(260)	119		
		C(267)-C(268)-C(261)	123		
		C(268)-C(261)-C(262)	123		
		C(256)-C(259)-H(263)	117		
		C(261)-C(262)-H(264)	120		
		C(259)-C(260)-C(265)	123		
		C(261)-C(260)-C(265)	118		
		C(260)-C(265)-C(266)	123		
		C(265)-C(266)-C(267)	119		
		C(271)-C(266)-C(267)	118		
		C(266)-C(267)-C(268)	119		
		C(272)-C(267)-C(268)	123		
		C(267)-C(268)-H(269)	120		
		C(260)-C(265)-H(270)	117		
		C(265)-C(266)-C(271)	123		
		C(266)-C(267)-C(272)	118		
		C(274)-C(273)-C(272)	119		
		C(267)-C(272)-C(273)	123		
		C(271)-C(274)-C(273)	119		
		C(279)-C(274)-C(273)	118		
		C(266)-C(271)-C(274)	123		
		C(266)-C(271)-H(275)	117		
		C(267)-C(272)-H(276)	117		
		C(272)-C(273)-C(277)	123		
		C(274)-C(273)-C(277)	118		
		C(280)-C(278)-C(277)	119		
		C(286)-C(278)-C(277)	123		
		C(273)-C(277)-C(278)	123		
		C(279)-C(280)-C(278)	120		
		C(285)-C(286)-C(278)	123		
		C(283)-C(280)-C(278)	118		
		C(271)-C(274)-C(279)	123		

TABLE 2.1   (Continued).

Bond lengths	R,A	Valent corners	Grad	Atom	Charge (by Milliken)
		C(274)-C(279)-C(280)	123		
		C(286)-C(278)-C(280)	118		
		C(274)-C(279)-H(281)	117		
		C(273)-C(277)-H(282)	117		
		C(279)-C(280)-C(283)	123		
		C(285)-C(284)-C(283)	120		
		C(280)-C(283)-C(284)	123		
		C(292)-C(285)-C(284)	118		
		C(286)-C(285)-C(284)	119		
		C(291)-C(292)-C(285)	123		
		C(292)-C(285)-C(286)	123		
		C(285)-C(286)-H(287)	120		
		C(280)-C(283)-H(288)	117		
		C(283)-C(284)-C(289)	123		
		C(285)-C(284)-C(289)	118		
		C(291)-C(290)-C(289)	120		
		C(284)-C(289)-C(290)	123		
		C(57)-C(291)-C(290)	118		
		C(292)-C(291)-C(290)	119		
		C(52)-C(57)-C(291)	123		
		C(57)-C(291)-C(292)	123		
		C(291)-C(292)-H(293)	120		
		C(284)-C(289)-H(294)	117		
		C(247)-C(248)-C(295)	123		
		C(297)-C(296)-C(295)	119		
		C(248)-C(295)-C(296)	122		
		C(298)-C(297)-C(296)	119		
		C(304)-C(297)-C(296)	119		
		C(249)-C(298)-C(297)	122		
		C(303)-C(304)-C(297)	122		
		C(248)-C(249)-C(298)	119		
		C(304)-C(297)-C(298)	123		
		C(248)-C(295)-H(299)	119		

TABLE 2.1   (Continued).

Bond lengths	R,A	Valent corners	Grad	Atom	Charge (by Milliken)
		C(249)-C(298)-H(300)	119		
		C(295)-C(296)-C(301)	123		
		C(297)-C(296)-C(301)	119		
		C(303)-C(302)-C(301)	119		
		C(296)-C(301)-C(302)	122		
		C(229)-C(303)-C(302)	118		
		C(304)-C(303)-C(302)	119		
		C(224)-C(229)-C(303)	122		
		C(229)-C(303)-C(304)	123		
		C(296)-C(301)-H(305)	118		
		C(303)-C(304)-H(306)	120		

TABLE 2.2   Optimized Bond Lengths, Valence Corners and Charges on Atoms of the Molecule of Hectacontacene

Bond lengths	R,A	Valent corners	Grad	Atom	Charge (by Milliken)
C(2)-C(1)	1.47	C(1)-C(2)-C(3)	119	C(1)	−0.04
C(3)-C(2)	1.38	C(10)-C(2)-C(3)	123	C(2)	−0.04
C(4)-C(3)	1.46	C(5)-C(4)-C(3)	118	C(3)	−0.01
C(4)-C(5)	1.47	C(2)-C(3)-C(4)	123	C(4)	−0.04
C(5)-C(6)	1.46	C(6)-C(5)-C(4)	118	C(5)	−0.04
C(6)-C(1)	1.38	C(54)-C(5)-C(4)	120	C(6)	−0.01
H(7)-C(3)	1.09	C(1)-C(6)-C(5)	123	H(7)	0.06
H(8)-C(6)	1.09	C(2)-C(1)-C(6)	120	H(8)	0.06
C(9)-C(1)	1.46	C(2)-C(3)-H(7)	120	C(9)	−0.01
C(10)-C(2)	1.46	C(1)-C(6)-H(8)	120	C(10)	−0.01
C(11)-C(10)	1.38	C(2)-C(1)-C(9)	118	C(11)	−0.04
C(11)-C(12)	1.47	C(1)-C(2)-C(10)	118	C(12)	−0.04
C(11)-C(18)	1.46	C(12)-C(11)-C(10)	119	H(13)	0.06
C(12)-C(9)	1.38	C(18)-C(11)-C(10)	123	H(14)	0.06
H(13)-C(10)	1.09	C(2)-C(10)-C(11)	123	C(15)	−0.01
H(14)-C(9)	1.09	C(9)-C(12)-C(11)	120	C(16)	−0.04

TABLE 2.2 (Continued).

Bond lengths	R,A	Valent corners	Grad	Atom	Charge (by Milliken)
C(15)-C(12)	1.46	C(17)-C(18)-C(11)	123	C(17)	−0.04
C(16)-C(15)	1.38	C(15)-C(12)-C(11)	118	C(18)	−0.01
C(16)-C(17)	1.47	C(1)-C(9)-C(12)	123	H(19)	0.06
C(17)-C(24)	1.46	C(18)-C(11)-C(12)	118	H(20)	0.06
C(18)-C(17)	1.38	C(2)-C(10)-H(13)	117	C(21)	−0.01
H(19)-C(15)	1.09	C(1)-C(9)-H(14)	117	C(22)	−0.04
H(20)-C(18)	1.09	C(9)-C(12)-C(15)	123	C(23)	−0.04
C(21)-C(16)	1.46	C(17)-C(16)-C(15)	120	C(24)	−0.01
C(22)-C(21)	1.38	C(12)-C(15)-C(16)	123	H(25)	0.06
C(23)-C(22)	1.47	C(24)-C(17)-C(16)	118	H(26)	0.06
C(24)-C(23)	1.38	C(18)-C(17)-C(16)	119	C(27)	−0.01
H(25)-C(24)	1.09	C(23)-C(24)-C(17)	123	C(28)	−0.01
H(26)-C(21)	1.09	C(24)-C(17)-C(18)	123	C(29)	−0.04
C(27)-C(22)	1.46	C(12)-C(15)-H(19)	117	C(30)	−0.04
C(28)-C(23)	1.46	C(17)-C(18)-H(20)	120	H(31)	0.06
C(29)-C(28)	1.38	C(15)-C(16)-C(21)	123	H(32)	0.06
C(29)-C(30)	1.47	C(17)-C(16)-C(21)	118	C(33)	−0.01
C(30)-C(27)	1.38	C(16)-C(21)-C(22)	123	C(34)	−0.04
H(31)-C(27)	1.09	C(21)-C(22)-C(23)	120	C(35)	−0.01
H(32)-C(28)	1.09	C(27)-C(22)-C(23)	118	C(36)	−0.04
C(33)-C(29)	1.46	C(22)-C(23)-C(24)	119	H(37)	0.06
C(34)-C(33)	1.38	C(28)-C(23)-C(24)	123	H(38)	0.06
C(34)-C(36)	1.47	C(23)-C(24)-H(25)	120	C(39)	−0.01
C(34)-C(42)	1.46	C(16)-C(21)-H(26)	117	C(40)	−0.04
C(35)-C(30)	1.46	C(21)-C(22)-C(27)	123	C(41)	−0.04
C(36)-C(35)	1.38	C(22)-C(23)-C(28)	118	C(42)	−0.01
H(37)-C(35)	1.09	C(30)-C(29)-C(28)	119	H(43)	0.06
H(38)-C(33)	1.09	C(23)-C(28)-C(29)	123	H(44)	0.06
C(39)-C(36)	1.46	C(27)-C(30)-C(29)	119	C(45)	−0.01
C(40)-C(39)	1.38	C(35)-C(30)-C(29)	118	C(46)	−0.04
C(40)-C(41)	1.47	C(22)-C(27)-C(30)	123	C(47)	−0.04
C(41)-C(48)	1.46	C(22)-C(27)-H(31)	117	C(48)	−0.01
C(42)-C(41)	1.38	C(23)-C(28)-H(32)	117	H(49)	0.06

TABLE 2.2   (Continued).

Bond lengths	R,A	Valent corners	Grad	Atom	Charge (by Milliken)
H(43)-C(42)	1.09	C(28)-C(29)-C(33)	123	H(50)	0.06
H(44)-C(39)	1.09	C(30)-C(29)-C(33)	118	C(51)	−0.01
C(45)-C(40)	1.46	C(36)-C(34)-C(33)	119	C(52)	−0.04
C(46)-C(45)	1.38	C(42)-C(34)-C(33)	123	C(53)	−0.04
C(46)-C(47)	1.47	C(29)-C(33)-C(34)	123	C(54)	−0.01
C(46)-C(118)	1.46	C(35)-C(36)-C(34)	119	H(55)	0.06
C(47)-C(117)	1.46	C(41)-C(42)-C(34)	123	H(56)	0.06
C(48)-C(47)	1.38	C(39)-C(36)-C(34)	118	C(57)	−0.01
H(49)-C(48)	1.09	C(27)-C(30)-C(35)	123	C(58)	−0.01
H(50)-C(45)	1.09	C(30)-C(35)-C(36)	123	H(59)	0.06
C(51)-C(4)	1.38	C(42)-C(34)-C(36)	118	H(60)	0.06
C(52)-C(51)	1.46	C(30)-C(35)-H(37)	117	C(61)	−0.04
C(52)-C(53)	1.47	C(29)-C(33)-H(38)	117	C(62)	−0.04
C(53)-C(54)	1.46	C(35)-C(36)-C(39)	123	C(63)	−0.01
C(54)-C(5)	1.38	C(41)-C(40)-C(39)	119	C(64)	−0.04
H(55)-C(51)	1.09	C(36)-C(39)-C(40)	123	C(65)	−0.04
H(56)-C(54)	1.09	C(48)-C(41)-C(40)	118	C(66)	−0.01
C(57)-C(52)	1.38	C(42)-C(41)-C(40)	119	H(67)	0.06
C(58)-C(53)	1.38	C(47)-C(48)-C(41)	123	H(68)	0.06
C(58)-C(288)	1.46	C(48)-C(41)-C(42)	123	C(69)	−0.01
H(59)-C(57)	1.09	C(41)-C(42)-H(43)	120	C(70)	−0.01
H(60)-C(58)	1.09	C(36)-C(39)-H(44)	117	C(71)	−0.04
C(61)-C(66)	1.38	C(39)-C(40)-C(45)	123	C(72)	−0.04
C(62)-C(61)	1.47	C(41)-C(40)-C(45)	118	H(73)	0.06
C(63)-C(62)	1.38	C(47)-C(46)-C(45)	119	H(74)	0.06
C(64)-C(63)	1.46	C(118)-C(46)-C(45)	123	C(75)	−0.01
C(65)-C(64)	1.47	C(40)-C(45)-C(46)	123	C(76)	−0.04
C(66)-C(65)	1.46	C(117)-C(47)-C(46)	118	C(77)	−0.04
H(67)-C(63)	1.09	C(113)-C(118)-C(46)	123	C(78)	−0.01
H(68)-C(66)	1.09	C(48)-C(47)-C(46)	119	H(79)	0.06
C(69)-C(61)	1.46	C(112)-C(117)-C(47)	123	H(80)	0.06
C(70)-C(62)	1.46	C(118)-C(46)-C(47)	118	C(81)	−0.02
C(71)-C(70)	1.38	C(117)-C(47)-C(48)	123	C(82)	−0.04

TABLE 2.2   (Continued).

Bond lengths	R,A	Valent corners	Grad	Atom	Charge (by Milliken)
C(71)-C(72)	1.47	C(47)-C(48)-H(49)	120	C(83)	−0.04
C(71)-C(78)	1.46	C(40)-C(45)-H(50)	117	C(84)	−0.02
C(72)-C(69)	1.38	C(3)-C(4)-C(51)	123	H(85)	0.06
H(73)-C(70)	1.09	C(5)-C(4)-C(51)	119	H(86)	0.06
H(74)-C(69)	1.09	C(53)-C(52)-C(51)	118	C(87)	−0.02
C(75)-C(72)	1.46	C(4)-C(51)-C(52)	123	C(88)	−0.02
C(76)-C(75)	1.38	C(54)-C(53)-C(52)	118	C(89)	−0.04
C(76)-C(77)	1.48	C(58)-C(53)-C(52)	120	C(90)	−0.04
C(77)-C(84)	1.46	C(5)-C(54)-C(53)	123	H(91)	0.06
C(78)-C(77)	1.38	C(288)-C(58)-C(53)	123	H(92)	0.06
H(79)-C(75)	1.09	C(6)-C(5)-C(54)	123	C(93)	−0.02
H(80)-C(78)	1.09	C(4)-C(51)-H(55)	120	C(94)	−0.04
C(81)-C(76)	1.46	C(5)-C(54)-H(56)	120	C(95)	−0.02
C(82)-C(81)	1.38	C(51)-C(52)-C(57)	123	C(96)	−0.04
C(83)-C(82)	1.48	C(53)-C(52)-C(57)	119	H(97)	0.05
C(84)-C(83)	1.38	C(54)-C(53)-C(58)	123	H(98)	0.05
H(85)-C(84)	1.09	C(287)-C(288)-C(58)	123	C(99)	−0.02
H(86)-C(81)	1.09	C(289)-C(288)-C(58)	118	C(100)	−0.04
C(87)-C(82)	1.46	C(52)-C(57)-H(59)	120	C(101)	−0.04
C(88)-C(83)	1.46	C(53)-C(58)-H(60)	120	C(102)	−0.02
C(89)-C(88)	1.38	C(288)-C(58)-H(60)	117	H(103)	0.05
C(89)-C(90)	1.48	C(65)-C(66)-C(61)	123	H(104)	0.05
C(90)-C(87)	1.38	C(66)-C(61)-C(62)	120	C(105)	−0.03
H(91)-C(87)	1.09	C(69)-C(61)-C(62)	118	C(106)	−0.04
H(92)-C(88)	1.09	C(61)-C(62)-C(63)	119	C(107)	−0.04
C(93)-C(89)	1.46	C(70)-C(62)-C(63)	123	C(108)	−0.03
C(94)-C(93)	1.37	C(62)-C(63)-C(64)	123	H(109)	0.05
C(94)-C(96)	1.48	C(63)-C(64)-C(65)	118	H(110)	0.05
C(94)-C(102)	1.46	C(111)-C(64)-C(65)	119	C(111)	−0.01
C(95)-C(90)	1.46	C(64)-C(65)-C(66)	118	C(112)	−0.04
C(96)-C(95)	1.37	C(114)-C(65)-C(66)	123	C(113)	−0.04
H(97)-C(95)	1.09	C(62)-C(63)-H(67)	120	C(114)	−0.01
H(98)-C(93)	1.09	C(65)-C(66)-H(68)	117	H(115)	0.06

TABLE 2.2   (Continued).

Bond lengths	R,A	Valent corners	Grad	Atom	Charge (by Milliken)
C(99)-C(96)	1.46	C(66)-C(61)-C(69)	123	H(116)	0.06
C(100)-C(99)	1.39	C(61)-C(62)-C(70)	118	C(117)	−0.01
C(101)-C(100)	1.49	C(72)-C(71)-C(70)	119	C(118)	−0.01
C(102)-C(101)	1.39	C(78)-C(71)-C(70)	123	H(119)	0.06
H(103)-C(102)	1.09	C(62)-C(70)-C(71)	123	H(120)	0.06
H(104)-C(99)	1.09	C(69)-C(72)-C(71)	120	C(121)	−0.04
C(105)-C(100)	1.43	C(77)-C(78)-C(71)	123	C(122)	−0.04
C(106)-C(105)	1.43	C(75)-C(72)-C(71)	118	C(123)	−0.02
C(106)-C(107)	1.49	C(61)-C(69)-C(72)	123	C(124)	−0.04
C(107)-C(108)	1.43	C(78)-C(71)-C(72)	118	C(125)	−0.04
C(108)-C(101)	1.43	C(62)-C(70)-H(73)	117	C(126)	−0.02
H(109)-C(108)	1.09	C(61)-C(69)-H(74)	117	H(127)	0.06
H(110)-C(105)	1.09	C(69)-C(72)-C(75)	123	H(128)	0.06
C(111)-C(64)	1.38	C(77)-C(76)-C(75)	120	C(129)	−0.02
C(112)-C(111)	1.46	C(72)-C(75)-C(76)	123	C(130)	−0.02
C(112)-C(113)	1.47	C(84)-C(77)-C(76)	118	C(131)	−0.04
C(113)-C(114)	1.46	C(78)-C(77)-C(76)	120	C(132)	−0.04
C(114)-C(65)	1.38	C(83)-C(84)-C(77)	123	H(133)	0.06
H(115)-C(111)	1.09	C(84)-C(77)-C(78)	123	H(134)	0.06
H(116)-C(114)	1.09	C(72)-C(75)-H(79)	117	C(135)	−0.01
C(117)-C(112)	1.38	C(77)-C(78)-H(80)	120	C(136)	−0.04
C(118)-C(113)	1.38	C(75)-C(76)-C(81)	123	C(137)	−0.04
H(119)-C(117)	1.09	C(77)-C(76)-C(81)	118	C(138)	−0.01
H(120)-C(118)	1.09	C(76)-C(81)-C(82)	123	H(139)	0.06
C(121)-C(129)	1.38	C(81)-C(82)-C(83)	120	H(140)	0.06
C(121)-C(122)	1.47	C(87)-C(82)-C(83)	118	C(141)	−0.01
C(121)-C(126)	1.46	C(82)-C(83)-C(84)	120	C(142)	−0.04
C(121)-C(391)	1.38	C(88)-C(83)-C(84)	123	C(143)	−0.04
C(121)-C(393)	1.47	C(83)-C(84)-H(85)	120	C(144)	−0.01
C(121)-C(399)	1.46	C(76)-C(81)-H(86)	117	H(145)	0.06
C(122)-C(392)	1.38	C(81)-C(82)-C(87)	123	H(146)	0.06
C(122)-C(123)	1.46	C(82)-C(83)-C(88)	118	C(147)	−0.01
C(122)-C(394)	1.48	C(90)-C(89)-C(88)	120	C(148)	−0.01

TABLE 2.2   (Continued).

Bond lengths	R,A	Valent corners	Grad	Atom	Charge (by Milliken)
C(122)-C(397)	1.46	C(83)-C(88)-C(89)	123	C(149)	−0.04
C(123)-C(124)	1.38	C(87)-C(90)-C(89)	120	C(150)	−0.04
C(123)-C(393)	1.46	C(95)-C(90)-C(89)	118	H(151)	0.06
C(123)-C(398)	1.38	C(82)-C(87)-C(90)	123	H(152)	0.06
C(124)-C(397)	1.37	C(82)-C(87)-H(91)	117	C(153)	−0.01
C(124)-C(125)	1.48	C(83)-C(88)-H(92)	117	C(154)	−0.04
C(124)-C(169)	1.46	C(88)-C(89)-C(93)	123	C(155)	−0.01
C(124)-C(400)	1.48	C(90)-C(89)-C(93)	118	C(156)	−0.04
C(124)-C(406)	1.46	C(96)-C(94)-C(93)	120	H(157)	0.06
C(125)-C(126)	1.38	C(102)-C(94)-C(93)	123	H(158)	0.06
C(125)-C(172)	1.46	C(89)-C(93)-C(94)	123	C(159)	−0.01
C(125)-C(398)	1.47	C(95)-C(96)-C(94)	120	C(160)	−0.04
C(125)-C(399)	1.38	C(101)-C(102)-C(94)	123	C(161)	−0.04
C(125)-C(403)	1.46	C(99)-C(96)-C(94)	118	C(162)	−0.01
C(126)-C(400)	1.38	C(87)-C(90)-C(95)	123	H(163)	0.06
C(126)-C(394)	1.46	C(90)-C(95)-C(96)	123	H(164)	0.06
H(127)-C(123)	1.09	C(102)-C(94)-C(96)	118	C(165)	−0.01
H(128)-C(399)	1.09	C(90)-C(95)-H(97)	117	C(166)	−0.01
C(129)-C(394)	1.38	C(89)-C(93)-H(98)	117	H(167)	0.06
C(129)-C(132)	1.46	C(95)-C(96)-C(99)	123	H(168)	0.06
C(129)-C(386)	1.46	C(96)-C(99)-C(100)	123	C(169)	−0.02
C(130)-C(122)	1.38	C(99)-C(100)-C(101)	119	C(170)	−0.04
C(130)-C(131)	1.46	C(105)-C(100)-C(101)	118	C(171)	−0.04
C(130)-C(387)	1.46	C(100)-C(101)-C(102)	119	C(172)	−0.02
C(130)-C(393)	1.38	C(108)-C(101)-C(102)	123	H(173)	0.05
C(131)-C(388)	1.38	C(101)-C(102)-H(103)	120	H(174)	0.05
C(131)-C(132)	1.47	C(96)-C(99)-H(104)	117	C(175)	−0.04
C(131)-C(386)	1.48	C(99)-C(100)-C(105)	123	C(176)	−0.06
C(131)-C(392)	1.46	C(107)-C(106)-C(105)	118	C(177)	−0.06
C(132)-C(385)	1.38	C(100)-C(105)-C(106)	124	C(178)	−0.04
C(132)-C(387)	1.48	C(108)-C(107)-C(106)	118	H(179)	0.06
C(132)-C(391)	1.46	C(481)-C(107)-C(106)	119	H(180)	0.06
H(133)-C(130)	1.09	C(101)-C(108)-C(107)	124	H(181)	0.06

TABLE 2.2 (Continued).

Bond lengths	R,A	Valent corners	Grad	Atom	Charge (by Milliken)
H(134)-C(391)	1.09	C(100)-C(101)-C(108)	118	H(182)	0.06
C(135)-C(132)	1.38	C(101)-C(108)-H(109)	118	C(183)	−0.04
C(135)-C(136)	1.46	C(100)-C(105)-H(110)	118	C(184)	−0.04
C(135)-C(380)	1.46	C(63)-C(64)-C(111)	123	C(185)	−0.01
C(135)-C(386)	1.38	C(113)-C(112)-C(111)	118	C(186)	−0.04
C(136)-C(379)	1.38	C(64)-C(111)-C(112)	123	C(187)	−0.04
C(136)-C(137)	1.47	C(114)-C(113)-C(112)	118	C(188)	−0.01
C(136)-C(381)	1.48	C(118)-C(113)-C(112)	120	H(189)	0.06
C(136)-C(385)	1.46	C(65)-C(114)-C(113)	123	H(190)	0.06
C(137)-C(382)	1.38	C(64)-C(65)-C(114)	120	C(191)	−0.01
C(137)-C(138)	1.46	C(64)-C(111)-H(115)	120	C(192)	−0.01
C(137)-C(380)	1.47	C(65)-C(114)-H(116)	120	C(193)	−0.04
C(137)-C(388)	1.46	C(111)-C(112)-C(117)	123	C(194)	−0.04
C(138)-C(131)	1.38	C(113)-C(112)-C(117)	119	H(195)	0.06
C(138)-C(381)	1.46	C(114)-C(113)-C(118)	123	H(196)	0.06
C(138)-C(387)	1.38	C(112)-C(117)-H(119)	120	C(197)	−0.01
H(139)-C(135)	1.09	C(113)-C(118)-H(120)	120	C(198)	−0.04
H(140)-C(388)	1.09	C(394)-C(129)-C(121)	1	C(199)	−0.04
C(141)-C(136)	1.38	C(132)-C(129)-C(121)	123	C(200)	−0.01
C(141)-C(142)	1.46	C(386)-C(129)-C(121)	123	H(201)	0.06
C(141)-C(376)	1.46	C(392)-C(122)-C(121)	120	H(202)	0.06
C(141)-C(380)	1.38	C(123)-C(122)-C(121)	118	C(203)	−0.01
C(142)-C(373)	1.38	C(394)-C(122)-C(121)	1	C(204)	−0.04
C(142)-C(143)	1.47	C(397)-C(122)-C(121)	118	C(205)	−0.04
C(142)-C(375)	1.48	C(400)-C(126)-C(121)	123	C(206)	−0.01
C(142)-C(379)	1.46	C(394)-C(126)-C(121)	1	H(207)	0.06
C(143)-C(148)	1.38	C(129)-C(391)-C(121)	89	H(208)	0.06
C(143)-C(144)	1.46	C(394)-C(391)-C(121)	1	C(209)	−0.01
C(143)-C(374)	1.38	C(122)-C(393)-C(121)	95	C(210)	−0.01
C(143)-C(376)	1.47	C(394)-C(393)-C(121)	1	C(211)	−0.04
C(143)-C(382)	1.46	C(397)-C(393)-C(121)	118	C(212)	−0.04
C(144)-C(137)	1.38	C(126)-C(399)-C(121)	60	H(213)	0.06
C(144)-C(375)	1.45	C(400)-C(399)-C(121)	123	H(214)	0.06

TABLE 2.2   (Continued).

Bond lengths	R,A	Valent corners	Grad	Atom	Charge (by Milliken)
C(144)-C(381)	1.38	C(130)-C(122)-C(121)	120	C(215)	−0.02
H(145)-C(382)	1.09	C(125)-C(126)-C(121)	123	C(216)	−0.04
H(146)-C(141)	1.09	C(132)-C(391)-C(121)	123	C(217)	−0.02
C(147)-C(142)	1.38	C(386)-C(391)-C(121)	123	C(218)	−0.04
C(147)-C(150)	1.46	C(123)-C(393)-C(121)	118	H(219)	0.06
C(147)-C(365)	1.46	C(130)-C(393)-C(121)	119	H(220)	0.06
C(147)-C(376)	1.38	C(392)-C(393)-C(121)	119	C(221)	−0.01
C(148)-C(375)	1.38	C(125)-C(399)-C(121)	123	C(222)	−0.04
C(148)-C(149)	1.46	C(394)-C(399)-C(121)	1	C(223)	−0.04
C(148)-C(366)	1.45	C(130)-C(392)-C(122)	94	C(224)	−0.01
C(149)-C(153)	1.38	C(393)-C(392)-C(122)	1	H(225)	0.06
C(149)-C(150)	1.47	C(124)-C(123)-C(122)	123	H(226)	0.06
C(149)-C(365)	1.47	C(393)-C(123)-C(122)	1	C(227)	−0.02
C(149)-C(367)	1.38	C(398)-C(123)-C(122)	123	C(228)	−0.02
C(149)-C(374)	1.46	C(121)-C(394)-C(122)	80	H(229)	0.06
C(150)-C(370)	1.37	C(399)-C(394)-C(122)	118	H(230)	0.06
C(150)-C(366)	1.48	C(123)-C(397)-C(122)	85	C(231)	−0.01
C(150)-C(373)	1.45	C(398)-C(397)-C(122)	123	C(232)	−0.04
H(151)-C(147)	1.09	C(126)-C(121)-C(122)	118	C(233)	−0.04
H(152)-C(374)	1.09	C(129)-C(121)-C(122)	120	C(234)	−0.01
C(153)-C(366)	1.38	C(391)-C(121)-C(122)	120	H(235)	0.06
C(153)-C(154)	1.46	C(393)-C(121)-C(122)	1	H(236)	0.06
C(153)-C(368)	1.45	C(399)-C(121)-C(122)	118	C(237)	−0.01
C(154)-C(162)	1.38	C(131)-C(130)-C(122)	123	C(238)	−0.04
C(154)-C(156)	1.47	C(387)-C(130)-C(122)	123	C(239)	−0.04
C(154)-C(367)	1.47	C(393)-C(130)-C(122)	1	C(240)	−0.01
C(154)-C(369)	1.46	C(131)-C(392)-C(122)	123	H(241)	0.06
C(154)-C(417)	1.38	C(387)-C(392)-C(122)	123	H(242)	0.06
C(155)-C(150)	1.38	C(126)-C(394)-C(122)	118	C(243)	−0.04
C(155)-C(156)	1.46	C(129)-C(394)-C(122)	119	C(244)	−0.04
C(155)-C(365)	1.38	C(391)-C(394)-C(122)	119	C(245)	−0.02
C(155)-C(369)	1.46	C(393)-C(394)-C(122)	1	C(246)	−0.04
C(156)-C(420)	1.37	C(124)-C(397)-C(122)	123	C(247)	−0.04

TABLE 2.2   (Continued).

Bond lengths	R,A	Valent corners	Grad	Atom	Charge (by Milliken)
C(156)-C(368)	1.49	C(393)-C(397)-C(122)	1	C(248)	−0.02
C(156)-C(370)	1.45	C(397)-C(124)-C(123)	1	H(249)	0.06
H(157)-C(155)	1.09	C(125)-C(124)-C(123)	120	H(250)	0.06
H(158)-C(367)	1.08	C(169)-C(124)-C(123)	123	C(251)	−0.02
C(159)-C(156)	1.38	C(400)-C(124)-C(123)	120	C(252)	−0.02
C(159)-C(160)	1.46	C(406)-C(124)-C(123)	123	C(253)	−0.04
C(159)-C(369)	1.39	C(122)-C(393)-C(123)	92	C(254)	−0.04
C(159)-C(419)	1.46	C(394)-C(393)-C(123)	118	H(255)	0.06
C(160)-C(424)	1.37	C(397)-C(393)-C(123)	1	H(256)	0.06
C(160)-C(161)	1.47	C(124)-C(398)-C(123)	36	C(257)	−0.02
C(160)-C(418)	1.49	C(400)-C(398)-C(123)	120	C(258)	−0.04
C(160)-C(420)	1.45	C(406)-C(398)-C(123)	123	C(259)	−0.04
C(161)-C(166)	1.38	C(392)-C(122)-C(123)	123	C(260)	−0.02
C(161)-C(162)	1.46	C(394)-C(122)-C(123)	118	H(261)	0.06
C(161)-C(417)	1.47	C(397)-C(122)-C(123)	1	H(262)	0.06
C(161)-C(419)	1.46	C(130)-C(393)-C(123)	123	C(263)	−0.01
C(161)-C(423)	1.38	C(392)-C(393)-C(123)	123	C(264)	−0.04
C(162)-C(368)	1.38	C(125)-C(398)-C(123)	120	C(265)	−0.04
C(162)-C(418)	1.45	C(169)-C(398)-C(123)	122	C(266)	−0.01
H(163)-C(417)	1.08	C(397)-C(398)-C(123)	1	H(267)	0.06
H(164)-C(159)	1.09	C(123)-C(397)-C(124)	106	H(268)	0.06
C(165)-C(160)	1.38	C(398)-C(397)-C(124)	0	C(269)	−0.01
C(165)-C(350)	1.47	C(126)-C(125)-C(124)	120	C(270)	−0.01
C(165)-C(419)	1.39	C(172)-C(125)-C(124)	118	C(271)	−0.04
C(166)-C(418)	1.37	C(398)-C(125)-C(124)	0	C(272)	−0.04
H(167)-C(423)	1.07	C(399)-C(125)-C(124)	120	H(273)	0.06
H(168)-C(165)	1.09	C(403)-C(125)-C(124)	118	H(274)	0.06
C(169)-C(170)	1.37	C(170)-C(169)-C(124)	123	C(275)	−0.01
C(169)-C(398)	1.46	C(398)-C(169)-C(124)	180	C(276)	−0.04
C(169)-C(405)	1.38	C(405)-C(169)-C(124)	123	C(277)	−0.01
C(170)-C(406)	1.37	C(125)-C(400)-C(124)	67	C(278)	−0.04
C(170)-C(171)	1.48	C(403)-C(400)-C(124)	118	H(279)	0.06
C(170)-C(175)	1.47	C(169)-C(406)-C(124)	89	H(280)	0.06

TABLE 2.2   (Continued).

Bond lengths	R,A	Valent corners	Grad	Atom	Charge (by Milliken)
C(170)-C(404)	1.49	C(393)-C(123)-C(124)	123	C(281)	−0.01
C(170)-C(412)	1.47	C(398)-C(123)-C(124)	0	C(282)	−0.04
C(171)-C(172)	1.37	C(393)-C(397)-C(124)	123	C(283)	−0.04
C(171)-C(178)	1.47	C(126)-C(400)-C(124)	119	C(284)	−0.01
C(171)-C(403)	1.38	C(172)-C(400)-C(124)	118	H(285)	0.06
C(171)-C(405)	1.47	C(398)-C(400)-C(124)	0	H(286)	0.06
C(171)-C(409)	1.47	C(399)-C(400)-C(124)	119	C(287)	−0.01
C(172)-C(404)	1.37	C(170)-C(406)-C(124)	123	C(288)	−0.04
C(172)-C(400)	1.46	C(398)-C(406)-C(124)	180	C(289)	−0.04
H(173)-C(169)	1.09	C(405)-C(406)-C(124)	123	C(290)	−0.01
H(174)-C(403)	1.09	C(400)-C(126)-C(125)	180	H(291)	0.06
C(175)-C(176)	1.36	C(394)-C(126)-C(125)	123	H(292)	0.06
C(175)-C(405)	1.47	C(404)-C(172)-C(125)	123	C(293)	−0.02
C(175)-C(411)	1.37	C(400)-C(172)-C(125)	180	C(294)	−0.04
C(176)-C(412)	1.36	C(124)-C(398)-C(125)	148	C(295)	−0.04
C(176)-C(177)	1.45	C(400)-C(398)-C(125)	180	C(296)	−0.02
C(176)-C(410)	1.46	C(406)-C(398)-C(125)	118	H(297)	0.06
C(177)-C(178)	1.36	C(126)-C(399)-C(125)	69	H(298)	0.06
C(177)-C(409)	1.37	C(400)-C(399)-C(125)	180	C(299)	−0.02
C(177)-C(411)	1.45	C(172)-C(403)-C(125)	84	C(300)	−0.04
C(178)-C(410)	1.36	C(404)-C(403)-C(125)	123	C(301)	−0.04
C(178)-C(404)	1.46	C(169)-C(124)-C(125)	118	C(302)	−0.02
H(179)-C(175)	1.09	C(397)-C(124)-C(125)	119	H(303)	0.06
H(180)-H(416)	0.07	C(400)-C(124)-C(125)	180	H(304)	0.06
H(181)-H(415)	0.09	C(406)-C(124)-C(125)	117	C(305)	−0.04
H(182)-H(414)	0.07	C(171)-C(172)-C(125)	123	C(306)	−0.04
C(183)-C(188)	1.46	C(169)-C(398)-C(125)	118	C(307)	−0.01
C(184)-C(183)	1.47	C(397)-C(398)-C(125)	119	C(308)	−0.04
C(185)-C(184)	1.46	C(394)-C(399)-C(125)	123	C(309)	−0.04
C(186)-C(185)	1.38	C(171)-C(403)-C(125)	122	C(310)	−0.01
C(186)-C(187)	1.47	C(400)-C(403)-C(125)	180	H(311)	0.06
C(187)-C(234)	1.46	C(125)-C(400)-C(126)	96	H(312)	0.06
C(188)-C(187)	1.38	C(403)-C(400)-C(126)	123	C(313)	−0.01

TABLE 2.2   (Continued).

Bond lengths	R,A	Valent corners	Grad	Atom	Charge (by Milliken)
H(189)-C(185)	1.09	C(121)-C(394)-C(126)	98	C(314)	−0.01
H(190)-C(188)	1.09	C(399)-C(394)-C(126)	0	C(315)	−0.04
C(191)-C(183)	1.38	C(129)-C(121)-C(126)	123	C(316)	−0.04
C(192)-C(184)	1.38	C(391)-C(121)-C(126)	123	H(317)	0.06
C(193)-C(192)	1.46	C(393)-C(121)-C(126)	118	H(318)	0.06
C(193)-C(194)	1.47	C(399)-C(121)-C(126)	0	C(319)	−0.01
C(194)-C(191)	1.46	C(172)-C(125)-C(126)	123	C(320)	−0.04
H(195)-C(192)	1.09	C(398)-C(125)-C(126)	120	C(321)	−0.04
H(196)-C(191)	1.09	C(399)-C(125)-C(126)	0	C(322)	−0.01
C(197)-C(194)	1.38	C(403)-C(125)-C(126)	123	H(323)	0.06
C(198)-C(197)	1.46	C(129)-C(394)-C(126)	123	H(324)	0.06
C(198)-C(199)	1.47	C(391)-C(394)-C(126)	123	C(325)	−0.01
C(199)-C(200)	1.46	C(393)-C(394)-C(126)	118	C(326)	−0.04
C(200)-C(193)	1.38	C(172)-C(400)-C(126)	123	C(327)	−0.04
H(201)-C(197)	1.09	C(398)-C(400)-C(126)	119	C(328)	−0.01
H(202)-C(200)	1.09	C(399)-C(400)-C(126)	0	H(329)	0.06
C(203)-C(198)	1.38	C(124)-C(123)-H(127)	120	H(330)	0.06
C(204)-C(203)	1.46	C(393)-C(123)-H(127)	117	C(331)	−0.01
C(204)-C(205)	1.47	C(398)-C(123)-H(127)	120	C(332)	−0.01
C(205)-C(206)	1.46	C(126)-C(399)-H(128)	152	C(333)	−0.04
C(206)-C(199)	1.38	C(400)-C(399)-H(128)	120	C(334)	−0.04
H(207)-C(206)	1.09	C(121)-C(394)-C(129)	90	H(335)	0.06
H(208)-C(203)	1.09	C(399)-C(394)-C(129)	123	H(336)	0.06
C(209)-C(204)	1.38	C(385)-C(132)-C(129)	123	C(337)	−0.01
C(210)-C(205)	1.38	C(387)-C(132)-C(129)	118	C(338)	−0.04
C(211)-C(210)	1.46	C(391)-C(132)-C(129)	1	C(339)	−0.01
C(211)-C(212)	1.47	C(132)-C(386)-C(129)	90	C(340)	−0.04
C(212)-C(209)	1.46	C(387)-C(386)-C(129)	118	H(341)	0.06
H(213)-C(209)	1.09	C(391)-C(386)-C(129)	1	H(342)	0.06
H(214)-C(210)	1.09	C(391)-C(121)-C(129)	1	C(343)	−0.01
C(215)-C(211)	1.38	C(393)-C(121)-C(129)	120	C(344)	−0.04
C(216)-C(215)	1.45	C(399)-C(121)-C(129)	123	C(345)	−0.04
C(216)-C(218)	1.47	C(131)-C(132)-C(129)	118	C(346)	−0.01

TABLE 2.2 (Continued).

Bond lengths	R,A	Valent corners	Grad	Atom	Charge (by Milliken)
C(217)-C(212)	1.38	C(135)-C(132)-C(129)	123	H(347)	0.06
C(218)-C(217)	1.45	C(131)-C(386)-C(129)	118	H(348)	0.06
H(219)-C(217)	1.09	C(135)-C(386)-C(129)	123	C(349)	−0.01
H(220)-C(215)	1.09	C(385)-C(386)-C(129)	123	C(350)	−0.04
C(221)-C(218)	1.38	C(391)-C(394)-C(129)	1	C(351)	−0.04
C(222)-C(221)	1.45	C(393)-C(394)-C(129)	119	C(352)	−0.01
C(222)-C(223)	1.47	C(392)-C(122)-C(130)	1	H(353)	0.06
C(223)-C(224)	1.45	C(123)-C(122)-C(130)	123	H(354)	0.06
C(224)-C(216)	1.38	C(394)-C(122)-C(130)	120	C(355)	−0.01
H(225)-C(224)	1.09	C(397)-C(122)-C(130)	123	C(356)	−0.04
H(226)-C(221)	1.09	C(388)-C(131)-C(130)	123	C(357)	−0.04
C(227)-C(222)	1.39	C(132)-C(131)-C(130)	118	C(358)	−0.01
C(228)-C(223)	1.39	C(386)-C(131)-C(130)	118	H(359)	0.06
C(228)-C(300)	1.44	C(392)-C(131)-C(130)	1	H(360)	0.06
H(229)-C(228)	1.09	C(131)-C(387)-C(130)	92	C(361)	−0.01
H(230)-C(227)	1.09	C(388)-C(387)-C(130)	123	C(362)	−0.01
C(231)-C(186)	1.46	C(392)-C(387)-C(130)	1	H(363)	0.06
C(232)-C(231)	1.38	C(122)-C(393)-C(130)	81	H(364)	0.06
C(232)-C(233)	1.47	C(394)-C(393)-C(130)	119	C(365)	−0.04
C(233)-C(240)	1.46	C(397)-C(393)-C(130)	123	C(366)	−0.04
C(234)-C(233)	1.38	C(138)-C(131)-C(130)	123	C(367)	−0.01
H(235)-C(231)	1.09	C(132)-C(387)-C(130)	118	C(368)	−0.04
H(236)-C(234)	1.09	C(138)-C(387)-C(130)	123	C(369)	−0.04
C(237)-C(232)	1.46	C(386)-C(387)-C(130)	118	C(370)	−0.01
C(238)-C(237)	1.38	C(392)-C(393)-C(130)	1	H(371)	0.06
C(239)-C(238)	1.47	C(138)-C(388)-C(131)	91	H(372)	0.06
C(240)-C(239)	1.38	C(385)-C(132)-C(131)	119	C(373)	−0.01
H(241)-C(237)	1.09	C(387)-C(132)-C(131)	1	C(374)	−0.01
H(242)-C(240)	1.09	C(391)-C(132)-C(131)	118	C(375)	−0.04
C(243)-C(248)	1.39	C(132)-C(386)-C(131)	89	C(376)	−0.04
C(244)-C(243)	1.46	C(387)-C(386)-C(131)	1	H(377)	0.06
C(245)-C(244)	1.39	C(391)-C(386)-C(131)	118	H(378)	0.06
C(246)-C(245)	1.44	C(130)-C(392)-C(131)	85	C(379)	−0.01

TABLE 2.2   (Continued).

Bond lengths	R,A	Valent corners	Grad	Atom	Charge (by Milliken)
C(247)-C(246)	1.45	C(393)-C(392)-C(131)	123	C(380)	−0.04
C(248)-C(247)	1.44	C(387)-C(130)-C(131)	1	C(381)	−0.04
H(249)-C(245)	1.09	C(393)-C(130)-C(131)	123	C(382)	−0.01
H(250)-C(248)	1.09	C(135)-C(132)-C(131)	119	H(383)	0.06
C(251)-C(243)	1.45	C(381)-C(138)-C(131)	123	H(384)	0.06
C(252)-C(244)	1.45	C(387)-C(138)-C(131)	1	C(385)	−0.01
C(253)-C(252)	1.39	C(135)-C(386)-C(131)	119	C(386)	−0.04
C(253)-C(254)	1.47	C(385)-C(386)-C(131)	119	C(387)	−0.04
C(253)-C(260)	1.45	C(137)-C(388)-C(131)	123	C(388)	−0.01
C(254)-C(251)	1.39	C(381)-C(388)-C(131)	123	H(389)	0.06
H(255)-C(252)	1.09	C(387)-C(388)-C(131)	1	H(390)	0.06
H(256)-C(251)	1.09	C(387)-C(392)-C(131)	1	C(391)	−0.02
C(257)-C(254)	1.45	C(135)-C(385)-C(132)	90	C(392)	−0.02
C(258)-C(257)	1.38	C(386)-C(385)-C(132)	1	C(393)	−0.04
C(258)-C(259)	1.47	C(131)-C(387)-C(132)	88	C(394)	−0.04
C(259)-C(266)	1.46	C(388)-C(387)-C(132)	119	H(395)	0.06
C(260)-C(259)	1.38	C(392)-C(387)-C(132)	118	H(396)	0.06
H(261)-C(257)	1.09	C(129)-C(391)-C(132)	87	C(397)	−0.02
H(262)-C(260)	1.09	C(394)-C(391)-C(132)	123	C(398)	−0.04
C(263)-C(258)	1.46	C(386)-C(129)-C(132)	1	C(399)	−0.02
C(264)-C(263)	1.38	C(394)-C(129)-C(132)	123	C(400)	−0.04
C(265)-C(264)	1.47	C(386)-C(131)-C(132)	1	H(401)	0.06
C(266)-C(265)	1.38	C(388)-C(131)-C(132)	120	H(402)	0.06
H(267)-C(266)	1.09	C(392)-C(131)-C(132)	118	C(403)	−0.02
H(268)-C(263)	1.09	C(136)-C(135)-C(132)	123	C(404)	−0.04
C(269)-C(264)	1.46	C(380)-C(135)-C(132)	123	C(405)	−0.04
C(270)-C(265)	1.46	C(386)-C(135)-C(132)	1	C(406)	−0.02
C(271)-C(270)	1.38	C(136)-C(385)-C(132)	123	H(407)	0.05
C(271)-C(272)	1.47	C(380)-C(385)-C(132)	123	H(408)	0.05
C(272)-C(269)	1.38	C(138)-C(387)-C(132)	119	C(409)	−0.04
H(273)-C(269)	1.09	C(386)-C(387)-C(132)	1	C(410)	−0.06
H(274)-C(270)	1.09	C(386)-C(391)-C(132)	1	C(411)	−0.06
C(275)-C(271)	1.46	C(122)-C(130)-H(133)	120	C(412)	−0.04

TABLE 2.2 (Continued).

Bond lengths	R,A	Valent corners	Grad	Atom	Charge (by Milliken)
C(276)-C(275)	1.38	C(131)-C(130)-H(133)	117	H(413)	0.06
C(276)-C(278)	1.47	C(387)-C(130)-H(133)	117	H(414)	0.06
C(276)-C(284)	1.46	C(393)-C(130)-H(133)	120	H(415)	0.06
C(277)-C(272)	1.46	C(129)-C(391)-H(134)	92	H(416)	0.06
C(278)-C(277)	1.38	C(394)-C(391)-H(134)	120	C(417)	−0.01
H(279)-C(277)	1.09	C(385)-C(132)-C(135)	1	C(418)	−0.04
H(280)-C(275)	1.09	C(387)-C(132)-C(135)	120	C(419)	−0.04
C(281)-C(278)	1.46	C(391)-C(132)-C(135)	123	C(420)	−0.01
C(282)-C(281)	1.38	C(379)-C(136)-C(135)	123	H(421)	0.06
C(282)-C(283)	1.47	C(137)-C(136)-C(135)	118	H(422)	0.06
C(283)-C(290)	1.46	C(381)-C(136)-C(135)	118	C(423)	−0.01
C(284)-C(283)	1.38	C(385)-C(136)-C(135)	1	C(424)	−0.01
H(285)-C(284)	1.09	C(136)-C(380)-C(135)	88	H(425)	0.06
H(286)-C(281)	1.09	C(381)-C(380)-C(135)	118	H(426)	0.06
C(287)-C(282)	1.46	C(385)-C(380)-C(135)	1	C(427)	−0.04
C(288)-C(287)	1.38	C(132)-C(386)-C(135)	87	C(428)	−0.04
C(288)-C(289)	1.47	C(387)-C(386)-C(135)	120	C(429)	−0.01
C(289)-C(57)	1.46	C(391)-C(386)-C(135)	123	C(430)	−0.04
C(290)-C(289)	1.38	C(141)-C(136)-C(135)	123	C(431)	−0.04
H(291)-C(290)	1.09	C(137)-C(380)-C(135)	118	C(432)	−0.01
H(292)-C(287)	1.09	C(141)-C(380)-C(135)	122	H(433)	0.06
C(293)-C(246)	1.41	C(379)-C(380)-C(135)	123	H(434)	0.06
C(294)-C(293)	1.42	C(385)-C(386)-C(135)	1	C(435)	−0.02
C(294)-C(295)	1.45	C(141)-C(379)-C(136)	92	C(436)	−0.02
C(295)-C(296)	1.42	C(380)-C(379)-C(136)	1	C(437)	−0.04
C(295)-C(302)	1.43	C(382)-C(137)-C(136)	120	C(438)	−0.04
C(296)-C(247)	1.41	C(138)-C(137)-C(136)	118	H(439)	0.06
H(297)-C(293)	1.09	C(380)-C(137)-C(136)	1	H(440)	0.06
H(298)-C(296)	1.09	C(388)-C(137)-C(136)	118	C(441)	−0.01
C(299)-C(294)	1.43	C(137)-C(381)-C(136)	85	C(442)	−0.04
C(300)-C(299)	1.4	C(382)-C(381)-C(136)	119	C(443)	−0.04
C(300)-C(301)	1.46	C(388)-C(381)-C(136)	118	C(444)	−0.01
C(301)-C(227)	1.44	C(135)-C(385)-C(136)	90	H(445)	0.06

TABLE 2.2   (Continued).

Bond lengths	R,A	Valent corners	Grad	Atom	Charge (by Milliken)
C(302)-C(301)	1.4	C(386)-C(385)-C(136)	123	H(446)	0.06
H(303)-C(299)	1.09	C(380)-C(135)-C(136)	1	C(447)	−0.01
H(304)-C(302)	1.09	C(386)-C(135)-C(136)	123	C(448)	−0.04
C(305)-C(310)	1.38	C(144)-C(137)-C(136)	119	C(449)	−0.04
C(306)-C(305)	1.47	C(142)-C(141)-C(136)	123	C(450)	−0.01
C(307)-C(306)	1.38	C(376)-C(141)-C(136)	123	H(451)	0.06
C(308)-C(307)	1.46	C(380)-C(141)-C(136)	1	H(452)	0.06
C(309)-C(308)	1.47	C(142)-C(379)-C(136)	123	C(453)	−0.01
C(310)-C(309)	1.46	C(376)-C(379)-C(136)	123	C(454)	−0.01
H(311)-C(307)	1.09	C(138)-C(381)-C(136)	118	C(455)	−0.04
H(312)-C(310)	1.09	C(144)-C(381)-C(136)	119	C(456)	−0.04
C(313)-C(305)	1.46	C(380)-C(381)-C(136)	1	H(457)	0.06
C(314)-C(306)	1.46	C(380)-C(385)-C(136)	1	H(458)	0.06
C(315)-C(314)	1.38	C(144)-C(382)-C(137)	89	C(459)	−0.01
C(315)-C(316)	1.47	C(131)-C(138)-C(137)	123	C(460)	−0.04
C(315)-C(322)	1.46	C(381)-C(138)-C(137)	2	C(461)	−0.01
C(316)-C(313)	1.38	C(387)-C(138)-C(137)	123	C(462)	−0.04
H(317)-C(314)	1.09	C(136)-C(380)-C(137)	93	H(463)	0.06
H(318)-C(313)	1.09	C(381)-C(380)-C(137)	2	H(464)	0.06
C(319)-C(316)	1.46	C(385)-C(380)-C(137)	118	C(465)	−0.01
C(320)-C(319)	1.38	C(138)-C(388)-C(137)	84	C(466)	−0.04
C(320)-C(321)	1.47	C(379)-C(136)-C(137)	119	C(467)	−0.04
C(321)-C(328)	1.46	C(381)-C(136)-C(137)	2	C(468)	−0.01
C(322)-C(321)	1.38	C(385)-C(136)-C(137)	118	H(469)	0.06
H(323)-C(319)	1.09	C(375)-C(144)-C(137)	123	H(470)	0.06
H(324)-C(322)	1.09	C(381)-C(144)-C(137)	2	C(471)	−0.01
C(325)-C(320)	1.46	C(141)-C(380)-C(137)	119	C(472)	−0.01
C(326)-C(325)	1.38	C(379)-C(380)-C(137)	119	H(473)	0.06
C(327)-C(326)	1.47	C(143)-C(382)-C(137)	122	H(474)	0.06
C(328)-C(327)	1.38	C(375)-C(382)-C(137)	123	C(475)	−0.02
H(329)-C(328)	1.09	C(381)-C(382)-C(137)	2	C(476)	−0.04
H(330)-C(325)	1.09	C(381)-C(388)-C(137)	2	C(477)	−0.04
C(331)-C(326)	1.46	C(387)-C(388)-C(137)	123	C(478)	−0.02

TABLE 2.2 (Continued).

Bond lengths	R,A	Valent corners	Grad	Atom	Charge (by Milliken)
C(332)-C(327)	1.46	C(388)-C(131)-C(138)	2	H(479)	0.05
C(333)-C(332)	1.38	C(132)-C(131)-C(138)	119	H(480)	0.05
C(333)-C(334)	1.47	C(386)-C(131)-C(138)	120	C(481)	−0.02
C(334)-C(331)	1.38	C(392)-C(131)-C(138)	123	C(482)	−0.02
H(335)-C(331)	1.09	C(137)-C(381)-C(138)	93	H(483)	0.05
H(336)-C(332)	1.09	C(382)-C(381)-C(138)	123	H(484)	0.05
C(337)-C(333)	1.46	C(388)-C(381)-C(138)	2	C(485)	−0.04
C(338)-C(337)	1.38	C(131)-C(387)-C(138)	85	C(486)	−0.04
C(338)-C(340)	1.47	C(388)-C(387)-C(138)	2	C(487)	−0.01
C(338)-C(346)	1.46	C(392)-C(387)-C(138)	123	C(488)	−0.04
C(339)-C(334)	1.46	C(380)-C(137)-C(138)	118	C(489)	−0.04
C(340)-C(339)	1.38	C(382)-C(137)-C(138)	123	C(490)	−0.01
H(341)-C(339)	1.09	C(388)-C(137)-C(138)	2	H(491)	0.06
H(342)-C(337)	1.09	C(144)-C(381)-C(138)	123	H(492)	0.06
C(343)-C(340)	1.46	C(380)-C(381)-C(138)	118	C(493)	−0.01
C(344)-C(343)	1.38	C(386)-C(387)-C(138)	119	C(494)	−0.01
C(344)-C(345)	1.47	C(132)-C(135)-H(139)	120	C(495)	−0.04
C(345)-C(352)	1.46	C(136)-C(135)-H(139)	117	C(496)	−0.04
C(346)-C(345)	1.38	C(380)-C(135)-H(139)	117	H(497)	0.06
H(347)-C(346)	1.09	C(386)-C(135)-H(139)	120	H(498)	0.06
H(348)-C(343)	1.09	C(138)-C(388)-H(140)	89	C(499)	−0.01
C(349)-C(344)	1.46	C(379)-C(136)-C(141)	1	C(500)	−0.04
C(350)-C(349)	1.38	C(137)-C(136)-C(141)	119	C(501)	−0.04
C(350)-C(351)	1.47	C(381)-C(136)-C(141)	120	C(502)	−0.01
C(350)-C(424)	1.46	C(385)-C(136)-C(141)	123	H(503)	0.06
C(351)-C(166)	1.45	C(373)-C(142)-C(141)	123	H(504)	0.06
C(351)-C(423)	1.46	C(143)-C(142)-C(141)	118	C(505)	−0.01
C(352)-C(351)	1.38	C(375)-C(142)-C(141)	118	C(506)	−0.04
H(353)-C(352)	1.09	C(379)-C(142)-C(141)	1	C(507)	−0.04
H(354)-C(349)	1.09	C(142)-C(376)-C(141)	85	C(508)	−0.01
C(355)-C(308)	1.38	C(379)-C(376)-C(141)	1	H(509)	0.06
C(356)-C(355)	1.46	C(136)-C(380)-C(141)	85	H(510)	0.06
C(356)-C(357)	1.47	C(381)-C(380)-C(141)	120	C(511)	−0.01

TABLE 2.2   (Continued).

Bond lengths	R,A	Valent corners	Grad	Atom	Charge (by Milliken)
C(357)-C(358)	1.46	C(385)-C(380)-C(141)	123	C(512)	−0.01
C(358)-C(309)	1.38	C(147)-C(142)-C(141)	123	C(513)	−0.04
H(359)-C(355)	1.09	C(143)-C(376)-C(141)	118	C(514)	−0.04
H(360)-C(358)	1.09	C(147)-C(376)-C(141)	122	H(515)	0.06
C(361)-C(356)	1.38	C(373)-C(376)-C(141)	123	H(516)	0.06
C(362)-C(357)	1.38	C(375)-C(376)-C(141)	118	C(517)	−0.02
C(362)-C(590)	1.46	C(379)-C(380)-C(141)	1	C(518)	−0.04
H(363)-C(361)	1.09	C(147)-C(373)-C(142)	95	C(519)	−0.02
H(364)-C(362)	1.09	C(376)-C(373)-C(142)	1	C(520)	−0.04
C(365)-C(150)	0.03	C(148)-C(143)-C(142)	119	H(521)	0.06
C(365)-C(366)	1.47	C(144)-C(143)-C(142)	118	H(522)	0.06
C(365)-C(370)	1.38	C(374)-C(143)-C(142)	120	C(523)	−0.02
C(365)-C(373)	1.46	C(376)-C(143)-C(142)	1	C(524)	−0.04
C(366)-C(149)	0.03	C(382)-C(143)-C(142)	118	C(525)	−0.04
C(366)-C(367)	1.38	C(143)-C(375)-C(142)	80	C(526)	−0.02
C(366)-C(374)	1.46	C(376)-C(375)-C(142)	1	H(527)	0.06
C(367)-C(153)	0.03	C(382)-C(375)-C(142)	118	H(528)	0.06
C(367)-C(368)	1.46	C(141)-C(379)-C(142)	92	C(529)	−0.02
C(368)-C(154)	0.02	C(380)-C(379)-C(142)	123	C(530)	−0.02
C(368)-C(369)	1.47	C(376)-C(141)-C(142)	1	H(531)	0.06
C(368)-C(417)	1.38	C(380)-C(141)-C(142)	123	H(532)	0.06
C(369)-C(156)	0.02	C(150)-C(147)-C(142)	123	C(533)	−0.01
C(369)-C(370)	1.46	C(365)-C(147)-C(142)	122	C(534)	−0.04
C(369)-C(420)	1.38	C(376)-C(147)-C(142)	1	C(535)	−0.04
C(370)-C(155)	0.02	C(150)-C(373)-C(142)	123	C(536)	−0.01
H(371)-C(367)	1.09	C(365)-C(373)-C(142)	123	H(537)	0.06
H(372)-C(155)	1.08	C(144)-C(375)-C(142)	118	H(538)	0.06
C(373)-C(147)	0.03	C(148)-C(375)-C(142)	119	C(539)	−0.01
C(373)-C(376)	1.38	C(374)-C(375)-C(142)	119	C(540)	−0.04
C(374)-C(148)	0.04	C(376)-C(379)-C(142)	1	C(541)	−0.04
C(374)-C(375)	1.38	C(375)-C(148)-C(143)	2	C(542)	−0.01
C(375)-C(143)	0.04	C(149)-C(148)-C(143)	123	H(543)	0.06
C(375)-C(376)	1.47	C(366)-C(148)-C(143)	123	H(544)	0.06

TABLE 2.2   (Continued).

Bond lengths	R,A	Valent corners	Grad	Atom	Charge (by Milliken)
C(375)-C(382)	1.46	C(137)-C(144)-C(143)	123	C(545)	−0.04
C(376)-C(142)	0.03	C(375)-C(144)-C(143)	1	C(546)	−0.04
C(376)-C(379)	1.46	C(381)-C(144)-C(143)	123	C(547)	−0.02
H(377)-C(374)	1.09	C(148)-C(374)-C(143)	87	C(548)	−0.04
H(378)-C(147)	1.09	C(375)-C(374)-C(143)	2	C(549)	−0.04
C(379)-C(141)	0.03	C(142)-C(376)-C(143)	98	C(550)	−0.02
C(379)-C(380)	1.38	C(379)-C(376)-C(143)	118	H(551)	0.06
C(380)-C(136)	0.03	C(144)-C(382)-C(143)	82	H(552)	0.06
C(380)-C(381)	1.47	C(373)-C(142)-C(143)	119	C(553)	−0.01
C(380)-C(385)	1.46	C(375)-C(142)-C(143)	1	C(554)	−0.01
C(381)-C(137)	0.04	C(379)-C(142)-C(143)	118	C(555)	−0.04
C(381)-C(382)	1.38	C(149)-C(374)-C(143)	122	C(556)	−0.04
C(381)-C(388)	1.46	C(366)-C(374)-C(143)	122	H(557)	0.06
C(382)-C(144)	0.04	C(147)-C(376)-C(143)	120	H(558)	0.06
H(383)-C(141)	1.09	C(373)-C(376)-C(143)	119	C(559)	−0.02
H(384)-C(382)	1.09	C(375)-C(376)-C(143)	1	C(560)	−0.04
C(385)-C(135)	0.03	C(375)-C(382)-C(143)	1	C(561)	−0.04
C(385)-C(386)	1.38	C(381)-C(382)-C(143)	123	C(562)	−0.02
C(386)-C(132)	0.03	C(382)-C(137)-C(144)	2	H(563)	0.06
C(386)-C(387)	1.47	C(138)-C(137)-C(144)	123	H(564)	0.06
C(386)-C(391)	1.46	C(380)-C(137)-C(144)	120	C(565)	−0.01
C(387)-C(131)	0.04	C(388)-C(137)-C(144)	123	C(566)	−0.04
C(387)-C(388)	1.38	C(143)-C(375)-C(144)	96	C(567)	−0.04
C(387)-C(392)	1.46	C(376)-C(375)-C(144)	118	C(568)	−0.01
C(388)-C(138)	0.04	C(382)-C(375)-C(144)	2	H(569)	0.06
H(389)-C(388)	1.09	C(137)-C(381)-C(144)	87	H(570)	0.06
H(390)-C(135)	1.09	C(382)-C(381)-C(144)	2	C(571)	−0.01
C(391)-C(129)	0.02	C(388)-C(381)-C(144)	123	C(572)	−0.01
C(391)-C(394)	1.38	C(148)-C(143)-C(144)	123	C(573)	−0.04
C(392)-C(130)	0.03	C(374)-C(143)-C(144)	122	C(574)	−0.04
C(392)-C(393)	1.38	C(376)-C(143)-C(144)	118	H(575)	0.06
C(393)-C(122)	0.02	C(382)-C(143)-C(144)	2	H(576)	0.06
C(393)-C(394)	1.47	C(148)-C(375)-C(144)	123	C(577)	−0.01

TABLE 2.2 (Continued).

Bond lengths	R,A	Valent corners	Grad	Atom	Charge (by Milliken)
C(393)-C(397)	1.46	C(374)-C(375)-C(144)	123	C(578)	−0.04
C(394)-C(121)	0.01	C(380)-C(381)-C(144)	119	C(579)	−0.01
C(394)-C(399)	1.46	C(144)-C(382)-H(145)	93	C(580)	−0.04
H(395)-C(391)	1.09	C(136)-C(141)-H(146)	120	H(581)	0.06
H(396)-C(130)	1.09	C(142)-C(141)-H(146)	117	H(582)	0.06
C(397)-C(123)	0.02	C(376)-C(141)-H(146)	117	C(583)	−0.01
C(397)-C(398)	1.38	C(380)-C(141)-H(146)	120	C(584)	−0.04
C(398)-C(124)	0.01	C(373)-C(142)-C(147)	1	C(585)	−0.04
C(398)-C(400)	1.48	C(143)-C(142)-C(147)	119	C(586)	−0.01
C(398)-C(406)	1.46	C(375)-C(142)-C(147)	120	H(587)	0.06
C(399)-C(126)	0	C(379)-C(142)-C(147)	123	H(588)	0.06
C(399)-C(400)	1.38	C(370)-C(150)-C(147)	123	C(589)	−0.01
C(400)-C(125)	0.01	C(366)-C(150)-C(147)	118	C(590)	−0.04
C(400)-C(403)	1.46	C(373)-C(150)-C(147)	1	C(591)	−0.04
H(401)-C(399)	1.09	C(150)-C(365)-C(147)	81	C(592)	−0.01
H(402)-C(123)	1.09	C(366)-C(365)-C(147)	118	H(593)	0.06
C(403)-C(172)	0.03	C(370)-C(365)-C(147)	123	H(594)	0.06
C(403)-C(404)	1.37	C(373)-C(365)-C(147)	1	C(595)	−0.02
C(404)-C(171)	0.04	C(142)-C(376)-C(147)	83	C(596)	−0.04
C(404)-C(405)	1.48	C(379)-C(376)-C(147)	122	C(597)	−0.04
C(404)-C(409)	1.47	C(149)-C(150)-C(147)	118	C(598)	−0.02
C(405)-C(170)	0.03	C(155)-C(150)-C(147)	123	H(599)	0.06
C(405)-C(406)	1.37	C(149)-C(365)-C(147)	118	H(600)	0.06
C(405)-C(412)	1.47	C(155)-C(365)-C(147)	122	C(601)	−0.02
C(406)-C(169)	0.01	C(373)-C(376)-C(147)	1	C(602)	−0.04
H(407)-C(169)	1.09	C(375)-C(376)-C(147)	120	C(603)	−0.04
H(408)-C(403)	1.09	C(143)-C(375)-C(148)	90	C(604)	−0.02
C(409)-C(178)	0.06	C(376)-C(375)-C(148)	119	H(605)	0.06
C(409)-C(410)	1.36	C(382)-C(375)-C(148)	123	H(606)	0.06
C(410)-C(177)	0.07	C(153)-C(149)-C(148)	123		
C(410)-C(411)	1.45	C(150)-C(149)-C(148)	118		
C(411)-C(176)	0.06	C(365)-C(149)-C(148)	118		
C(411)-C(412)	1.36	C(367)-C(149)-C(148)	122		

TABLE 2.2 (Continued).

Bond lengths	R,A	Valent corners	Grad	Atom	Charge (by Milliken)
C(412)-C(175)	0.04	C(374)-C(149)-C(148)	1		
H(413)-C(175)	1.08	C(149)-C(366)-C(148)	101		
H(414)-H(182)	0.07	C(367)-C(366)-C(148)	123		
H(415)-H(181)	0.09	C(374)-C(366)-C(148)	1		
H(416)-H(180)	0.07	C(374)-C(143)-C(148)	1		
C(417)-C(162)	0.02	C(376)-C(143)-C(148)	119		
C(417)-C(418)	1.46	C(382)-C(143)-C(148)	123		
C(418)-C(161)	0.02	C(150)-C(366)-C(148)	118		
C(418)-C(419)	1.47	C(153)-C(366)-C(148)	123		
C(418)-C(423)	1.38	C(365)-C(366)-C(148)	118		
C(419)-C(160)	0.02	C(374)-C(375)-C(148)	1		
C(419)-C(420)	1.46	C(366)-C(153)-C(149)	1		
C(419)-C(424)	1.38	C(154)-C(153)-C(149)	123		
C(420)-C(159)	0.02	C(368)-C(153)-C(149)	123		
H(421)-C(417)	1.09	C(370)-C(150)-C(149)	119		
H(422)-C(159)	1.08	C(366)-C(150)-C(149)	1		
C(423)-C(166)	0.02	C(373)-C(150)-C(149)	118		
C(424)-C(165)	0.02	C(150)-C(365)-C(149)	108		
H(425)-C(423)	1.09	C(366)-C(365)-C(149)	1		
H(426)-C(165)	1.07	C(370)-C(365)-C(149)	119		
C(427)-C(432)	1.46	C(373)-C(365)-C(149)	118		
C(428)-C(427)	1.48	C(153)-C(367)-C(149)	84		
C(429)-C(428)	1.46	C(368)-C(367)-C(149)	122		
C(430)-C(429)	1.38	C(148)-C(374)-C(149)	79		
C(430)-C(431)	1.48	C(375)-C(374)-C(149)	122		
C(431)-C(478)	1.46	C(366)-C(148)-C(149)	1		
C(432)-C(431)	1.38	C(375)-C(148)-C(149)	123		
H(433)-C(429)	1.09	C(155)-C(150)-C(149)	119		
H(434)-C(432)	1.09	C(155)-C(365)-C(149)	120		
C(435)-C(427)	1.38	C(154)-C(367)-C(149)	122		
C(436)-C(428)	1.38	C(366)-C(367)-C(149)	1		
C(437)-C(436)	1.46	C(366)-C(374)-C(149)	1		
C(437)-C(438)	1.48	C(155)-C(370)-C(150)	101		

TABLE 2.2   (Continued).

Bond lengths	R,A	Valent corners	Grad	Atom	Charge (by Milliken)
C(438)-C(435)	1.46	C(149)-C(366)-C(150)	72		
H(439)-C(436)	1.09	C(367)-C(366)-C(150)	119		
H(440)-C(435)	1.09	C(374)-C(366)-C(150)	118		
C(441)-C(438)	1.38	C(147)-C(373)-C(150)	96		
C(442)-C(441)	1.46	C(376)-C(373)-C(150)	123		
C(442)-C(443)	1.47	C(365)-C(147)-C(150)	1		
C(443)-C(444)	1.46	C(376)-C(147)-C(150)	123		
C(444)-C(437)	1.38	C(153)-C(149)-C(150)	119		
H(445)-C(441)	1.09	C(365)-C(149)-C(150)	1		
H(446)-C(444)	1.09	C(367)-C(149)-C(150)	120		
C(447)-C(442)	1.38	C(374)-C(149)-C(150)	118		
C(448)-C(447)	1.46	C(156)-C(155)-C(150)	123		
C(448)-C(449)	1.47	C(365)-C(155)-C(150)	1		
C(449)-C(450)	1.46	C(369)-C(155)-C(150)	122		
C(450)-C(443)	1.38	C(153)-C(366)-C(150)	119		
H(451)-C(450)	1.09	C(365)-C(366)-C(150)	1		
H(452)-C(447)	1.09	C(156)-C(370)-C(150)	123		
C(453)-C(448)	1.38	C(365)-C(370)-C(150)	1		
C(454)-C(449)	1.38	C(369)-C(370)-C(150)	123		
C(455)-C(454)	1.46	C(365)-C(373)-C(150)	1		
C(455)-C(456)	1.47	C(142)-C(147)-H(151)	120		
C(456)-C(453)	1.46	C(150)-C(147)-H(151)	117		
H(457)-C(453)	1.09	C(365)-C(147)-H(151)	117		
H(458)-C(454)	1.09	C(376)-C(147)-H(151)	120		
C(459)-C(455)	1.38	C(148)-C(374)-H(152)	99		
C(460)-C(459)	1.46	C(375)-C(374)-H(152)	120		
C(460)-C(462)	1.47	C(149)-C(366)-C(153)	93		
C(461)-C(456)	1.38	C(367)-C(366)-C(153)	1		
C(462)-C(461)	1.46	C(374)-C(366)-C(153)	123		
H(463)-C(461)	1.09	C(162)-C(154)-C(153)	123		
H(464)-C(459)	1.09	C(156)-C(154)-C(153)	118		
C(465)-C(462)	1.38	C(367)-C(154)-C(153)	1		
C(466)-C(465)	1.46	C(369)-C(154)-C(153)	118		

TABLE 2.2 (Continued).

Bond lengths	R,A	Valent corners	Grad	Atom	Charge (by Milliken)
C(466)-C(467)	1.47	C(417)-C(154)-C(153)	122		
C(467)-C(468)	1.46	C(154)-C(368)-C(153)	112		
C(468)-C(460)	1.38	C(369)-C(368)-C(153)	117		
H(469)-C(468)	1.09	C(417)-C(368)-C(153)	123		
H(470)-C(465)	1.09	C(365)-C(149)-C(153)	119		
C(471)-C(466)	1.38	C(367)-C(149)-C(153)	1		
C(472)-C(467)	1.38	C(374)-C(149)-C(153)	123		
C(472)-C(540)	1.46	C(365)-C(366)-C(153)	119		
H(473)-C(472)	1.09	C(156)-C(368)-C(153)	117		
H(474)-C(471)	1.09	C(162)-C(368)-C(153)	124		
C(475)-C(430)	1.46	C(367)-C(368)-C(153)	1		
C(476)-C(475)	1.37	C(368)-C(162)-C(154)	1		
C(476)-C(477)	1.48	C(418)-C(162)-C(154)	123		
C(476)-C(481)	1.46	C(420)-C(156)-C(154)	119		
C(477)-C(482)	1.46	C(368)-C(156)-C(154)	1		
C(478)-C(477)	1.37	C(370)-C(156)-C(154)	117		
H(479)-C(475)	1.09	C(153)-C(367)-C(154)	72		
H(480)-C(478)	1.09	C(368)-C(367)-C(154)	1		
C(481)-C(107)	1.39	C(156)-C(369)-C(154)	129		
C(482)-C(106)	1.39	C(370)-C(369)-C(154)	118		
H(483)-C(481)	1.09	C(420)-C(369)-C(154)	119		
H(484)-C(482)	1.09	C(162)-C(417)-C(154)	77		
C(485)-C(490)	1.46	C(418)-C(417)-C(154)	122		
C(486)-C(485)	1.47	C(366)-C(153)-C(154)	123		
C(487)-C(486)	1.46	C(368)-C(153)-C(154)	1		
C(488)-C(487)	1.38	C(155)-C(156)-C(154)	118		
C(488)-C(489)	1.47	C(159)-C(156)-C(154)	120		
C(489)-C(536)	1.46	C(161)-C(162)-C(154)	123		
C(490)-C(489)	1.38	C(366)-C(367)-C(154)	122		
H(491)-C(487)	1.09	C(155)-C(369)-C(154)	118		
H(492)-C(490)	1.09	C(159)-C(369)-C(154)	120		
C(493)-C(485)	1.38	C(368)-C(369)-C(154)	1		
C(494)-C(486)	1.38	C(161)-C(417)-C(154)	122		

TABLE 2.2   (Continued).

Bond lengths	R,A	Valent corners	Grad	Atom	Charge (by Milliken)
C(495)-C(494)	1.46	C(368)-C(417)-C(154)	1		
C(495)-C(496)	1.47	C(370)-C(150)-C(155)	1		
C(496)-C(493)	1.46	C(366)-C(150)-C(155)	120		
H(497)-C(494)	1.09	C(373)-C(150)-C(155)	123		
H(498)-C(493)	1.09	C(420)-C(156)-C(155)	123		
C(499)-C(496)	1.38	C(368)-C(156)-C(155)	118		
C(500)-C(499)	1.46	C(370)-C(156)-C(155)	1		
C(500)-C(501)	1.47	C(150)-C(365)-C(155)	79		
C(501)-C(502)	1.46	C(366)-C(365)-C(155)	120		
C(502)-C(495)	1.38	C(370)-C(365)-C(155)	1		
H(503)-C(499)	1.09	C(373)-C(365)-C(155)	122		
H(504)-C(502)	1.09	C(156)-C(369)-C(155)	73		
C(505)-C(500)	1.38	C(370)-C(369)-C(155)	1		
C(506)-C(505)	1.46	C(420)-C(369)-C(155)	122		
C(506)-C(507)	1.47	C(159)-C(156)-C(155)	123		
C(507)-C(508)	1.46	C(159)-C(369)-C(155)	122		
C(508)-C(501)	1.38	C(368)-C(369)-C(155)	118		
H(509)-C(508)	1.09	C(159)-C(420)-C(156)	112		
H(510)-C(505)	1.09	C(154)-C(368)-C(156)	53		
C(511)-C(506)	1.38	C(369)-C(368)-C(156)	1		
C(512)-C(507)	1.38	C(417)-C(368)-C(156)	119		
C(513)-C(512)	1.46	C(155)-C(370)-C(156)	103		
C(513)-C(514)	1.47	C(162)-C(154)-C(156)	119		
C(514)-C(511)	1.46	C(367)-C(154)-C(156)	118		
H(515)-C(511)	1.09	C(369)-C(154)-C(156)	1		
H(516)-C(512)	1.09	C(417)-C(154)-C(156)	120		
C(517)-C(513)	1.38	C(365)-C(155)-C(156)	122		
C(518)-C(517)	1.45	C(369)-C(155)-C(156)	1		
C(518)-C(520)	1.47	C(160)-C(159)-C(156)	123		
C(519)-C(514)	1.38	C(369)-C(159)-C(156)	1		
C(520)-C(519)	1.45	C(419)-C(159)-C(156)	122		
H(521)-C(519)	1.09	C(162)-C(368)-C(156)	119		
H(522)-C(517)	1.09	C(367)-C(368)-C(156)	118		

TABLE 2.2　(Continued).

Bond lengths	R,A	Valent corners	Grad	Atom	Charge (by Milliken)
C(523)-C(520)	1.39	C(365)-C(370)-C(156)	123		
C(524)-C(523)	1.45	C(369)-C(370)-C(156)	1		
C(524)-C(525)	1.46	C(160)-C(420)-C(156)	124		
C(525)-C(526)	1.45	C(369)-C(420)-C(156)	1		
C(526)-C(518)	1.39	C(419)-C(420)-C(156)	123		
H(527)-C(526)	1.09	C(150)-C(155)-H(157)	120		
H(528)-C(523)	1.09	C(156)-C(155)-H(157)	117		
C(529)-C(524)	1.39	C(365)-C(155)-H(157)	120		
C(530)-C(525)	1.39	C(369)-C(155)-H(157)	117		
C(530)-C(602)	1.44	C(153)-C(367)-H(158)	112		
H(531)-C(530)	1.09	C(368)-C(367)-H(158)	117		
H(532)-C(529)	1.09	C(420)-C(156)-C(159)	1		
C(533)-C(488)	1.46	C(368)-C(156)-C(159)	120		
C(534)-C(533)	1.38	C(370)-C(156)-C(159)	123		
C(534)-C(535)	1.47	C(424)-C(160)-C(159)	123		
C(535)-C(542)	1.46	C(161)-C(160)-C(159)	118		
C(536)-C(535)	1.38	C(418)-C(160)-C(159)	118		
H(537)-C(533)	1.09	C(420)-C(160)-C(159)	1		
H(538)-C(536)	1.09	C(156)-C(369)-C(159)	70		
C(539)-C(534)	1.46	C(370)-C(369)-C(159)	122		
C(540)-C(539)	1.38	C(420)-C(369)-C(159)	1		
C(540)-C(541)	1.47	C(160)-C(419)-C(159)	63		
C(541)-C(471)	1.46	C(420)-C(419)-C(159)	1		
C(542)-C(541)	1.38	C(424)-C(419)-C(159)	122		
H(543)-C(539)	1.09	C(165)-C(160)-C(159)	123		
H(544)-C(542)	1.09	C(368)-C(369)-C(159)	120		
C(545)-C(550)	1.39	C(161)-C(419)-C(159)	118		
C(546)-C(545)	1.46	C(165)-C(419)-C(159)	122		
C(547)-C(546)	1.39	C(418)-C(419)-C(159)	118		
C(548)-C(547)	1.44	C(165)-C(424)-C(160)	116		
C(549)-C(548)	1.46	C(166)-C(161)-C(160)	119		
C(550)-C(549)	1.44	C(162)-C(161)-C(160)	118		
H(551)-C(547)	1.09	C(417)-C(161)-C(160)	118		

TABLE 2.2   (Continued).

Bond lengths	R,A	Valent corners	Grad	Atom	Charge (by Milliken)
H(552)-C(550)	1.09	C(419)-C(161)-C(160)	0		
C(553)-C(545)	1.45	C(423)-C(161)-C(160)	120		
C(554)-C(546)	1.45	C(161)-C(418)-C(160)	17		
C(555)-C(554)	1.38	C(419)-C(418)-C(160)	0		
C(555)-C(556)	1.47	C(423)-C(418)-C(160)	119		
C(555)-C(562)	1.45	C(159)-C(420)-C(160)	115		
C(556)-C(553)	1.38	C(369)-C(159)-C(160)	122		
H(557)-C(554)	1.09	C(419)-C(159)-C(160)	1		
H(558)-C(553)	1.09	C(350)-C(165)-C(160)	122		
C(559)-C(556)	1.45	C(419)-C(165)-C(160)	1		
C(560)-C(559)	1.38	C(162)-C(418)-C(160)	117		
C(560)-C(561)	1.47	C(166)-C(418)-C(160)	119		
C(561)-C(568)	1.46	C(417)-C(418)-C(160)	118		
C(562)-C(561)	1.38	C(369)-C(420)-C(160)	123		
H(563)-C(559)	1.09	C(419)-C(420)-C(160)	1		
H(564)-C(562)	1.09	C(350)-C(424)-C(160)	123		
C(565)-C(560)	1.46	C(419)-C(424)-C(160)	1		
C(566)-C(565)	1.38	C(418)-C(166)-C(161)	1		
C(567)-C(566)	1.47	C(368)-C(162)-C(161)	123		
C(568)-C(567)	1.38	C(418)-C(162)-C(161)	1		
H(569)-C(568)	1.09	C(162)-C(417)-C(161)	57		
H(570)-C(565)	1.09	C(418)-C(417)-C(161)	1		
C(571)-C(566)	1.46	C(160)-C(419)-C(161)	166		
C(572)-C(567)	1.46	C(420)-C(419)-C(161)	118		
C(573)-C(572)	1.38	C(424)-C(419)-C(161)	119		
C(573)-C(574)	1.47	C(166)-C(423)-C(161)	71		
C(574)-C(571)	1.38	C(418)-C(160)-C(161)	180		
H(575)-C(571)	1.09	C(420)-C(160)-C(161)	117		
H(576)-C(572)	1.09	C(424)-C(160)-C(161)	119		
C(577)-C(573)	1.46	C(351)-C(166)-C(161)	123		
C(578)-C(577)	1.38	C(368)-C(417)-C(161)	122		
C(578)-C(580)	1.47	C(165)-C(419)-C(161)	120		
C(578)-C(586)	1.46	C(418)-C(419)-C(161)	180		

TABLE 2.2   (Continued).

Bond lengths	R,A	Valent corners	Grad	Atom	Charge (by Milliken)
C(579)-C(574)	1.46	C(351)-C(423)-C(161)	122		
C(580)-C(579)	1.38	C(418)-C(423)-C(161)	1		
H(581)-C(579)	1.09	C(154)-C(368)-C(162)	100		
H(582)-C(577)	1.09	C(369)-C(368)-C(162)	119		
C(583)-C(580)	1.46	C(417)-C(368)-C(162)	1		
C(584)-C(583)	1.38	C(161)-C(418)-C(162)	125		
C(584)-C(585)	1.47	C(419)-C(418)-C(162)	117		
C(585)-C(592)	1.46	C(423)-C(418)-C(162)	123		
C(586)-C(585)	1.38	C(367)-C(154)-C(162)	122		
H(587)-C(586)	1.09	C(369)-C(154)-C(162)	119		
H(588)-C(583)	1.09	C(417)-C(154)-C(162)	1		
C(589)-C(584)	1.46	C(166)-C(161)-C(162)	123		
C(590)-C(589)	1.38	C(417)-C(161)-C(162)	1		
C(590)-C(591)	1.47	C(419)-C(161)-C(162)	118		
C(591)-C(361)	1.46	C(423)-C(161)-C(162)	122		
C(592)-C(591)	1.38	C(367)-C(368)-C(162)	123		
H(593)-C(592)	1.09	C(166)-C(418)-C(162)	124		
H(594)-C(589)	1.09	C(417)-C(418)-C(162)	1		
C(595)-C(548)	1.4	C(162)-C(417)-H(163)	140		
C(596)-C(595)	1.43	C(418)-C(417)-H(163)	117		
C(596)-C(597)	1.45	C(156)-C(159)-H(164)	120		
C(597)-C(604)	1.42	C(160)-C(159)-H(164)	117		
C(597)-C(598)	1.43	C(369)-C(159)-H(164)	121		
C(598)-C(549)	1.4	C(419)-C(159)-H(164)	118		
H(599)-C(595)	1.09	C(424)-C(160)-C(165)	1		
H(600)-C(598)	1.09	C(161)-C(160)-C(165)	120		
C(601)-C(596)	1.42	C(418)-C(160)-C(165)	120		
C(602)-C(601)	1.41	C(420)-C(160)-C(165)	123		
C(602)-C(603)	1.46	C(349)-C(350)-C(165)	122		
C(603)-C(529)	1.44	C(351)-C(350)-C(165)	118		
C(604)-C(603)	1.41	C(424)-C(350)-C(165)	1		
H(605)-C(601)	1.09	C(160)-C(419)-C(165)	60		
H(606)-C(604)	1.09	C(420)-C(419)-C(165)	122		

TABLE 2.2   (Continued).

Bond lengths	R,A	Valent corners	Grad	Atom	Charge (by Milliken)
		C(424)-C(419)-C(165)	1		
		C(418)-C(419)-C(165)	120		
		C(161)-C(418)-C(166)	109		
		C(419)-C(418)-C(166)	119		
		C(423)-C(418)-C(166)	1		
		C(417)-C(161)-C(166)	122		
		C(419)-C(161)-C(166)	119		
		C(423)-C(161)-C(166)	1		
		C(423)-C(351)-C(166)	1		
		C(417)-C(418)-C(166)	123		
		C(166)-C(423)-H(167)	167		
		C(160)-C(165)-H(168)	120		
		C(350)-C(165)-H(168)	118		
		C(419)-C(165)-H(168)	121		
		C(406)-C(170)-C(169)	180		
		C(171)-C(170)-C(169)	120		
		C(175)-C(170)-C(169)	123		
		C(404)-C(170)-C(169)	120		
		C(412)-C(170)-C(169)	123		
		C(124)-C(398)-C(169)	90		
		C(400)-C(398)-C(169)	118		
		C(406)-C(398)-C(169)	1		
		C(170)-C(405)-C(169)	80		
		C(406)-C(405)-C(169)	180		
		C(412)-C(405)-C(169)	123		
		C(397)-C(124)-C(169)	123		
		C(400)-C(124)-C(169)	118		
		C(406)-C(124)-C(169)	1		
		C(397)-C(398)-C(169)	123		
		C(171)-C(405)-C(169)	120		
		C(175)-C(405)-C(169)	122		
		C(404)-C(405)-C(169)	120		
		C(169)-C(406)-C(170)	113		

TABLE 2.2   (Continued).

Bond lengths	R,A	Valent corners	Grad	Atom	Charge (by Milliken)
		C(172)-C(171)-C(170)	120		
		C(178)-C(171)-C(170)	118		
		C(403)-C(171)-C(170)	120		
		C(405)-C(171)-C(170)	1		
		C(409)-C(171)-C(170)	118		
		C(176)-C(175)-C(170)	122		
		C(405)-C(175)-C(170)	1		
		C(411)-C(175)-C(170)	121		
		C(171)-C(404)-C(170)	79		
		C(405)-C(404)-C(170)	1		
		C(409)-C(404)-C(170)	118		
		C(175)-C(412)-C(170)	90		
		C(398)-C(169)-C(170)	123		
		C(405)-C(169)-C(170)	1		
		C(172)-C(404)-C(170)	119		
		C(178)-C(404)-C(170)	117		
		C(403)-C(404)-C(170)	119		
		C(398)-C(406)-C(170)	123		
		C(405)-C(406)-C(170)	1		
		C(176)-C(412)-C(170)	122		
		C(405)-C(412)-C(170)	1		
		C(411)-C(412)-C(170)	122		
		C(404)-C(172)-C(171)	2		
		C(400)-C(172)-C(171)	123		
		C(410)-C(178)-C(171)	122		
		C(404)-C(178)-C(171)	2		
		C(172)-C(403)-C(171)	84		
		C(404)-C(403)-C(171)	2		
		C(170)-C(405)-C(171)	101		
		C(406)-C(405)-C(171)	119		
		C(412)-C(405)-C(171)	118		
		C(178)-C(409)-C(171)	84		
		C(410)-C(409)-C(171)	121		

TABLE 2.2   (Continued).

Bond lengths	R,A	Valent corners	Grad	Atom	Charge (by Milliken)
		C(175)-C(170)-C(171)	118		
		C(404)-C(170)-C(171)	2		
		C(406)-C(170)-C(171)	119		
		C(412)-C(170)-C(171)	117		
		C(177)-C(178)-C(171)	122		
		C(409)-C(178)-C(171)	93		
		C(400)-C(403)-C(171)	122		
		C(175)-C(405)-C(171)	118		
		C(404)-C(405)-C(171)	2		
		C(177)-C(409)-C(171)	121		
		C(404)-C(409)-C(171)	2		
		C(171)-C(404)-C(172)	92		
		C(405)-C(404)-C(172)	119		
		C(409)-C(404)-C(172)	123		
		C(125)-C(400)-C(172)	104		
		C(403)-C(400)-C(172)	1		
		C(398)-C(125)-C(172)	118		
		C(399)-C(125)-C(172)	123		
		C(403)-C(125)-C(172)	1		
		C(178)-C(171)-C(172)	123		
		C(403)-C(171)-C(172)	1		
		C(405)-C(171)-C(172)	120		
		C(409)-C(171)-C(172)	123		
		C(398)-C(400)-C(172)	118		
		C(399)-C(400)-C(172)	123		
		C(178)-C(404)-C(172)	123		
		C(403)-C(404)-C(172)	1		
		C(170)-C(169)-H(173)	120		
		C(398)-C(169)-H(173)	117		
		C(405)-C(169)-H(173)	120		
		C(172)-C(403)-H(174)	99		
		C(404)-C(403)-H(174)	120		
		C(412)-C(176)-C(175)	2		

TABLE 2.2 (Continued).

Bond lengths	R,A	Valent corners	Grad	Atom	Charge (by Milliken)
		C(177)-C(176)-C(175)	121		
		C(410)-C(176)-C(175)	121		
		C(170)-C(405)-C(175)	87		
		C(406)-C(405)-C(175)	122		
		C(412)-C(405)-C(175)	2		
		C(176)-C(411)-C(175)	82		
		C(412)-C(411)-C(175)	2		
		C(404)-C(170)-C(175)	118		
		C(406)-C(170)-C(175)	123		
		C(412)-C(170)-C(175)	2		
		C(411)-C(176)-C(175)	95		
		C(404)-C(405)-C(175)	118		
		C(177)-C(411)-C(175)	121		
		C(410)-C(411)-C(175)	121		
		C(175)-C(412)-C(176)	96		
		C(178)-C(177)-C(176)	121		
		C(409)-C(177)-C(176)	121		
		C(411)-C(177)-C(176)	2		
		C(177)-C(410)-C(176)	80		
		C(411)-C(410)-C(176)	2		
		C(405)-C(175)-C(176)	121		
		C(411)-C(175)-C(176)	3		
		C(410)-C(177)-C(176)	97		
		C(178)-C(410)-C(176)	120		
		C(409)-C(410)-C(176)	120		
		C(412)-C(411)-C(176)	84		
		C(405)-C(412)-C(176)	122		
		C(411)-C(412)-C(176)	3		
		C(410)-C(178)-C(177)	3		
		C(404)-C(178)-C(177)	122		
		C(178)-C(409)-C(177)	84		
		C(410)-C(409)-C(177)	3		
		C(176)-C(411)-C(177)	95		

TABLE 2.2   (Continued).

Bond lengths	R,A	Valent corners	Grad	Atom	Charge (by Milliken)
		C(412)-C(411)-C(177)	121		
		C(410)-C(176)-C(177)	3		
		C(412)-C(176)-C(177)	120		
		C(409)-C(178)-C(177)	94		
		C(404)-C(409)-C(177)	121		
		C(411)-C(410)-C(177)	83		
		C(410)-C(411)-C(177)	3		
		C(177)-C(410)-C(178)	90		
		C(411)-C(410)-C(178)	120		
		C(171)-C(404)-C(178)	94		
		C(405)-C(404)-C(178)	117		
		C(409)-C(404)-C(178)	2		
		C(403)-C(171)-C(178)	123		
		C(405)-C(171)-C(178)	118		
		C(409)-C(171)-C(178)	2		
		C(409)-C(177)-C(178)	3		
		C(411)-C(177)-C(178)	121		
		C(403)-C(404)-C(178)	123		
		C(410)-C(409)-C(178)	87		
		C(409)-C(410)-C(178)	3		
		C(176)-C(175)-H(179)	120		
		C(405)-C(175)-H(179)	118		
		C(411)-C(175)-H(179)	121		
		C(187)-C(188)-C(183)	123		
		C(188)-C(183)-C(184)	118		
		C(191)-C(183)-C(184)	120		
		C(183)-C(184)-C(185)	118		
		C(192)-C(184)-C(185)	123		
		C(187)-C(186)-C(185)	119		
		C(184)-C(185)-C(186)	123		
		C(234)-C(187)-C(186)	118		
		C(188)-C(187)-C(186)	120		
		C(233)-C(234)-C(187)	123		

TABLE 2.2   (Continued).

Bond lengths	R,A	Valent corners	Grad	Atom	Charge (by Milliken)
		C(234)-C(187)-C(188)	123		
		C(184)-C(185)-H(189)	117		
		C(187)-C(188)-H(190)	120		
		C(188)-C(183)-C(191)	123		
		C(183)-C(184)-C(192)	119		
		C(194)-C(193)-C(192)	118		
		C(184)-C(192)-C(193)	123		
		C(191)-C(194)-C(193)	118		
		C(197)-C(194)-C(193)	119		
		C(183)-C(191)-C(194)	123		
		C(184)-C(192)-H(195)	120		
		C(183)-C(191)-H(196)	120		
		C(191)-C(194)-C(197)	123		
		C(199)-C(198)-C(197)	118		
		C(194)-C(197)-C(198)	123		
		C(200)-C(199)-C(198)	118		
		C(206)-C(199)-C(198)	119		
		C(193)-C(200)-C(199)	123		
		C(192)-C(193)-C(200)	123		
		C(194)-C(193)-C(200)	119		
		C(194)-C(197)-H(201)	120		
		C(193)-C(200)-H(202)	120		
		C(197)-C(198)-C(203)	123		
		C(199)-C(198)-C(203)	119		
		C(205)-C(204)-C(203)	118		
		C(198)-C(203)-C(204)	123		
		C(206)-C(205)-C(204)	118		
		C(210)-C(205)-C(204)	119		
		C(199)-C(206)-C(205)	123		
		C(200)-C(199)-C(206)	123		
		C(199)-C(206)-H(207)	120		
		C(198)-C(203)-H(208)	120		
		C(203)-C(204)-C(209)	123		

TABLE 2.2   (Continued).

Bond lengths	R,A	Valent corners	Grad	Atom	Charge (by Milliken)
		C(205)-C(204)-C(209)	119		
		C(206)-C(205)-C(210)	123		
		C(212)-C(211)-C(210)	118		
		C(205)-C(210)-C(211)	123		
		C(209)-C(212)-C(211)	118		
		C(217)-C(212)-C(211)	119		
		C(204)-C(209)-C(212)	123		
		C(204)-C(209)-H(213)	120		
		C(205)-C(210)-H(214)	120		
		C(210)-C(211)-C(215)	123		
		C(212)-C(211)-C(215)	119		
		C(218)-C(216)-C(215)	118		
		C(211)-C(215)-C(216)	123		
		C(217)-C(218)-C(216)	118		
		C(221)-C(218)-C(216)	119		
		C(209)-C(212)-C(217)	123		
		C(212)-C(217)-C(218)	123		
		C(212)-C(217)-H(219)	120		
		C(211)-C(215)-H(220)	120		
		C(217)-C(218)-C(221)	123		
		C(223)-C(222)-C(221)	118		
		C(218)-C(221)-C(222)	123		
		C(224)-C(223)-C(222)	118		
		C(228)-C(223)-C(222)	119		
		C(216)-C(224)-C(223)	123		
		C(300)-C(228)-C(223)	122		
		C(215)-C(216)-C(224)	123		
		C(218)-C(216)-C(224)	119		
		C(216)-C(224)-H(225)	120		
		C(218)-C(221)-H(226)	120		
		C(221)-C(222)-C(227)	123		
		C(223)-C(222)-C(227)	119		
		C(224)-C(223)-C(228)	123		

TABLE 2.2 (Continued).

Bond lengths	R,A	Valent corners	Grad	Atom	Charge (by Milliken)
		C(299)-C(300)-C(228)	123		
		C(301)-C(300)-C(228)	118		
		C(223)-C(228)-H(229)	120		
		C(300)-C(228)-H(229)	118		
		C(222)-C(227)-H(230)	120		
		C(185)-C(186)-C(231)	123		
		C(187)-C(186)-C(231)	118		
		C(233)-C(232)-C(231)	119		
		C(186)-C(231)-C(232)	123		
		C(240)-C(233)-C(232)	118		
		C(234)-C(233)-C(232)	120		
		C(239)-C(240)-C(233)	123		
		C(240)-C(233)-C(234)	123		
		C(186)-C(231)-H(235)	117		
		C(233)-C(234)-H(236)	120		
		C(231)-C(232)-C(237)	123		
		C(233)-C(232)-C(237)	118		
		C(232)-C(237)-C(238)	123		
		C(237)-C(238)-C(239)	119		
		C(238)-C(239)-C(240)	120		
		C(232)-C(237)-H(241)	117		
		C(239)-C(240)-H(242)	120		
		C(247)-C(248)-C(243)	122		
		C(248)-C(243)-C(244)	119		
		C(251)-C(243)-C(244)	118		
		C(243)-C(244)-C(245)	119		
		C(252)-C(244)-C(245)	123		
		C(244)-C(245)-C(246)	122		
		C(245)-C(246)-C(247)	118		
		C(293)-C(246)-C(247)	119		
		C(246)-C(247)-C(248)	118		
		C(296)-C(247)-C(248)	123		
		C(244)-C(245)-H(249)	120		

TABLE 2.2 (Continued).

Bond lengths	R,A	Valent corners	Grad	Atom	Charge (by Milliken)
		C(247)-C(248)-H(250)	118		
		C(248)-C(243)-C(251)	123		
		C(243)-C(244)-C(252)	118		
		C(254)-C(253)-C(252)	119		
		C(260)-C(253)-C(252)	123		
		C(244)-C(252)-C(253)	123		
		C(251)-C(254)-C(253)	119		
		C(259)-C(260)-C(253)	123		
		C(257)-C(254)-C(253)	118		
		C(243)-C(251)-C(254)	123		
		C(260)-C(253)-C(254)	118		
		C(244)-C(252)-H(255)	118		
		C(243)-C(251)-H(256)	117		
		C(251)-C(254)-C(257)	123		
		C(259)-C(258)-C(257)	119		
		C(254)-C(257)-C(258)	123		
		C(266)-C(259)-C(258)	118		
		C(260)-C(259)-C(258)	119		
		C(265)-C(266)-C(259)	123		
		C(266)-C(259)-C(260)	123		
		C(254)-C(257)-H(261)	117		
		C(259)-C(260)-H(262)	120		
		C(257)-C(258)-C(263)	123		
		C(259)-C(258)-C(263)	118		
		C(258)-C(263)-C(264)	123		
		C(263)-C(264)-C(265)	119		
		C(269)-C(264)-C(265)	118		
		C(264)-C(265)-C(266)	119		
		C(270)-C(265)-C(266)	123		
		C(265)-C(266)-H(267)	120		
		C(258)-C(263)-H(268)	117		
		C(263)-C(264)-C(269)	123		
		C(264)-C(265)-C(270)	118		

TABLE 2.2 (Continued).

Bond lengths	R,A	Valent corners	Grad	Atom	Charge (by Milliken)
		C(272)-C(271)-C(270)	119		
		C(265)-C(270)-C(271)	123		
		C(269)-C(272)-C(271)	119		
		C(277)-C(272)-C(271)	118		
		C(264)-C(269)-C(272)	123		
		C(264)-C(269)-H(273)	117		
		C(265)-C(270)-H(274)	117		
		C(270)-C(271)-C(275)	123		
		C(272)-C(271)-C(275)	118		
		C(278)-C(276)-C(275)	119		
		C(284)-C(276)-C(275)	123		
		C(271)-C(275)-C(276)	123		
		C(277)-C(278)-C(276)	119		
		C(283)-C(284)-C(276)	123		
		C(281)-C(278)-C(276)	118		
		C(269)-C(272)-C(277)	123		
		C(272)-C(277)-C(278)	123		
		C(284)-C(276)-C(278)	118		
		C(272)-C(277)-H(279)	117		
		C(271)-C(275)-H(280)	117		
		C(277)-C(278)-C(281)	123		
		C(283)-C(282)-C(281)	120		
		C(278)-C(281)-C(282)	123		
		C(290)-C(283)-C(282)	118		
		C(284)-C(283)-C(282)	119		
		C(289)-C(290)-C(283)	123		
		C(290)-C(283)-C(284)	123		
		C(283)-C(284)-H(285)	120		
		C(278)-C(281)-H(286)	117		
		C(281)-C(282)-C(287)	123		
		C(283)-C(282)-C(287)	118		
		C(289)-C(288)-C(287)	120		
		C(282)-C(287)-C(288)	123		

TABLE 2.2   (Continued).

Bond lengths	R,A	Valent corners	Grad	Atom	Charge (by Milliken)
		C(57)-C(289)-C(288)	118		
		C(290)-C(289)-C(288)	119		
		C(52)-C(57)-C(289)	123		
		C(57)-C(289)-C(290)	123		
		C(289)-C(290)-H(291)	120		
		C(282)-C(287)-H(292)	117		
		C(245)-C(246)-C(293)	123		
		C(295)-C(294)-C(293)	119		
		C(246)-C(293)-C(294)	122		
		C(296)-C(295)-C(294)	119		
		C(302)-C(295)-C(294)	119		
		C(247)-C(296)-C(295)	122		
		C(301)-C(302)-C(295)	122		
		C(246)-C(247)-C(296)	119		
		C(302)-C(295)-C(296)	123		
		C(246)-C(293)-H(297)	119		
		C(247)-C(296)-H(298)	119		
		C(293)-C(294)-C(299)	123		
		C(295)-C(294)-C(299)	119		
		C(301)-C(300)-C(299)	119		
		C(294)-C(299)-C(300)	122		
		C(227)-C(301)-C(300)	118		
		C(302)-C(301)-C(300)	119		
		C(222)-C(227)-C(301)	122		
		C(227)-C(301)-C(302)	123		
		C(294)-C(299)-H(303)	118		
		C(301)-C(302)-H(304)	119		
		C(309)-C(310)-C(305)	123		
		C(310)-C(305)-C(306)	120		
		C(313)-C(305)-C(306)	118		
		C(305)-C(306)-C(307)	119		
		C(314)-C(306)-C(307)	123		
		C(306)-C(307)-C(308)	123		

TABLE 2.2   (Continued).

Bond lengths	R,A	Valent corners	Grad	Atom	Charge (by Milliken)
		C(307)-C(308)-C(309)	118		
		C(355)-C(308)-C(309)	119		
		C(308)-C(309)-C(310)	118		
		C(358)-C(309)-C(310)	123		
		C(306)-C(307)-H(311)	120		
		C(309)-C(310)-H(312)	117		
		C(310)-C(305)-C(313)	123		
		C(305)-C(306)-C(314)	118		
		C(316)-C(315)-C(314)	119		
		C(322)-C(315)-C(314)	123		
		C(306)-C(314)-C(315)	123		
		C(313)-C(316)-C(315)	120		
		C(321)-C(322)-C(315)	123		
		C(319)-C(316)-C(315)	118		
		C(305)-C(313)-C(316)	123		
		C(322)-C(315)-C(316)	118		
		C(306)-C(314)-H(317)	117		
		C(305)-C(313)-H(318)	117		
		C(313)-C(316)-C(319)	123		
		C(321)-C(320)-C(319)	120		
		C(316)-C(319)-C(320)	123		
		C(328)-C(321)-C(320)	118		
		C(322)-C(321)-C(320)	119		
		C(327)-C(328)-C(321)	123		
		C(328)-C(321)-C(322)	123		
		C(316)-C(319)-H(323)	117		
		C(321)-C(322)-H(324)	120		
		C(319)-C(320)-C(325)	123		
		C(321)-C(320)-C(325)	118		
		C(320)-C(325)-C(326)	123		
		C(325)-C(326)-C(327)	120		
		C(331)-C(326)-C(327)	118		
		C(326)-C(327)-C(328)	119		

TABLE 2.2   (Continued).

Bond lengths	R,A	Valent corners	Grad	Atom	Charge (by Milliken)
		C(332)-C(327)-C(328)	123		
		C(327)-C(328)-H(329)	120		
		C(320)-C(325)-H(330)	117		
		C(325)-C(326)-C(331)	123		
		C(326)-C(327)-C(332)	118		
		C(334)-C(333)-C(332)	119		
		C(327)-C(332)-C(333)	123		
		C(331)-C(334)-C(333)	120		
		C(339)-C(334)-C(333)	118		
		C(326)-C(331)-C(334)	123		
		C(326)-C(331)-H(335)	117		
		C(327)-C(332)-H(336)	117		
		C(332)-C(333)-C(337)	123		
		C(334)-C(333)-C(337)	118		
		C(340)-C(338)-C(337)	119		
		C(346)-C(338)-C(337)	123		
		C(333)-C(337)-C(338)	123		
		C(339)-C(340)-C(338)	120		
		C(345)-C(346)-C(338)	123		
		C(343)-C(340)-C(338)	118		
		C(331)-C(334)-C(339)	123		
		C(334)-C(339)-C(340)	123		
		C(346)-C(338)-C(340)	118		
		C(334)-C(339)-H(341)	117		
		C(333)-C(337)-H(342)	117		
		C(339)-C(340)-C(343)	123		
		C(345)-C(344)-C(343)	119		
		C(340)-C(343)-C(344)	123		
		C(352)-C(345)-C(344)	118		
		C(346)-C(345)-C(344)	119		
		C(351)-C(352)-C(345)	123		
		C(352)-C(345)-C(346)	123		
		C(345)-C(346)-H(347)	120		

TABLE 2.2   (Continued).

Bond lengths	R,A	Valent corners	Grad	Atom	Charge (by Milliken)
		C(340)-C(343)-H(348)	117		
		C(343)-C(344)-C(349)	123		
		C(345)-C(344)-C(349)	118		
		C(351)-C(350)-C(349)	119		
		C(424)-C(350)-C(349)	123		
		C(344)-C(349)-C(350)	123		
		C(166)-C(351)-C(350)	117		
		C(423)-C(351)-C(350)	118		
		C(165)-C(424)-C(350)	121		
		C(419)-C(165)-C(350)	121		
		C(352)-C(351)-C(350)	119		
		C(419)-C(424)-C(350)	123		
		C(418)-C(166)-C(351)	124		
		C(166)-C(423)-C(351)	51		
		C(424)-C(350)-C(351)	118		
		C(418)-C(423)-C(351)	123		
		C(166)-C(351)-C(352)	123		
		C(423)-C(351)-C(352)	123		
		C(351)-C(352)-H(353)	120		
		C(344)-C(349)-H(354)	117		
		C(307)-C(308)-C(355)	123		
		C(357)-C(356)-C(355)	118		
		C(308)-C(355)-C(356)	123		
		C(358)-C(357)-C(356)	118		
		C(362)-C(357)-C(356)	120		
		C(309)-C(358)-C(357)	123		
		C(590)-C(362)-C(357)	123		
		C(308)-C(309)-C(358)	120		
		C(308)-C(355)-H(359)	120		
		C(309)-C(358)-H(360)	120		
		C(355)-C(356)-C(361)	123		
		C(357)-C(356)-C(361)	119		
		C(358)-C(357)-C(362)	123		

TABLE 2.2 (Continued).

Bond lengths	R,A	Valent corners	Grad	Atom	Charge (by Milliken)
		C(589)-C(590)-C(362)	123		
		C(591)-C(590)-C(362)	118		
		C(356)-C(361)-H(363)	120		
		C(357)-C(362)-H(364)	120		
		C(590)-C(362)-H(364)	117		
		C(370)-C(150)-C(365)	99		
		C(366)-C(150)-C(365)	70		
		C(373)-C(150)-C(365)	97		
		C(149)-C(366)-C(365)	73		
		C(367)-C(366)-C(365)	119		
		C(374)-C(366)-C(365)	118		
		C(155)-C(370)-C(365)	102		
		C(147)-C(373)-C(365)	97		
		C(376)-C(373)-C(365)	123		
		C(376)-C(147)-C(365)	122		
		C(367)-C(149)-C(365)	120		
		C(374)-C(149)-C(365)	118		
		C(369)-C(155)-C(365)	122		
		C(369)-C(370)-C(365)	123		
		C(153)-C(149)-C(366)	86		
		C(150)-C(149)-C(366)	107		
		C(365)-C(149)-C(366)	106		
		C(367)-C(149)-C(366)	85		
		C(374)-C(149)-C(366)	77		
		C(153)-C(367)-C(366)	85		
		C(368)-C(367)-C(366)	123		
		C(148)-C(374)-C(366)	80		
		C(375)-C(374)-C(366)	123		
		C(375)-C(148)-C(366)	123		
		C(370)-C(150)-C(366)	119		
		C(373)-C(150)-C(366)	117		
		C(368)-C(153)-C(366)	123		
		C(370)-C(365)-C(366)	119		

TABLE 2.2 (Continued).

Bond lengths	R,A	Valent corners	Grad	Atom	Charge (by Milliken)
		C(373)-C(365)-C(366)	118		
		C(366)-C(153)-C(367)	94		
		C(154)-C(153)-C(367)	107		
		C(368)-C(153)-C(367)	107		
		C(154)-C(368)-C(367)	113		
		C(369)-C(368)-C(367)	118		
		C(417)-C(368)-C(367)	123		
		C(374)-C(149)-C(367)	122		
		C(369)-C(154)-C(367)	118		
		C(417)-C(154)-C(367)	122		
		C(374)-C(366)-C(367)	123		
		C(162)-C(154)-C(368)	79		
		C(156)-C(154)-C(368)	127		
		C(367)-C(154)-C(368)	67		
		C(369)-C(154)-C(368)	126		
		C(417)-C(154)-C(368)	78		
		C(156)-C(369)-C(368)	130		
		C(370)-C(369)-C(368)	118		
		C(420)-C(369)-C(368)	119		
		C(162)-C(417)-C(368)	78		
		C(418)-C(417)-C(368)	123		
		C(370)-C(156)-C(368)	117		
		C(420)-C(156)-C(368)	119		
		C(418)-C(162)-C(368)	124		
		C(420)-C(156)-C(369)	109		
		C(368)-C(156)-C(369)	50		
		C(370)-C(156)-C(369)	106		
		C(155)-C(370)-C(369)	104		
		C(159)-C(420)-C(369)	112		
		C(417)-C(154)-C(369)	120		
		C(419)-C(159)-C(369)	122		
		C(417)-C(368)-C(369)	119		
		C(419)-C(420)-C(369)	123		

TABLE 2.2   (Continued).

Bond lengths	R,A	Valent corners	Grad	Atom	Charge (by Milliken)
		C(150)-C(155)-C(370)	78		
		C(156)-C(155)-C(370)	76		
		C(365)-C(155)-C(370)	77		
		C(369)-C(155)-C(370)	75		
		C(373)-C(150)-C(370)	123		
		C(420)-C(156)-C(370)	124		
		C(373)-C(365)-C(370)	123		
		C(420)-C(369)-C(370)	123		
		C(153)-C(367)-H(371)	114		
		C(368)-C(367)-H(371)	117		
		C(150)-C(155)-H(372)	120		
		C(156)-C(155)-H(372)	117		
		C(365)-C(155)-H(372)	121		
		C(369)-C(155)-H(372)	117		
		C(142)-C(147)-C(373)	84		
		C(150)-C(147)-C(373)	83		
		C(365)-C(147)-C(373)	82		
		C(376)-C(147)-C(373)	83		
		C(142)-C(376)-C(373)	85		
		C(379)-C(376)-C(373)	123		
		C(375)-C(142)-C(373)	119		
		C(379)-C(142)-C(373)	123		
		C(375)-C(376)-C(373)	119		
		C(375)-C(148)-C(374)	90		
		C(149)-C(148)-C(374)	100		
		C(366)-C(148)-C(374)	99		
		C(143)-C(375)-C(374)	91		
		C(376)-C(375)-C(374)	119		
		C(382)-C(375)-C(374)	123		
		C(376)-C(143)-C(374)	120		
		C(382)-C(143)-C(374)	122		
		C(148)-C(143)-C(375)	89		
		C(144)-C(143)-C(375)	83		

TABLE 2.2   (Continued).

Bond lengths	R,A	Valent corners	Grad	Atom	Charge (by Milliken)
		C(374)-C(143)-C(375)	87		
		C(376)-C(143)-C(375)	97		
		C(382)-C(143)-C(375)	81		
		C(142)-C(376)-C(375)	100		
		C(379)-C(376)-C(375)	118		
		C(144)-C(382)-C(375)	84		
		C(379)-C(142)-C(375)	117		
		C(381)-C(144)-C(375)	123		
		C(381)-C(382)-C(375)	123		
		C(373)-C(142)-C(376)	94		
		C(143)-C(142)-C(376)	81		
		C(375)-C(142)-C(376)	79		
		C(379)-C(142)-C(376)	92		
		C(141)-C(379)-C(376)	94		
		C(380)-C(379)-C(376)	123		
		C(380)-C(141)-C(376)	122		
		C(382)-C(143)-C(376)	118		
		C(382)-C(375)-C(376)	118		
		C(148)-C(374)-H(377)	101		
		C(375)-C(374)-H(377)	120		
		C(142)-C(147)-H(378)	120		
		C(150)-C(147)-H(378)	117		
		C(365)-C(147)-H(378)	117		
		C(376)-C(147)-H(378)	120		
		C(136)-C(141)-C(379)	87		
		C(142)-C(141)-C(379)	86		
		C(376)-C(141)-C(379)	85		
		C(380)-C(141)-C(379)	85		
		C(136)-C(380)-C(379)	87		
		C(381)-C(380)-C(379)	119		
		C(385)-C(380)-C(379)	123		
		C(381)-C(136)-C(379)	119		
		C(385)-C(136)-C(379)	123		

TABLE 2.2   (Continued).

Bond lengths	R,A	Valent corners	Grad	Atom	Charge (by Milliken)
		C(379)-C(136)-C(380)	92		
		C(137)-C(136)-C(380)	86		
		C(381)-C(136)-C(380)	84		
		C(385)-C(136)-C(380)	90		
		C(137)-C(381)-C(380)	86		
		C(382)-C(381)-C(380)	119		
		C(388)-C(381)-C(380)	118		
		C(135)-C(385)-C(380)	91		
		C(386)-C(385)-C(380)	123		
		C(386)-C(135)-C(380)	123		
		C(382)-C(137)-C(380)	120		
		C(388)-C(137)-C(380)	118		
		C(382)-C(137)-C(381)	89		
		C(138)-C(137)-C(381)	85		
		C(380)-C(137)-C(381)	92		
		C(388)-C(137)-C(381)	84		
		C(144)-C(382)-C(381)	91		
		C(138)-C(388)-C(381)	86		
		C(385)-C(136)-C(381)	118		
		C(387)-C(138)-C(381)	123		
		C(385)-C(380)-C(381)	118		
		C(387)-C(388)-C(381)	123		
		C(137)-C(144)-C(382)	89		
		C(375)-C(144)-C(382)	95		
		C(381)-C(144)-C(382)	87		
		C(388)-C(137)-C(382)	122		
		C(388)-C(381)-C(382)	123		
		C(136)-C(141)-H(383)	120		
		C(142)-C(141)-H(383)	117		
		C(376)-C(141)-H(383)	117		
		C(380)-C(141)-H(383)	120		
		C(144)-C(382)-H(384)	96		
		C(132)-C(135)-C(385)	88		

TABLE 2.2   (Continued).

Bond lengths	R,A	Valent corners	Grad	Atom	Charge (by Milliken)
		C(136)-C(135)-C(385)	89		
		C(380)-C(135)-C(385)	88		
		C(386)-C(135)-C(385)	87		
		C(132)-C(386)-C(385)	88		
		C(387)-C(386)-C(385)	119		
		C(391)-C(386)-C(385)	123		
		C(387)-C(132)-C(385)	119		
		C(391)-C(132)-C(385)	123		
		C(385)-C(132)-C(386)	90		
		C(387)-C(132)-C(386)	88		
		C(391)-C(132)-C(386)	88		
		C(131)-C(387)-C(386)	90		
		C(388)-C(387)-C(386)	119		
		C(392)-C(387)-C(386)	118		
		C(129)-C(391)-C(386)	88		
		C(394)-C(391)-C(386)	123		
		C(394)-C(129)-C(386)	123		
		C(388)-C(131)-C(386)	120		
		C(392)-C(131)-C(386)	118		
		C(388)-C(131)-C(387)	92		
		C(132)-C(131)-C(387)	90		
		C(386)-C(131)-C(387)	89		
		C(392)-C(131)-C(387)	85		
		C(138)-C(388)-C(387)	93		
		C(130)-C(392)-C(387)	86		
		C(393)-C(392)-C(387)	123		
		C(393)-C(130)-C(387)	123		
		C(391)-C(132)-C(387)	118		
		C(391)-C(386)-C(387)	118		
		C(131)-C(138)-C(388)	87		
		C(381)-C(138)-C(388)	93		
		C(387)-C(138)-C(388)	86		
		C(392)-C(131)-C(388)	123		

TABLE 2.2   (Continued).

Bond lengths	R,A	Valent corners	Grad	Atom	Charge (by Milliken)
		C(392)-C(387)-C(388)	123		
		C(138)-C(388)-H(389)	92		
		C(132)-C(135)-H(390)	120		
		C(136)-C(135)-H(390)	117		
		C(380)-C(135)-H(390)	117		
		C(386)-C(135)-H(390)	120		
		C(394)-C(129)-C(391)	90		
		C(132)-C(129)-C(391)	92		
		C(386)-C(129)-C(391)	91		
		C(121)-C(394)-C(391)	91		
		C(399)-C(394)-C(391)	123		
		C(393)-C(121)-C(391)	120		
		C(399)-C(121)-C(391)	123		
		C(393)-C(394)-C(391)	120		
		C(122)-C(130)-C(392)	84		
		C(131)-C(130)-C(392)	94		
		C(387)-C(130)-C(392)	92		
		C(393)-C(130)-C(392)	83		
		C(122)-C(393)-C(392)	82		
		C(394)-C(393)-C(392)	120		
		C(397)-C(393)-C(392)	123		
		C(394)-C(122)-C(392)	120		
		C(397)-C(122)-C(392)	123		
		C(392)-C(122)-C(393)	97		
		C(123)-C(122)-C(393)	87		
		C(394)-C(122)-C(393)	84		
		C(397)-C(122)-C(393)	86		
		C(121)-C(394)-C(393)	81		
		C(399)-C(394)-C(393)	118		
		C(123)-C(397)-C(393)	86		
		C(398)-C(397)-C(393)	123		
		C(399)-C(121)-C(393)	118		
		C(398)-C(123)-C(393)	123		

TABLE 2.2 (Continued).

Bond lengths	R,A	Valent corners	Grad	Atom	Charge (by Milliken)
		C(129)-C(121)-C(394)	89		
		C(122)-C(121)-C(394)	99		
		C(126)-C(121)-C(394)	82		
		C(391)-C(121)-C(394)	88		
		C(393)-C(121)-C(394)	98		
		C(399)-C(121)-C(394)	82		
		C(126)-C(399)-C(394)	60		
		C(400)-C(399)-C(394)	123		
		C(397)-C(122)-C(394)	118		
		C(400)-C(126)-C(394)	123		
		C(397)-C(393)-C(394)	118		
		C(129)-C(391)-H(395)	93		
		C(394)-C(391)-H(395)	120		
		C(122)-C(130)-H(396)	120		
		C(131)-C(130)-H(396)	117		
		C(387)-C(130)-H(396)	117		
		C(393)-C(130)-H(396)	120		
		C(124)-C(123)-C(397)	74		
		C(393)-C(123)-C(397)	93		
		C(398)-C(123)-C(397)	74		
		C(124)-C(398)-C(397)	36		
		C(400)-C(398)-C(397)	120		
		C(406)-C(398)-C(397)	123		
		C(400)-C(124)-C(397)	120		
		C(406)-C(124)-C(397)	123		
		C(397)-C(124)-C(398)	144		
		C(125)-C(124)-C(398)	32		
		C(169)-C(124)-C(398)	90		
		C(400)-C(124)-C(398)	32		
		C(406)-C(124)-C(398)	89		
		C(125)-C(400)-C(398)	67		
		C(403)-C(400)-C(398)	118		
		C(169)-C(406)-C(398)	90		

TABLE 2.2   (Continued).

Bond lengths	R,A	Valent corners	Grad	Atom	Charge (by Milliken)
		C(399)-C(125)-C(398)	120		
		C(403)-C(125)-C(398)	118		
		C(405)-C(169)-C(398)	122		
		C(399)-C(400)-C(398)	120		
		C(405)-C(406)-C(398)	123		
		C(400)-C(126)-C(399)	110		
		C(394)-C(126)-C(399)	120		
		C(125)-C(400)-C(399)	96		
		C(403)-C(400)-C(399)	123		
		C(403)-C(125)-C(399)	122		
		C(126)-C(125)-C(400)	83		
		C(172)-C(125)-C(400)	75		
		C(398)-C(125)-C(400)	112		
		C(399)-C(125)-C(400)	83		
		C(403)-C(125)-C(400)	74		
		C(172)-C(403)-C(400)	84		
		C(404)-C(403)-C(400)	123		
		C(406)-C(124)-C(400)	117		
		C(404)-C(172)-C(400)	123		
		C(406)-C(398)-C(400)	118		
		C(126)-C(399)-H(401)	153		
		C(400)-C(399)-H(401)	120		
		C(124)-C(123)-H(402)	120		
		C(393)-C(123)-H(402)	117		
		C(398)-C(123)-H(402)	120		
		C(404)-C(172)-C(403)	93		
		C(400)-C(172)-C(403)	95		
		C(171)-C(404)-C(403)	93		
		C(405)-C(404)-C(403)	120		
		C(409)-C(404)-C(403)	123		
		C(405)-C(171)-C(403)	120		
		C(409)-C(171)-C(403)	122		
		C(172)-C(171)-C(404)	86		

TABLE 2.2 (Continued).

Bond lengths	R,A	Valent corners	Grad	Atom	Charge (by Milliken)
		C(178)-C(171)-C(404)	85		
		C(403)-C(171)-C(404)	85		
		C(405)-C(171)-C(404)	98		
		C(409)-C(171)-C(404)	83		
		C(170)-C(405)-C(404)	102		
		C(406)-C(405)-C(404)	120		
		C(412)-C(405)-C(404)	118		
		C(178)-C(409)-C(404)	86		
		C(410)-C(409)-C(404)	122		
		C(406)-C(170)-C(404)	119		
		C(412)-C(170)-C(404)	117		
		C(410)-C(178)-C(404)	122		
		C(406)-C(170)-C(405)	99		
		C(171)-C(170)-C(405)	78		
		C(175)-C(170)-C(405)	92		
		C(404)-C(170)-C(405)	77		
		C(412)-C(170)-C(405)	90		
		C(169)-C(406)-C(405)	115		
		C(175)-C(412)-C(405)	91		
		C(409)-C(171)-C(405)	118		
		C(411)-C(175)-C(405)	121		
		C(409)-C(404)-C(405)	118		
		C(411)-C(412)-C(405)	122		
		C(170)-C(169)-C(406)	66		
		C(398)-C(169)-C(406)	90		
		C(405)-C(169)-C(406)	65		
		C(412)-C(170)-C(406)	123		
		C(412)-C(405)-C(406)	123		
		C(170)-C(169)-H(407)	120		
		C(398)-C(169)-H(407)	117		
		C(405)-C(169)-H(407)	120		
		C(172)-C(403)-H(408)	101		
		C(404)-C(403)-H(408)	120		

TABLE 2.2   (Continued).

Bond lengths	R,A	Valent corners	Grad	Atom	Charge (by Milliken)
		C(410)-C(178)-C(409)	91		
		C(404)-C(178)-C(409)	92		
		C(177)-C(410)-C(409)	93		
		C(411)-C(410)-C(409)	121		
		C(411)-C(177)-C(409)	121		
		C(178)-C(177)-C(410)	87		
		C(409)-C(177)-C(410)	84		
		C(411)-C(177)-C(410)	94		
		C(176)-C(411)-C(410)	98		
		C(412)-C(411)-C(410)	121		
		C(412)-C(176)-C(410)	120		
		C(412)-C(176)-C(411)	94		
		C(177)-C(176)-C(411)	82		
		C(410)-C(176)-C(411)	79		
		C(175)-C(412)-C(411)	99		
		C(176)-C(175)-C(412)	82		
		C(405)-C(175)-C(412)	87		
		C(411)-C(175)-C(412)	79		
		C(176)-C(175)-H(413)	120		
		C(405)-C(175)-H(413)	118		
		C(411)-C(175)-H(413)	120		
		C(368)-C(162)-C(417)	101		
		C(418)-C(162)-C(417)	122		
		C(161)-C(418)-C(417)	125		
		C(419)-C(418)-C(417)	118		
		C(423)-C(418)-C(417)	123		
		C(419)-C(161)-C(417)	118		
		C(423)-C(161)-C(417)	122		
		C(166)-C(161)-C(418)	70		
		C(162)-C(161)-C(418)	55		
		C(417)-C(161)-C(418)	54		
		C(419)-C(161)-C(418)	163		
		C(423)-C(161)-C(418)	70		

TABLE 2.2 (Continued).

Bond lengths	R,A	Valent corners	Grad	Atom	Charge (by Milliken)
		C(160)-C(419)-C(418)	166		
		C(420)-C(419)-C(418)	118		
		C(424)-C(419)-C(418)	119		
		C(166)-C(423)-C(418)	72		
		C(420)-C(160)-C(418)	117		
		C(424)-C(160)-C(418)	119		
		C(424)-C(160)-C(419)	119		
		C(161)-C(160)-C(419)	14		
		C(418)-C(160)-C(419)	14		
		C(420)-C(160)-C(419)	115		
		C(159)-C(420)-C(419)	116		
		C(165)-C(424)-C(419)	117		
		C(423)-C(161)-C(419)	120		
		C(423)-C(418)-C(419)	119		
		C(156)-C(159)-C(420)	68		
		C(160)-C(159)-C(420)	64		
		C(369)-C(159)-C(420)	67		
		C(419)-C(159)-C(420)	64		
		C(424)-C(160)-C(420)	124		
		C(424)-C(419)-C(420)	123		
		C(162)-C(417)-H(421)	141		
		C(418)-C(417)-H(421)	117		
		C(156)-C(159)-H(422)	120		
		C(160)-C(159)-H(422)	117		
		C(369)-C(159)-H(422)	121		
		C(419)-C(159)-H(422)	117		
		C(418)-C(166)-C(423)	107		
		C(160)-C(165)-C(424)	63		
		C(350)-C(165)-C(424)	59		
		C(419)-C(165)-C(424)	63		
		C(166)-C(423)-H(425)	168		
		C(160)-C(165)-H(426)	120		
		C(350)-C(165)-H(426)	118		

TABLE 2.2 (Continued).

Bond lengths	R,A	Valent corners	Grad	Atom	Charge (by Milliken)
		C(419)-C(165)-H(426)	121		
		C(431)-C(432)-C(427)	123		
		C(432)-C(427)-C(428)	118		
		C(435)-C(427)-C(428)	120		
		C(427)-C(428)-C(429)	118		
		C(436)-C(428)-C(429)	123		
		C(431)-C(430)-C(429)	120		
		C(428)-C(429)-C(430)	123		
		C(478)-C(431)-C(430)	118		
		C(432)-C(431)-C(430)	120		
		C(477)-C(478)-C(431)	123		
		C(478)-C(431)-C(432)	123		
		C(428)-C(429)-H(433)	117		
		C(431)-C(432)-H(434)	120		
		C(432)-C(427)-C(435)	123		
		C(427)-C(428)-C(436)	120		
		C(438)-C(437)-C(436)	118		
		C(428)-C(436)-C(437)	123		
		C(435)-C(438)-C(437)	118		
		C(441)-C(438)-C(437)	120		
		C(427)-C(435)-C(438)	123		
		C(428)-C(436)-H(439)	120		
		C(427)-C(435)-H(440)	120		
		C(435)-C(438)-C(441)	123		
		C(443)-C(442)-C(441)	118		
		C(438)-C(441)-C(442)	123		
		C(444)-C(443)-C(442)	118		
		C(450)-C(443)-C(442)	119		
		C(437)-C(444)-C(443)	123		
		C(436)-C(437)-C(444)	123		
		C(438)-C(437)-C(444)	120		
		C(438)-C(441)-H(445)	120		
		C(437)-C(444)-H(446)	120		

TABLE 2.2   (Continued).

Bond lengths	R,A	Valent corners	Grad	Atom	Charge (by Milliken)
		C(441)-C(442)-C(447)	123		
		C(443)-C(442)-C(447)	120		
		C(449)-C(448)-C(447)	118		
		C(442)-C(447)-C(448)	123		
		C(450)-C(449)-C(448)	118		
		C(454)-C(449)-C(448)	119		
		C(443)-C(450)-C(449)	123		
		C(444)-C(443)-C(450)	123		
		C(443)-C(450)-H(451)	120		
		C(442)-C(447)-H(452)	120		
		C(447)-C(448)-C(453)	123		
		C(449)-C(448)-C(453)	119		
		C(450)-C(449)-C(454)	123		
		C(456)-C(455)-C(454)	118		
		C(449)-C(454)-C(455)	123		
		C(453)-C(456)-C(455)	118		
		C(461)-C(456)-C(455)	119		
		C(448)-C(453)-C(456)	123		
		C(448)-C(453)-H(457)	120		
		C(449)-C(454)-H(458)	120		
		C(454)-C(455)-C(459)	123		
		C(456)-C(455)-C(459)	120		
		C(462)-C(460)-C(459)	118		
		C(455)-C(459)-C(460)	123		
		C(461)-C(462)-C(460)	118		
		C(465)-C(462)-C(460)	119		
		C(453)-C(456)-C(461)	123		
		C(456)-C(461)-C(462)	123		
		C(456)-C(461)-H(463)	120		
		C(455)-C(459)-H(464)	120		
		C(461)-C(462)-C(465)	123		
		C(467)-C(466)-C(465)	118		
		C(462)-C(465)-C(466)	123		

TABLE 2.2   (Continued).

Bond lengths	R,A	Valent corners	Grad	Atom	Charge (by Milliken)
		C(468)-C(467)-C(466)	118		
		C(472)-C(467)-C(466)	119		
		C(460)-C(468)-C(467)	123		
		C(540)-C(472)-C(467)	123		
		C(459)-C(460)-C(468)	123		
		C(462)-C(460)-C(468)	120		
		C(460)-C(468)-H(469)	120		
		C(462)-C(465)-H(470)	120		
		C(465)-C(466)-C(471)	123		
		C(467)-C(466)-C(471)	119		
		C(468)-C(467)-C(472)	123		
		C(539)-C(540)-C(472)	123		
		C(541)-C(540)-C(472)	118		
		C(467)-C(472)-H(473)	120		
		C(540)-C(472)-H(473)	117		
		C(466)-C(471)-H(474)	120		
		C(429)-C(430)-C(475)	123		
		C(431)-C(430)-C(475)	118		
		C(477)-C(476)-C(475)	120		
		C(481)-C(476)-C(475)	123		
		C(430)-C(475)-C(476)	123		
		C(482)-C(477)-C(476)	118		
		C(107)-C(481)-C(476)	123		
		C(478)-C(477)-C(476)	120		
		C(106)-C(482)-C(477)	123		
		C(481)-C(476)-C(477)	118		
		C(482)-C(477)-C(478)	123		
		C(430)-C(475)-H(479)	117		
		C(477)-C(478)-H(480)	120		
		C(108)-C(107)-C(481)	123		
		C(105)-C(106)-C(482)	123		
		C(107)-C(106)-C(482)	119		
		C(107)-C(481)-H(483)	120		

TABLE 2.2 (Continued).

Bond lengths	R,A	Valent corners	Grad	Atom	Charge (by Milliken)
		C(106)-C(482)-H(484)	120		
		C(489)-C(490)-C(485)	123		
		C(490)-C(485)-C(486)	118		
		C(493)-C(485)-C(486)	120		
		C(485)-C(486)-C(487)	118		
		C(494)-C(486)-C(487)	123		
		C(489)-C(488)-C(487)	119		
		C(486)-C(487)-C(488)	123		
		C(536)-C(489)-C(488)	118		
		C(490)-C(489)-C(488)	120		
		C(535)-C(536)-C(489)	123		
		C(536)-C(489)-C(490)	123		
		C(486)-C(487)-H(491)	117		
		C(489)-C(490)-H(492)	120		
		C(490)-C(485)-C(493)	123		
		C(485)-C(486)-C(494)	119		
		C(496)-C(495)-C(494)	118		
		C(486)-C(494)-C(495)	123		
		C(493)-C(496)-C(495)	118		
		C(499)-C(496)-C(495)	119		
		C(485)-C(493)-C(496)	123		
		C(486)-C(494)-H(497)	120		
		C(485)-C(493)-H(498)	120		
		C(493)-C(496)-C(499)	123		
		C(501)-C(500)-C(499)	118		
		C(496)-C(499)-C(500)	123		
		C(502)-C(501)-C(500)	118		
		C(508)-C(501)-C(500)	119		
		C(495)-C(502)-C(501)	123		
		C(494)-C(495)-C(502)	123		
		C(496)-C(495)-C(502)	119		
		C(496)-C(499)-H(503)	120		
		C(495)-C(502)-H(504)	120		

TABLE 2.2   (Continued).

Bond lengths	R,A	Valent corners	Grad	Atom	Charge (by Milliken)
		C(499)-C(500)-C(505)	123		
		C(501)-C(500)-C(505)	119		
		C(507)-C(506)-C(505)	118		
		C(500)-C(505)-C(506)	123		
		C(508)-C(507)-C(506)	118		
		C(512)-C(507)-C(506)	119		
		C(501)-C(508)-C(507)	123		
		C(502)-C(501)-C(508)	123		
		C(501)-C(508)-H(509)	120		
		C(500)-C(505)-H(510)	120		
		C(505)-C(506)-C(511)	123		
		C(507)-C(506)-C(511)	119		
		C(508)-C(507)-C(512)	123		
		C(514)-C(513)-C(512)	118		
		C(507)-C(512)-C(513)	123		
		C(511)-C(514)-C(513)	118		
		C(519)-C(514)-C(513)	119		
		C(506)-C(511)-C(514)	123		
		C(506)-C(511)-H(515)	120		
		C(507)-C(512)-H(516)	120		
		C(512)-C(513)-C(517)	123		
		C(514)-C(513)-C(517)	119		
		C(520)-C(518)-C(517)	118		
		C(513)-C(517)-C(518)	123		
		C(519)-C(520)-C(518)	118		
		C(523)-C(520)-C(518)	119		
		C(511)-C(514)-C(519)	123		
		C(514)-C(519)-C(520)	123		
		C(514)-C(519)-H(521)	120		
		C(513)-C(517)-H(522)	120		
		C(519)-C(520)-C(523)	123		
		C(525)-C(524)-C(523)	118		
		C(520)-C(523)-C(524)	123		

TABLE 2.2 (Continued).

Bond lengths	R,A	Valent corners	Grad	Atom	Charge (by Milliken)
		C(526)-C(525)-C(524)	118		
		C(530)-C(525)-C(524)	119		
		C(518)-C(526)-C(525)	123		
		C(602)-C(530)-C(525)	122		
		C(517)-C(518)-C(526)	123		
		C(520)-C(518)-C(526)	119		
		C(518)-C(526)-H(527)	120		
		C(520)-C(523)-H(528)	120		
		C(523)-C(524)-C(529)	123		
		C(525)-C(524)-C(529)	119		
		C(526)-C(525)-C(530)	123		
		C(601)-C(602)-C(530)	123		
		C(603)-C(602)-C(530)	118		
		C(525)-C(530)-H(531)	120		
		C(602)-C(530)-H(531)	118		
		C(524)-C(529)-H(532)	120		
		C(487)-C(488)-C(533)	123		
		C(489)-C(488)-C(533)	118		
		C(535)-C(534)-C(533)	119		
		C(488)-C(533)-C(534)	123		
		C(542)-C(535)-C(534)	118		
		C(536)-C(535)-C(534)	120		
		C(541)-C(542)-C(535)	123		
		C(542)-C(535)-C(536)	123		
		C(488)-C(533)-H(537)	117		
		C(535)-C(536)-H(538)	120		
		C(533)-C(534)-C(539)	123		
		C(535)-C(534)-C(539)	118		
		C(541)-C(540)-C(539)	119		
		C(534)-C(539)-C(540)	123		
		C(471)-C(541)-C(540)	118		
		C(542)-C(541)-C(540)	120		
		C(466)-C(471)-C(541)	123		

TABLE 2.2 (Continued).

Bond lengths	R,A	Valent corners	Grad	Atom	Charge (by Milliken)
		C(471)-C(541)-C(542)	123		
		C(534)-C(539)-H(543)	117		
		C(541)-C(542)-H(544)	120		
		C(549)-C(550)-C(545)	122		
		C(550)-C(545)-C(546)	119		
		C(553)-C(545)-C(546)	118		
		C(545)-C(546)-C(547)	119		
		C(554)-C(546)-C(547)	123		
		C(546)-C(547)-C(548)	122		
		C(547)-C(548)-C(549)	118		
		C(595)-C(548)-C(549)	119		
		C(548)-C(549)-C(550)	118		
		C(598)-C(549)-C(550)	123		
		C(546)-C(547)-H(551)	120		
		C(549)-C(550)-H(552)	118		
		C(550)-C(545)-C(553)	123		
		C(545)-C(546)-C(554)	118		
		C(556)-C(555)-C(554)	119		
		C(562)-C(555)-C(554)	123		
		C(546)-C(554)-C(555)	123		
		C(553)-C(556)-C(555)	119		
		C(561)-C(562)-C(555)	123		
		C(559)-C(556)-C(555)	118		
		C(545)-C(553)-C(556)	123		
		C(562)-C(555)-C(556)	118		
		C(546)-C(554)-H(557)	117		
		C(545)-C(553)-H(558)	117		
		C(553)-C(556)-C(559)	123		
		C(561)-C(560)-C(559)	119		
		C(556)-C(559)-C(560)	123		
		C(568)-C(561)-C(560)	118		
		C(562)-C(561)-C(560)	119		
		C(567)-C(568)-C(561)	123		

TABLE 2.2   (Continued).

Bond lengths	R,A	Valent corners	Grad	Atom	Charge (by Milliken)
		C(568)-C(561)-C(562)	123		
		C(556)-C(559)-H(563)	117		
		C(561)-C(562)-H(564)	120		
		C(559)-C(560)-C(565)	123		
		C(561)-C(560)-C(565)	118		
		C(560)-C(565)-C(566)	123		
		C(565)-C(566)-C(567)	119		
		C(571)-C(566)-C(567)	118		
		C(566)-C(567)-C(568)	119		
		C(572)-C(567)-C(568)	123		
		C(567)-C(568)-H(569)	120		
		C(560)-C(565)-H(570)	117		
		C(565)-C(566)-C(571)	123		
		C(566)-C(567)-C(572)	118		
		C(574)-C(573)-C(572)	119		
		C(567)-C(572)-C(573)	123		
		C(571)-C(574)-C(573)	119		
		C(579)-C(574)-C(573)	118		
		C(566)-C(571)-C(574)	123		
		C(566)-C(571)-H(575)	117		
		C(567)-C(572)-H(576)	117		
		C(572)-C(573)-C(577)	123		
		C(574)-C(573)-C(577)	118		
		C(580)-C(578)-C(577)	119		
		C(586)-C(578)-C(577)	123		
		C(573)-C(577)-C(578)	123		
		C(579)-C(580)-C(578)	119		
		C(585)-C(586)-C(578)	123		
		C(583)-C(580)-C(578)	118		
		C(571)-C(574)-C(579)	123		
		C(574)-C(579)-C(580)	123		
		C(586)-C(578)-C(580)	118		
		C(574)-C(579)-H(581)	117		

TABLE 2.2   (Continued).

Bond lengths	R,A	Valent corners	Grad	Atom	Charge (by Milliken)
		C(573)-C(577)-H(582)	117		
		C(579)-C(580)-C(583)	123		
		C(585)-C(584)-C(583)	120		
		C(580)-C(583)-C(584)	123		
		C(592)-C(585)-C(584)	118		
		C(586)-C(585)-C(584)	119		
		C(591)-C(592)-C(585)	123		
		C(592)-C(585)-C(586)	123		
		C(585)-C(586)-H(587)	120		
		C(580)-C(583)-H(588)	117		
		C(583)-C(584)-C(589)	123		
		C(585)-C(584)-C(589)	118		
		C(591)-C(590)-C(589)	120		
		C(584)-C(589)-C(590)	123		
		C(361)-C(591)-C(590)	118		
		C(592)-C(591)-C(590)	119		
		C(356)-C(361)-C(591)	123		
		C(361)-C(591)-C(592)	123		
		C(591)-C(592)-H(593)	120		
		C(584)-C(589)-H(594)	117		
		C(547)-C(548)-C(595)	123		
		C(597)-C(596)-C(595)	119		
		C(548)-C(595)-C(596)	122		
		C(604)-C(597)-C(596)	119		
		C(598)-C(597)-C(596)	119		
		C(603)-C(604)-C(597)	122		
		C(549)-C(598)-C(597)	122		
		C(548)-C(549)-C(598)	119		
		C(604)-C(597)-C(598)	123		
		C(548)-C(595)-H(599)	119		
		C(549)-C(598)-H(600)	119		
		C(595)-C(596)-C(601)	123		
		C(597)-C(596)-C(601)	119		

TABLE 2.2 (Continued).

Bond lengths	R,A	Valent corners	Grad	Atom	Charge (by Milliken)
		C(603)-C(602)-C(601)	119		
		C(596)-C(601)-C(602)	122		
		C(529)-C(603)-C(602)	118		
		C(604)-C(603)-C(602)	119		
		C(524)-C(529)-C(603)	122		
		C(529)-C(603)-C(604)	123		
		C(596)-C(601)-H(605)	119		
		C(603)-C(604)-H(606)	119		

TABLE 2.3 Total Energy (E_0), Electronic Energy ($E_{эл}$), Maximal Charge on a Hydrogene Atom (q_{max}^{H+}), Universal Acidity (pKa) of Molecules Pentacontacene and Hectacontacene

S. No.	Molecule	$-E_{0k}$Dg/mol	q_{max}^{H+}	pKa
1	Pentacontacene	–2,629,630	+ 0.06	33
2	Hectacontacene	–5,228,911	+ 0.06	33

KEYWORDS

- acid strength
- hectacontacene
- MNDO method
- pentacontacene
- quantum chemical calculation

REFERENCES

1. K. S. Novoselov, et al. Electric Field Effect in Atomically Thin Carbon Films, Science 306, 666 (2004);DOI:10.1126/science.1102896.
2. M.W.Shmidt, K.K.Baldrosge, J.A. Elbert, M.S. Gordon, J.H. Enseh, S.Koseki, N.Matsvnaga., K.A. Nguyen, S. J. Su, And Anothers. J. Comput. Chem.14, 1347–1363, (1993).

3. Bode, B. M. and Gordon, M. S. *J. Mol. Graphics Mod., 16*, 1998, 133–138.
4. Babkin V.A., Fedunov R.G., Minsker K.S. and anothers. Oxidation communication, 2002, №1, 25, 21–47.
5. V.A. Babkin, A.V. Ignatov, A.N. Ignatov, M.N. Gulyukin, V.Yu. Dmitriev, O.V. Stoyanov, G.E. Zaikov. Quantum-chemical Calculation of some molecules of triboratols. Kazan. "Vestnik" of Kazan State Technological University. 2013., Vol. 16, N2, p. 15–17.
6. V.A. Babkin, A.V. Ignatov, O.V. Stoyanov, G.E. Zaikov. Quantum-chemical Calculation of some monomers of cationic polymerization with small cycles. Kazan. "Vestnik" of Kazan State Technological University. 2013., Vol. 16, N 4, p.21–22
7. V.A. Babkin, V.V. Trifonov B.B., N.G. Lebedev, V.Yu. Dmitriev, D.S. Andreev, O.V. Stoyanov, G.E. Zaikov. Quantum-chemical Calculation of tetracene and pentacene by MNDO in approximation of the linear molecular model of graphene. Kazan. "Vestnik" of Kazan State Technological University. 2013., Vol. 16, N 7, p.16–18.

CHAPTER 3

A NEW CONCEPT OF PHOTOSYNTHESIS

GENNADY G. KOMISSAROV

N.N. Semenov Institute for Chemical Physics, Russian Academy of Sciences, Kosygin St. 4, Moscow 119991, Russia,
E-mail: komiss@chph.ras.ru; gkomiss@yandex.ru

CONTENTS

ABSTRACT

A history of the formation of a new concept of photosynthesis proposed by the author is considered ranging from 1966 to 2013. Its essence is as follows: the photosynthetic oxygen (hydrogen) source is not water, but exo- and endogenous hydrogen peroxide; thermal energy is a necessary party of the photosynthetic process; along with the carbon dioxide the air (oxygen, inert gases) is included in the photosynthetic equation. It is briefly touched the mechanism of the photovoltaic (Becquerel) effect in films of chlorophyll and its synthetic analog – phthalocyanine. There are presented works on artificial photosynthesis performed in the laboratory of Photobionics of the Semenov Institute of Chemical Physics RAS.

3.1   INTRODUCTION

Photosynthesis is the GLOBAL, FUNDAMENTAL and UNIQUE biological process. Its mechanism study is one of the central tasks of the modern natural science. Investigations into photosynthesis started in 1771, when the outstanding English chemist Joseph Priestley discovered the capacity of plants to 'repair air, distorted by the burning of candles,' that is, release oxygen. A significant contribution to the development of the concept was introduced by Ingen-Housz, who showed the necessity of solar light for the occurrence of photosynthesis: the release of oxygen occurs only if the plants are illuminated (in darkness, they lose this capacity). He also established that photosynthesis is accompanied by the buildup of organic products. The experiments carried out by J. Senebier and N.Th. Saussure revealed initial substances of photosynthesis (carbon dioxide and water). The energy aspect of the problem was discussed for the first time by J.R. Mayer. In 1941, A.P. Vinogradov and R.V Teis in USSR and S. Ruben with his colleagues in the USA established that oxygen is released from the water and not from the carbon dioxide [1]. During the last 70 years the main equation of photosynthesis does not change and is written is following:

$$CO_2 + H_2O \xrightarrow{\text{Light}} \text{carbohydrates} + O_2 \qquad (1)$$

During breathing (reaction (1) taking place from the left to the right), the energy stored in the final products by the plants is released. It should also be mentioned that all living substances are constructed from molecules, which initially formed in plants. Examination of the unique biological process of storage of solar energy has been continuing for more than two centuries, but the final mechanism has not as yet been completely explained.

3.2   FUNCTIONAL MODELING OF PHOTOSYNTHESIS

Regardless of the complexity of photosynthesis, it is possible to define two main stages in this process: the light stage, whose occurrence requires the direct effect of light, and the dark stage, which follows the light stage. It is assumed that the first, light stage, is characterized by the occurrence of photosplitting of water with the generation of molecular oxygen which is injected into the atmosphere as a secondary product:

$$H_2O \xrightarrow[\text{Chlorophyll}]{\text{Light}} [H] + O_2 \tag{2}$$

The resultant hydrogen subsequently enters to the thermal cycle of the fixation of carbon dioxide, which is completed, with the formation of carbohydrates, conventionally denoted by $\{CH_2O\}$:

$$[H] + CO_2 \xrightarrow{\text{Ferments}} \{CH_2O\} \tag{3}$$

This stage, including a large number of fermentation reactions, has been examined in detail in studies by the outstanding American chemists M. Calvin who was awarded the Nobel prize in 1961.

Thus, the PHOTOSYNTHESIS consists of two main stages: the PHOTOlysis of water and dark SYNTHESIS of carbohydrates. In the second stage, there are no specific PHOTOsynthesic process, the fixation of carbon dioxide in the darkness may be carried out by the liver cells of a rat, if a suitable donor of hydrogen is available.

In this article, we shall approach the problem of photosynthesis from the physical–chemical position. Can artificial physical-chemical systems reproduced the light stage of photosynthesis?

The sequence of transformation of energy during photosynthesis, as described in Refs. [1–3], may be written in the following form:

$$E_L \rightarrow E_E \rightarrow E_C \tag{4}$$

where E is the energy with the appropriate indexes 'light,' 'electrical,' 'chemical.' The equation (4) greatly simplified the problem of functional modeling of photosynthesis because it makes it possible to solve the problem in two independent stages:

$$E_L \rightarrow E_E \tag{5},$$

$$E_E \rightarrow E_C \tag{6}$$

and transformation Eq. (6) (electrolysis of water with the release of gaseous oxygen) has been solved a long time ago. On the industrial scale, the process is carried out with high efficiency (current efficiency is 95–99%). Thus, the initial task is to find a device capable of generating electrical energy under the effect of light.

In 1839, Becquerel [4] described the first of sometimes-detected photoeffects. The principle of the phenomenon, subsequently referred to as the Becquerel effect or photovoltaic effect, is the formation of a difference of the potentials between two metallic electrodes placed in an electrolyte; one of the electrodes is coated with a layer of a light-sensitive substance.

In 1874, Becquerel observed that the illumination of a silver-coated platinum electrode, carrying a chlorophyll film, with red light, changes its potential [5]. We believe that this date should be regarded as the basic date in the physical–chemical modeling of photosynthesis, although Becquerel never mentioned this approach. His work was directed to confirm the phenomenon of optical sensitization in photography, discovered by Vogel [6].

Prior to starting modeling experiments, we regarded it as essential to evaluate (at least approximately) the intensity of the photoflux in chloroplast. Using the Faraday law, it can easily be shown that the total current in chloroplasts, containing 0.1 mg of chlorophyll, is equal to 10^{-5} A. However, in a single chloroplast, the current is obviously many times smaller (10^{-13} A).

In 1966, as a functional model of chloroplast, we proposed a photovoltaic battery [1–3]. Its first in the world variant was realized at the Institute of Chemical Physics of the Academy of Sciences of the USSR in 1968 [1,

7, 8]. Then a similar battery was built in the USA [9]. The battery consisted of 4 electrodes coated with a synthetic analog of chlorophyll, that is, phthalocyanine. The battery was characterized by the following parameters [8]: the light potential reached 2.4 V, which is fully sufficient for electrolysis of water with the generation of molecular oxygen (the dissociation potential of water 1.23 V); the current, taken from the battery was 5.6×10^{-5} A so that it was possible to record oxygen by the conventional methods. The quantum yield of the photocurrent was not high, 0.01–0.1%. *The proposed device is capable of releasing oxygen from water under the effect of visible light, that is, reproduce one of the main functions of the chloroplast, and the process, taking place in the battery, is of the photocatalytic nature.*

3.3   THE BEQUREL EFFECT IN PHTALOCIANINE FILMS

The maximum value of the quantum yield of the photocurrent is detected in the case of monomolecular coating of the electrode with the pigment [10] (Fig. 3.1). In the chloroplast, the membranes are also covered on average by a monomolecular layer of chlorophyll.

To increase the quantum yield, it was necessary to carry out a systematic examination of the mechanism of the Becquerel effect. This was carried out in our laboratory. As a result, the quantum yield was increased to units and, subsequently, to tens of percent [1, 10–14]. In other words, it was possible to show that as regards efficiency, the modern photovoltaic systems are not inferior to photosynthetic structures.

Examination of the dependence of the magnitude of current and potential of pigmented electrodes on the pH value of the electrolyte gave an unexpected result. It was observed that at a fixed value of pH in darkness these values are almost identical with the values in platinum electrodes in the absence of the pigment in them. This indicates that the pigment film was penetrated by a large number of pores (Fig. 3.2) [1, 13].

The methods of the preparation of films, the values of current, potentials and the dynamics of these values unambiguously show that they used porous films, but the interpretation of the observed relationships was based on the considerations regarding the monolithic pigmented layer of

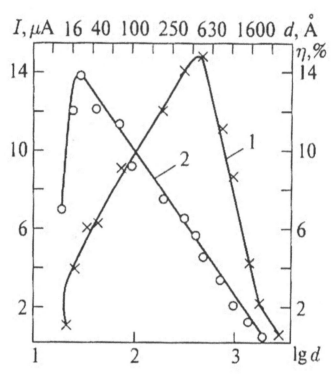

FIGURE 3.1 Dependence of the short circuit current (curve1) and the quantum yield (curve 2) on the thickness of the pigment film.

the electrode. This was followed by the development of methods of production of porefree films and at the present time it is possible to specify the main types of photovoltaic effects in the films of semiconductor pigments contacting, on the one side, with the electrolyte and, on the other side with the metal [1, 12–16] (Fig. 3.3).

The nature of the generation of photoresponse in the films differs principally. The values of current and their kinetic parameters differ tens, hundred or more times. The previously described considerations enabled us to start investigations in the area of structural-functional modeling, the development of artificial systems reproducing the structure (composition) and function of natural photosynthetic formations. The analysis of literature data on the generation of oxygen from water in physical–chemical systems (electrolysis, photolysis and radiolysis of water) and also the

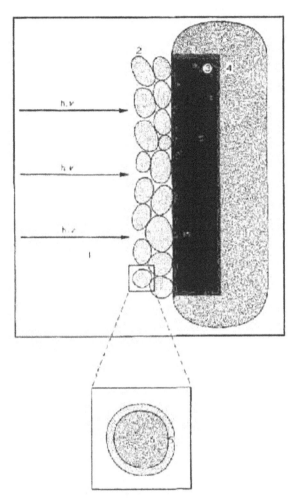

FIGURE 3.2   Scheme of an electrode coated with a layer of pigment: 1 – electrolyte, 2 – pigment layer, 3 – platinum electrode, 4 – insulating layer, 5 – thickness of layer penetrated by incident light, 6 – main body of the pigment suspension.

data obtained in our laboratory on the generation of oxygen in natural and modeling systems have made it possible to propose an original schema of photosynthetic release of oxygen [1, 12]:

According to the schema, the generation of a single molecule of oxygen requires at least four light quanta, each of which generates an

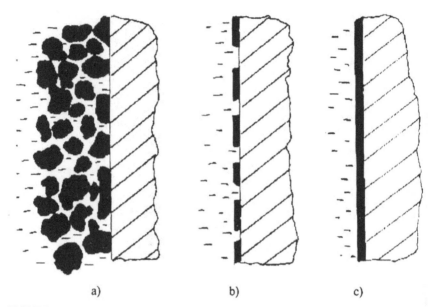

a)   b)   c)

FIGURE 3.3 Schematic of three types of pigment films (photovoltaic effects). Dark areas show the pigment, dashes indicate electrolyte, the metal is shaded by solid inclined lines. (a) A thick porous film (light does not reach the pigment-metal contact). Light-induced potential aries due to a change in concentration of potential-governing ions at the electrode surface ($\varphi = A(B\text{-ln } C)$; φ is light-induced potential; $A = RT/nF$ R, T, n and F have the commonly accepted meaning). The time needed to reach the steady photo-induced potential is of oder minutes, the dark current $\approx 10^{-6}$ A/cm^3. (b) a thin nonporous film (light penetration depth exceeds the film thickness). The response to illumination is a result of semiconductor processes in the bulk of the film and its boundaries. The time taken to reach the steady photo-induced potential is a fraction of second, the dark current is 10^{-12} A/cm^3. (c) intermediate type.

'electron–hole' couple, and the electron is used in the reaction $H^+ + e \rightarrow H$, required for subsequent fixation of CO_2. The surface of lamellae (sheets from which chloroplasts is produced) is a unique photo-electrode, consisting of alternating anodic and cathodic microregions (Fig. 3.4) [1, 12, 13]. On the whole, the proposed schema of the dissociation of water has made it possible to explain several relationships in the photosynthetic generation of oxygen.

The experimental confirmation required many years of work resulting in the development of the new concept of photosynthesis.

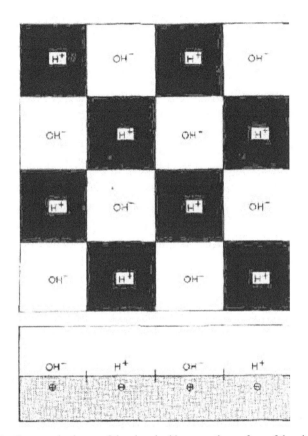

FIGURE 3.4   Structural scheme of the absorbed layer on the surface of the pigment (view from above and from side).

3.4   HYDROGEN PEROXIDE – SOURCE OF PHOTOSYNTHETIC OXYGEN (HYDROGEN)

3.4.1   THE EFFECT OF HYDROGEN PEROXIDE ON THE KINETICS OF GENERATION OF OXYGEN DURING PHOTOSYNTHESIS

The sequence of dissociation of water during photosynthesis, proposed by us in 1973, may be represented in the form: $H_2O \rightarrow H_2O_2 \rightarrow HO_2 \rightarrow O_2$, that is, water is oxidizes to hydrogen peroxide, which, in the final analysis, results in the generation of molecular oxygen (see the previous chapter). At present, there is a large number of literature data indicating the participation of hydrogen peroxide as an intermediate product in the course of formation of oxygen. We attempted to use the kinetic methods for confirming the participation of hydrogen peroxide in the course of photosynthetic generation of oxygen. For this purpose, we proposed a kinetic model based on the representation of vector algebra and projection geometry [1, 17]. The results of kinetic analyzes and the experimentally obtained changes of the kinetics of photosynthetic generation of oxygen in the presence of exogenous hydrogen peroxide indicate that it is capable of penetrating in light into the oxygen-generating complex of growth instead of water and act as an independent source of electrons and an independent source of generation of oxygen situated outside the four stage oxygen cycle.

3.4.2   HYDROGEN PEROXIDE – A SINGLE SOURCE OF PHOTOSYNTHETIC OXYGEN

As already reported, the main method of overcoming the problem of photosynthesis is based on the light stage of photosynthesis (the stage of photodissociation of water). The work of current investigators, concerned with the examination of the photosynthesis mechanism, has been and is still directed to finding approaches capable of explaining the mechanism by which the chlorophyll (the photosynthetic oxygen-generating reactions center) is capable of storing the energy of several light quanta in order to use this energy for the formation of molecular oxygen. Many original studies have been published in the course of these investigations.

Already in our first study, concerned with the justification of the photoelectrochemical hypothesis of photosynthesis [3], we paid attention to the need to take into account the changes of the properties of water in the chloroplast where it is situated between the lamellae with the distance between them not exceeding 100 Å. It is well known that the properties of structured liquid (in particular, and/or situated at the boundary with the solid phase) greatly differ from the properties in the volume. The viscosity of water in capillaries is an order of magnitude higher than the viscosity of water in the volume, and the heat conductivity of water in the layers increases tens of times, dielectric permittivity decreases from 81 (water in the volume) to 3–4 (in the interlayers with a thickness of 0.5–0.6 nm). Similar examples make it possible to assume that the puzzle of biological oxidation of water with the generation of oxygen is 'hidden not only in the properties of the chloroplast but also of the water itself' [3].

The extensive and long-term examination of the physical–chemical properties of water resulted in an unambiguous conclusion: *in the nature there is no pure water, and the water always contains an impurity, that is, hydrogen peroxide.* In a thri-distilate, the concentration of hydrogen peroxide is 10^{-9} M [18], and in the water of natural reservoirs (seas, rivers, lakes) it reaches 10^{-6} M, in rainwater is 10^{-5} M [1,19]. The above makes it possible to supplement the equation of biological oxidation of water [2] by another term, that is, hydrogen peroxide, which was discovered in 1818 by L.J. Thernard and referred to as 'oxidized water' [20]:

$$H_2O_2, H_2O \xrightarrow[\text{Chlorophyll}]{\text{Light}} [H] + O_2 \qquad (7)$$

At first sight, it may appear that in this case we are talking about some negligibly small impurity, with no relationship to photosynthesis. However, this is not so. By evaporating the aqueous solution of the peroxide, it is possible to increase is concentration tens of times, because its volatility is considerably smaller than in the case of water. The method of concentration of the aqueous solutions of hydrogen peroxide by evaporation of water has been used in the chemical practice for a long time now [20]. Transpiration (evaporation of water by plants) evidently plays the same function in addition to the protection of plants against overheating. For each kg of water, absorbed by the roots from soil, only 1 g (1/1000 part of) is used by the plant for the construction of tissue. Thus, the green leaves may be regarded as a unique concentrator of hydrogen peroxide. It should also be mentioned that the second initial substance in photosynthesis is CO_2 whose content in air is only 0.03% (less than the content of inert gases). Its concentration is higher than the concentration of hydrogen peroxide in the initial water.

However, the solubility of hydrogen peroxide in water is 7 (!) orders of magnitude higher than that for CO_2 (2×10^5 and 4.5×10^{-2} mole/L atm, respectively). This result in a serious consequence. Regardless of the area of formation of H_2O_2 (in air, in air bubbles in soil), its concentration in contacting water will be higher than in CO_2. Naturally, in this case it is necessary to take into account the initial concentration of these substances. In addition to exogenous hydrogen peroxide, the photosynthesized cell also contains endogenous hydrogen peroxide. Its source in the cytoplasma is the mitochondrin (in high-intensity photosynthesis, it converges to chloroplasts), peroxisoms, etc. For example, up to 40% of hydrogen peroxide, generated in peroxisoms, are transferred into cytosol. In other words, the hydrogen peroxide in vivo is completely sufficient for explaining the observed intensity of the generation of molecular oxygen from this hydrogen peroxide [1].

The data, obtained in [21], are most interesting from the viewpoint of examining the role of hydrogen peroxide in the generation of photosynthetic oxygen. The authors carried out mass spectrometric examination of photosynthetic generation of oxygen using hydrogen peroxide, marked with respect to oxygen (H_2 $^{18}O_2$). The results show that H_2 $^{18}O_2$ is the source of the entire amount of generated oxygen. It is well known [18] that the hydrogen peroxide in chemical systems rapidly exchanges hydrogen (deuterium) with water. This was already established in 1934. However, when using the oxide marked with respect to oxygen, the situation is completely different. As reported in the monograph [20], 'the oxygen of water, used as a solvent, does not take part in the dissociation or reaction of hydrogen peroxide; no exchange was found between the water and the resultant molecular oxygen or oxidized products.' Consequently, if the photosynthesized systems contain water and hydrogen peroxide, it is evident that the dissociation with the generation of oxygen explained case of the latter compound.

In the photoelectrochemical mechanism of formation of oxygen in vivo from two possible sources (hydrogen peroxide and water), the first mechanism is obviously preferred. In our laboratory, using the method of a spinning disk electrode with a ring (the spinning speed up to 3000 rpm) we showed the catalase activity of chlorophyll films, which were in contact with the aqueous solution of hydrogen peroxide. Under the effect of light, the rate of dissociation of hydrogen peroxide increased 2–3 times in comparison with the darkness values [1].

The results obtained in our laboratory and also the critical analysis of the literature data on the photosynthetic generation of oxygen have made it possible to propose a completely new viewpoint. According to the viewpoint, the source of photosynthetic oxygen (hydrogen) is not water but the exogenous and endogenous hydrogen peroxide [1, 22–26].

Naturally, it is difficult for a conventional photosynthetist to understand our position, because in the current literature the water is regarded as a source of oxygen. It is useful to mention that up to the studies carried out by Vinogradov [27] and Ruben [28], which appeared 70 years ago, the majority of investigators of photosynthesis had assumed that CO_2 is the source of oxygen in photosynthesis.

At the Fifth International Biochemical Congress (Moscow, 1961), A.P. Vinogradov and V.M. Kutyurin [29] attempted to evaluate the variants of the methods of dehydration of water during photosynthesis. We believe that A.P. Vinogradov had experimental justification for proposing the hydrogen peroxide as a source of oxygen in photosynthesis. We shall discuss the results. The studies [27, 28] show convincingly that CO_2 cannot be the source of photosynthetic oxygen as assumed at that time by the majority of researchers. This completed the revolutionary break in the examination of the photosynthesis mechanism. Since during the experiments it was not possible to achieve the equality of the isotope composition of the oxygen generated during photosynthesis and the oxygen in water, it would be necessary to introduce an assumption on the effect of breathing on the investigated process [29]. The situation existing in 1961 in this problem was characterized by R. Wurmser [30] as: 'it is almost evident (bold face by me, *G.K.*) that the generated oxygen comes from water.' In 1975, H.J. Metzner [31] published a large article: 'The dissociation of water during photosynthesis? Critical review.' Analyzing the literature data and his own results, the author concluded that they reject the hypothesis on the oxidation of water in photosynthesis. The study ended with the words: 'If we take together the data of published isotope experiments as a confirmation against the splitting of O–H bonds, we should postulate the rapture of another oxygen-containing bond, that is, or C–O bond or O–O bond in the peroxide precursor of oxygen' (bold face by me, *G.K.*).

We have assumed (and we shall remain on these positions) that the pigment system of the chloroplast is a highly autonomous structure, designed for the generation of protons and molecular oxygen.

3.4.3 THE ROLE OF THERMAL ENERGY IN PHOTOSYNTHESIS AND CORRECTION OF THE FUNDAMENTAL PHOTOSYNTHESIS EQUATION

It is generally accepted (see any textbook on photosynthesis) that the thermal energy is a waste production of the photosynthetic process. In the case of a high-intensity process, only 0.5–5% of light energy is used up for photosynthesis, whereas ~95% of energy 'degrades into heat.' It is difficult to assume that in billions of years of their existence of the earth, the plants have not adapted themselves to a more efficient utilization of light energy. In 1973, the author assumed that the thermal energy is not only an important but also essential participant of the photosynthetic process [1, 12, 32]. This viewpoint is presented in the most complete form in the study 'Photosynthesis as a thermal process' [33]. According to this concept, in the regions with the size of the order of the chlorophyll molecule, the local temperature may greatly exceed (by several tens of degrees) the temperature of the surrounding medium. According to our estimates, this temperature reaches 70°C as a result of the recombination of charge carriers in the reaction center in which the adsorption of initial substances has not been completed at the given moment. Increase of the temperature inside the chloroplasts accelerates at the diffusion of both the products of photosynthesis and initial substances. An increase of temperature also facilitates the transport of ions through the membrane. According to calculations, the energy required for the transport of anion from the electrolyte into a lipid membrane, is 250 kJ/mole. However, the energy of transport of the ion (for example, sodium, potassium) through a membrane channel is considerably lower (approximately 20 kJ/mole) [34].

Already at the start of previous century, Timiryazev [35] proposed a hypothesis regarding the effect of thermal energy in photosynthesis. He assumed that the heating of chloroplasts, caused by sunlight, may be sufficient for the occurrence, in the chloroplasts, of a process thermodynamically reversed in relation to combustion, and, consequently, this may be used to explain the principle of photosynthesis. This viewpoint is at present of only historic interest because it was criticized. However, in the light of our considerations regarding hydrogen peroxide as a source of molecular oxygen in photosynthesis and local heating of the chloroplasts, it is possible that this hypothesis will be developed further.

It may be assumed that in photosynthesis together with the photoelectrochemical mechanism, there is also the possibility of the thermal dissociation of hydrogen peroxide as the release of molecular oxygen. The thermal stability of water is incomparably higher than that of hydrogen peroxide [1, 20].

An additional confirmation of the possibility of thermal dissociation of H_2O_2 in photosynthesis may be the data of modeling systems – the films of phthalocyanine (synthetic analog of chlorophyll) on a Pt electrode, which is in contact with the electrolyte, where the dependence of the photopotential of the films of phthalocyanine on heating of the electrolyte may be examined. At a temperature higher than 80°C, the photoresponse cannot be recorded. This can be naturally explained by the thermal dissociation of hydrogen peroxide formed on the surface of the platinum electrode [1, 22].

It should be mentioned that the assumptions regarding the local heating of chloroplast, introduced by us in 1973, were initially met with excessive criticism, but in recent years these concepts in the sphere of biophysics do not lead to any dispute (see, for example, [36]). In addition, a number of studies have been published recently on polymers, the theory of the photographic process, media for optical memory, glasses, where the assumption on the local temperature is used successfully for explaining the observed relationships. In the monograph by Timashev [34] 'Physics and Chemistry of Membrane Processes,' the role of the local heating is treated in a special section.

On the basis of all these considerations it may be assumed that of the local equation of photosynthesis should include not only light but also thermal energy [1, 22–26, 33]. The point is that in the implicit form, the thermal energy, from our viewpoint, has been included in the general equation of photosynthesis for a long time. We shall pay attention to the thermodynamic potential of dissociation of water. The minimum difference of the potentials, required for the electrolysis of water at 25°C,

1 atm, and on the condition of the supply of thermal energy, is 1.23 V. The thermoneutral potential of dissociation of water at 25°C is 1.47 V. In all investigations into photosynthesis, only the first magnitude (1.23 V) has been used, but the necessity for supplying thermal energy in this case is simply not mentioned.

Thus, the previously presented data on the exo- and endogenous hydrogen peroxide as a source of oxygen (hydrogen) in photosynthesis and the role of thermal energy make it possible to write the basic equation of photosynthesis in the following form [1, 22–26]:

$$CO_2 \text{ (air)} + H_2O_2 \text{ (water)} \xrightarrow[\text{Thermal energy } (+/-)]{\text{Light energy}} \text{carbohydrates} + O_2 \qquad (8)$$

The main difference between this equation and Eq. (1) is the replacement of H_2O by H_2O_2, and the water plays the role of the reaction medium for CO_2 and H_2O_2. It should be mentioned that the identical situation was recorded previously with the CO_2. Initially, it was assumed that 'air' (my italics, the author) damaged by the combustion of candles takes place in photosynthesis, and, subsequently, the modification CO_2 was introduced. The concentration of CO_2 in air is only 0.03%. It is also possible that in future new components of both air and natural water, taking part in photosynthesis, will be detected. For example, reports have appeared according to which the inert gases affect the rate of splitting of cells. Therefore, we regard it as useful to write in the main photosynthesis equation air in addition to CO_2 and water together with H_2O_2 and, naturally, photosynthesis, like any other life processes, is not possible without water.

Two words about thermal energy. The sign (+/–) in Eq. (8) indicates that at high densities of the light the leaf (chloroplast) transfers the energy to the surrounding medium, and at low densities, it takes the energy from the surrounding medium. In the latter case, the coefficient of transformation of solar energy may be higher than 100% because the contribution of thermal energy is not taken into account. A similar situation was found in the early stages of examination of the efficiency of operation of fuel elements.

Thus, we shall make a conclusion. A new concept, according to which the source of oxygen (hydrogen) in photosynthesis is the exogenous and endogenous hydrogen peroxide, and not water, has been proposed. The dissociation of hydrogen peroxide with the generation of molecular oxygen is possible either by photoelectrochemical and/or thermochemical mechanism. The thermal energy is not a reject of photosynthesis, and becomes an essential participant of this process. What consequences result from these considerations?

3.4.4 SOME CONSEQUENCES RESULTING FROM THE NEW CONCEPT OF PHOTOSYNTHESIS

3.4.4.1 The Effect of Hydrogen Peroxide on the Growth of Plants

Within the framework of the proposed concept it was reasonable to propose that the variation of the concentration of peroxide in water has a significant effect on the rate of growth of plants because photosynthesis in particular determines the rate of the physiological processes. The authors of Ref. [37] presented results of experimental verification of this assumption. Naturally, the effect of different amounts of exogenous hydrogen peroxide may have an appropriate effect also on the rate of generation of endogenous peroxide. The mechanism of participation of hydrogen peroxide in the physiological processes of growth is being studied intensively at the present time. Peroxide determines the intensity of photophosphorylation, photobreathing, fungitotoxicity of the surface of leaves, etc.

In order to record of the growth of plants in the presence of hydrogen peroxide, in addition to the traditional biological approach, we have also used the recently proposed method of laser interference auxanometry [1, 24]. It should be mentioned that the problem of the growth of plants is one of the central problems in current phytophysiology.

Analysis of the literature shows that the stimulating effect of hydrogen peroxide on the growth and development of plants has been known for a long period of time. Already at the beginning of this century, it was known that the solutions of hydrogen peroxide stimulates the growth of seeds. The introduction of hydrogen peroxide into soil increases the yield of corn, soya beans, accelerates the growth of seeds, but the reason for this effect, as reported in the above investigations, was not clear [20].

3.4.4.2 Using Plants For the Generation of Air in a Closed Volume

If we compare the traditional form of the main equation of photosynthesis (1) with the Eq. (8) proposed by us, it may be seen that in addition to the differences mentioned previously, there is one important consequence in our view. The classic equation of photosynthesis assumes that there is complete agreement with respect to the initial and final products between photosynthesis and breathing (Eq. (1) read from right to left, is the equation of breathing). In the concepts proposed by us, the relationship between the fundamental biological process and breathing is complicated because the final product in breathing is water, which, in our view, does not split during photosynthesis. This is illustrated by the diagram of formation of oxygen and its usage during breathing (Fig. 3.5).

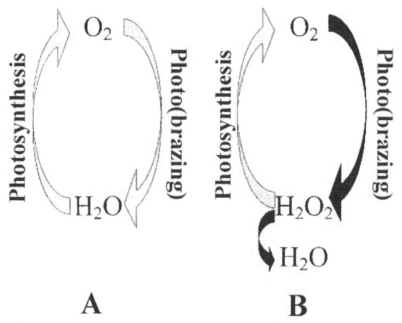

FIGURE 3.5 The relationship between photosynthesis and (photo) breathing within the framework of the conventional considerations regarding photosynthesis (A) and in accordance with the concept proposed by the author of the article (B).

It is well known that in addition to photosynthesis, the leaf is characterized by the occurrence of an opposite process stimulated by light, that is, the generation of carbon dioxide and absorption of oxygen (photobreathing). In the dark conditions, the intensity of the process, opposite photosynthesis (dark breathing), is 5–7% of photosynthetic gas exchange (plants, like living creatures, require oxygen). However, the intensity of photobreathing is an order of magnitude higher in comparison with dark breathing and equals ~50% of photosynthetic gas exchange [38, 39]. Since the photobreathing greatly reduces the rate of real photosynthesis and, consequently, the productivity of the plants, a larger number of attempts have been made to liquidate this process (or, at least, reduce its intensity). These attempts have not been successful. According to Chikov [38], 'none of the directions in the search for the methods of decreasing the intensity of photobreathing have resulted in a positive result for increasing the productivity of the plants. In addition, it is believed that the conditions, suppressing photobreathing, also decrease the productivity of the plants.'

Previously, it was mentioned that breathing is accompanied by the formation of endogenous hydrogen peroxide. In accordance with the proposed concept of photosynthesis, the endogenous hydrogen peroxide is a source of not only oxygen, injected into the atmosphere, but also of hydrogen used in the synthetic processes of growth. It should be mentioned that photobreathing is most active when the plant is supplied with abundance of mineral nitrogen, that is, favorable conditions are created for the growth of biomass. The stimulating effect of the exogenous hydrogen peroxide (at the optimum concentration of the latter) was examined in the previous section.

Thus, it may be assumed that the activity of synthetic processes in the plants is determined in approximately equal parts by the intensity of photosynthesis and photobreathing. Naturally, photobreathing uses the products stored as a result of photosynthesis plus atmospheric oxygen. A further examination of this process by physiologists of plants would enable, in our opinion, the development of the optimum technology cultivating of agricultural plants.

The considerations described here cast doubts on the exclusive role of plants in the formation of the current oxygen atmosphere of the earth, widely cited in the biological literature. We believe that in the geochemical investigations, which, in our view, are most important in the given problem, it is easy to find an opposite viewpoint.

Explanation of the mechanism of generation of photosynthesing oxygen is very important in the current period because of the development of space technology, the development of essential conditions for the life of man in a closed space. If the plants ensure the supply of oxygen to our atmosphere, it is natural that it should be attempted to use them for the generation of air in a closed system, for example, in a space station. Our concept of photosynthesis is directly related to this problem.

In September of 1991, eight volunteers started two-year tests in the Biosphere-2 – complex, isolated from the outer world. The complex is a prototype of future cosmic stations on the planets of the solar system [40]. The Biosphere-2 is a large experimental system generated for the examination of ecological processes taking place on the earth, and also for the development of the conditions of life activity in future cosmic stations which will be constructed mainly on Mars. The experiment was planned for 2 years. However, already in June 1992, the oxygen content inside the complex was greatly reduced (from 20.94 to 16.4%). The subsequent decrease of the oxygen concentration was 0.25–0.3% per month and it was therefore necessary to supply oxygen into the complex [41]. A special commission has been formed for investigating the reasons resulted in a decrease of the oxygen concentration.

We believe that one of the reasons resulting in the unsatisfactory dynamics of oxygen in the Biosphere-2 is associated with the fact that the developers of the object used the conventional assumptions according to which the water is a source of oxygen in photosynthesis. In accordance with our concept of photosynthesis (see Eq. (8) and Fig. 3.5), for the normal functioning of a hermetically sealed complex, containing plants, it is necessary to ensure the occurrence in the complex of the processes resulting in the generation of hydrogen peroxide. On the earth, the hydrogen peroxide forms as a result of storm discharges, the radiolysis of underground water, etc.

3.4.4.3 Artificial Photosynthesis and a Problem of the Life Origin

The attempts to model natural photosynthesis have been made since long ago. Back in the early twentieth century, there were studies aimed to reproduce the fundamental biological process or its separate steps by means of simple photochemical systems [1]. Currently, the number of publications dealing with this topic increases in the avalanche-like fashion (Fig. 3.6). However, the attempts made did not result in the design of artificial physicochemical systems able to form organic products from carbon dioxide and water with simultaneous evolution of molecular oxygen under the action of visible light. It was not until 1969 that only one of the key functions of chloroplast was reproduced for this first time, in particular, photocatalytic evolution of molecular oxygen from water under the action of light absorbed by synthetic analogs of chlorophyll [8, 9].

What is the main obstacle that precluded implementation of artificial photosynthesis despite numerous attempts? Probably, the traditional equa-

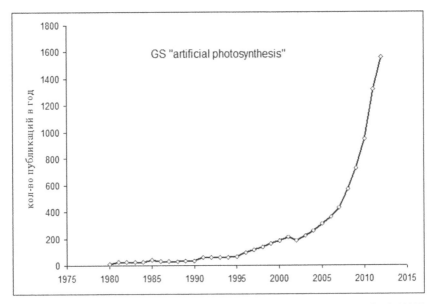

FIGURE 3.6 Annual number of publications mentioning artificial photosynthesis (1980–2012).

tion of photosynthesis bears some error, which prevents implementation of the process in vitro.

In 1993, a new photosynthesis equation was proposed at the Institute of Chemical Physics of the RAS and substantiated in detail in a monograph [1]. According to this equation, exo- and endogenic hydrogen peroxide rather than water serves as the source of oxygen (hydrogen) in the photosynthesis. In chloroplast, there is one hydrogen peroxide molecule per chlorophyll molecule [1]. As shown by quantum chemical calculations, the bond energy of peroxide with chlorophyll dimers is higher than that for water. Depending on the configuration of the diametric complex, the difference may reach 3.8 kcal/mol.

In 1994 we proposed to organize the international project "Artificial photosynthesis" [42]. It should be noted that the developed concept allows to se a new approach to the problem artificial photosynthesis. Thermodynamic estimates of the enthalpy ($\Delta H°$) and the Gibbs energy ($\Delta G°$) for reactions of CO_2 both with water and with hydrogen peroxide to give formaldehyde, formic acid, glucose, and other products were reported [43]. The results demonstrated unambiguously that syntheses of organic compounds from CO_2 and H_2O_2 requires less energy than the corresponding reactions using water. Indeed, replacement of H_2O by H_2O_2 decreases the G° of product formation by 30% for formaldehyde, by 34% for methanol, and by 42% for formic acid.

On the basis of the above, a successful attempt was made in 2004 to detect the formation of organic compounds upon the action of light on an aqueous suspension of adsorbed phthalocyanines, which were used as synthetic analogs of chlorophyll [43, 44]. The suspension contained 0.2 mol/L of hydrogen peroxide and 0.4 mol/L of $NaHCO_3$. By means of chemical analysis and spectrophotometry, photocatalytic formation of formaldehyde was detected (Fig. 3.7).

Analysis of the IR spectra of the irradiated reaction mixture attested to possible formation of organic products of various nature in such systems [43–45].

The work of Ref. [46] deals with a GC/MS detection of photogenerated organic products in an aqueous suspension of adsorbed aluminum phthalocyanine, H_2O_2, and HCO_3 and in the system containing H_2O_2 and CO_2, which we considered as a new step towards artificial photosynthesis.

The photophysical and photochemical properties of phthalocyanine are similar to those of chlorophyll but the stability against destructive impacts is much higher. This is why phthalocyanine was chosen for the experiment [1]. We used spectral grade chlorinated aluminum phthalocyanine (Kodak). Since chlorophyll is linked in vivo to a support (protein), we used phthalocyanine also in the adsorbed state. As a support with developed surface, we took silica gel L 40/100, which allowed easy separation of the pigment from the reaction products by mere centrifugation. Aluminum phthalocyanine was adsorbed from a solution in DMF. The adsorbed amount was measured as the decrease in the pigment concentration in solution by spectrophotometry on a DR/4000 V instrument (HACH-Lange, USA). As found from the adsorption isotherm, a monomolecular layer of the phthalocyanine metal complex is formed on the support surface when the phthalocyanine concentration in the initial solution is in the range. 10–4 mol/L [45]. The preliminary experiments showed that the highest photocatalytic activity of the adsorbed aluminum phthalocyanine in decomposition of H2O2 is observed in the region of premonomolecular coverage of the support surface. Considering these data,

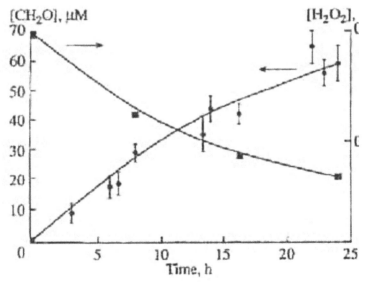

FIGURE 3.7 Kinetics of phormaldehyde accumulation and hydrogen peroxide consumption in the photocatalitic reduction bicarbonate anion ($[NaHCO_3]_0 = 0.4$ M, 20°C, photocatalist is aluminum phtalocyanine) (time, hours).

the catalysts were prepared by adding 10 mL of a 5×10^{-4} M solution of aluminum phthalocyanine to 1 g of silica gel. The adsorbates thus obtained contained 5.4×10^{-7} mol of the pigment per g of silica gel.

Kinetic experiments were carried out in 10 mL of a reaction mixture containing H_2O_2 (0.2 mol/L), special purity grade $NaHCO_3$ (0.4 mol/L) (Reakhim), and supported catalyst (200 mg) in distilled water. The suspension was distilled in a visible light with intense stirring. A halogen lamp (150 W) fitted with lenses, a condenser, and a KS-13 light filter cutting off the radiation below 630 nm was used as the source of visible light. The light flux power was 10 mW/cm^2. A quantitative determination of formaldehyde showed that its concentration reaches 10^{-5} mol/L after 24 h of irradiation. By this moment, more than 70% of the initial amount of H2O2 has been consumed, mainly via (photo)catalytic disproportionation. Under these conditions, we did not observe destruction or poisoning of the catalyst. The conclusion about the stability of the metal complex is also confirmed by the results of control runs carried out without $NaHCO_3$ in which no formaldehyde was detected. This implies that only CO_2 (HCO_3^-) can serve as the source of CH_2O.

Other products were analyzed by the GC/MS method. The GC/MS facility used to analyze the samples comprised a Thermo Focus GC gas chromatograph and a Thermo DSQ II mass spectrometer with a 60 m. 0.25 mm capillary glass column (0.25.m thick 100% dimethtylpolysiloxane as the stationary phase). The temperature program of the chromatograph included 10-min heating from 35 to 80°C (heating rate 1°C/min), then to 110°C (heating rate 5°C/min), and to 210°C (heating rate 10°C/min). The injector temperature was 200°C and the interface temperature was 250°C. Electron impact ionization was used, the electron energy being 70 eV and the mass spectrum being scanned in the range of 20–270 amu. Helium served as the carrier gas, the flow rate was 1 mL/min. Compounds were identified using the NIST library of mass spectra. Under the conditions used, the GC/MS method revealed organic compounds of alcohol and ketone classes in the reaction mixture (Fig. 3.8). If one of the components (hydrogen peroxide, hydrocarbonate, or phthalocyanine) is removed from the reaction mixture, organic products cannot be detected.

In the next experiment, $NaHCO_3$ was replaced by gaseous CO_2, which was passed continuously through the suspension during the irradiation (24 h).

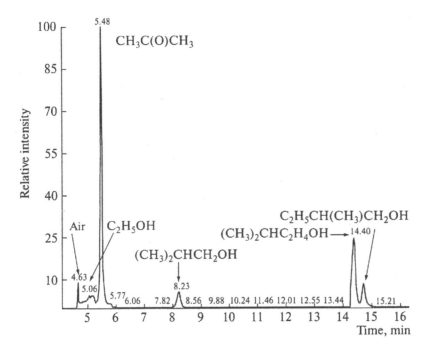

FIGURE 3.8 Composition of the organic products in the reaction mixture containing H_2O_2, NaHCO$_3$, and adsorbed aluminum phthalocyanine after 24 h of irradiation according to GC/MS data.

All other conditions were the same as in the previous experiment. The GC/MS analysis of the reaction mixture after irradiation showed the presence of formic acid (Fig. 3.9).

Note that in this run, too, GC/MS analysis did not detect organic compounds when either hydrogen peroxide or carbon dioxide was missing.

Positive results of experiments using hydrogen peroxide are of considerable interest. They open up attractive prospects: selection of the most appropriate pigment (variation of the ligands and the central atom), support, and reactant ratio, which may provide significant results. At the current stage, we can say with confidence that a new step was made towards artificial photosynthesis in a purely abiogenic system. The value of these results is beyond the framework of photosynthesis. Particular paths to biofuel from carbon dioxide and hydrogen peroxide have been outlined, which is of paramount importance for modern ecology. According to

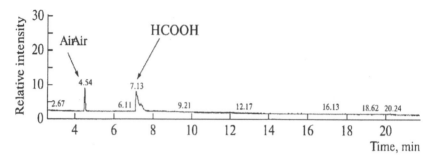

FIGURE 3.9   Formation of formic acid in the reaction mixture containing H_2O_2 and adsorbed aluminum phthalocyanine after 24 h of irradiation with continuous purging with CO_2 according to GC/MS data.

various estimates, the period when the mankind will face the problem of exhaustion of the resources of fossil combustibles is near at hand.

The obtained results can be used also in considering the problem of cosmic origin of life, as hydrogen peroxide was detected in space objects [47–50].

3.5   CONCLUSION

In the justification of the new concept of photosynthesis resulting from all our previous investigations, carried out in the laboratory, we use the following results. Photosynthesis (or, more accurately, its light stage – the generation of oxygen) is basically a relatively simple physical–chemical process-taking place in a highly complicated biological system. The pigments of the chloroplasts represented a highly autonomous system whose function is mainly the absorption of light (therefore, the intensive color of the pigments) and transformation of the light to chemical energy. The effect of light in the chloroplasts results in the formation of an excess number of protons ensuring the possibility of occurrence of fermentation reactions leading to the synthesis of hydrocarbons. The results of our literature analysis and the data obtained in our laboratory convincingly indicate that the source of oxygen (hydrogen) in photosynthesis is the exogenous and endogenous hydrogen peroxide. The water plays the role given to the solvent in conventional chemical reactions. Naturally, the water protons may take part in the formation of endogenous hydrogen peroxide. However, the problem of dissociation (oxidation) of water is not so important for us at the moment because the amount of hydrogen peroxide ('oxidized water,' the initial name of hydrogen peroxide) in the chloroplasts (cytoplasma) is such that it is possible to explain the experimental detected intensity of generation of oxygen (for each molecule of chlorophyll, including the molecule of the antennae, there is one molecule of hydrogen peroxide). Evidently, if the system contains water and hydrogen peroxide, only hydrogen peroxide will undergo photodissociation under the effect of visible light. Naturally, the development of detailed schema of the dissociation of hydrogen peroxide in photosynthesis requires a certain period of time. As an example, in current catalysis, hydrogen peroxide is used widely in heavy and light organic synthesis. This is one of the oldest variants of catalysis. However, according to the descriptive expression by Academician I. I. Moiseev "Hydrogen peroxide is widely used in organic synthesis industry, although the mechanism of its action is not elucidated fully" (Report of the seminar on Catalysis in ICP RAS).

The participation of hydrogen peroxide in the photosynthetic generation oxygen is far more likely than the participation of water, not only from the physical–chemical viewpoint (comparison of the dissociation potentials, thermal stability, etc.). It makes it possible to explain the existence of a large number of physiological processes accompanying photosynthesis. For example, in the case of high-intensity photosynthesis of mitochondria (generators of hydrogen per oxide) and chloroplast 'converge.' In the literature on physiology of plants, transpiration is treated as an 'unavoidable evil,' with attempts being made to eliminate it, although these attempts have not been successful. Within the framework of our considerations, transpiration is essential for the concentration of exogenous hydrogen peroxide. The identical situation is also characteristics of the process of photobreathing with which the physiologists 'fight' without success and cannot get rid of it. Our considerations, presented in this article, indicate that the plants accumulate not only light but also thermal energy (unique thermal pump). The significance of the latter is especially large in the plant associations where the illumination of leaves because of mutual shading is relatively low (in the individually standing trees, the mutual screening of the leaves is also high). This makes it possible to use a new approach to explaining the fact that the intensity of photosynthesis in the case of slight illumination is almost an order of magnitude higher in comparison with high-intensity illumination.

In conclusion, it should be mentioned that the long-term investigations into modeling and examination of the photosynthesis mechanism, described in the article, have been used by us in the formulation and computer analysis of the model in which the process of search for the solution of a scientific problem by the investigator is examined in the generalized form [51, 52].

3.6 ACKNOWLEDGEMENTS

I am very grateful to the colleagues and graduates of the Laboratory of Photobionics of the Institute of Chemical Physics. I am also grateful to a large number of undergraduates and graduates (in most cases, of the Physical Department of the Moscow State University and Moscow Physico-Technical Institute) for solving the given problems.

I am especially grateful to Academician A.L. Buchachenko, the Head of the Department of the Dynamics of Chemical and Biological Processes of the Institute of Chemical Physics of the Russian Academy of Sciences, for supporting investigations carried out in our laboratory. The experiments in recent years have been supported by continuing financial support of the Russian Fund of Fundamental investigations to which we are grateful (Grants No. 94–02–04972a, No. 95–03- 08982a, No. 96–0334064a, No. 98–0332061a, № 04–03- 32890a and No. 00–15–97404a, 08–03–00875a); the Presidential program "Leading Scientific Schools (NSh – 2003–2013) – Coordinator Academician A.L. Buchachenko.

2005–2007 Basic Research Program of the Presidium of RAS "Organic and hybrid nanostructured materials for photonics" – Program Coordinator Academician M.V. Alfimov.

2005–2006 Program of Presidium of RAS # 7P-05 "Hydrogen Energy" – Program Coordinator Academician I.I. Moiseev.

2006–2014 Program of the Presidium of RAS № 18 "The problem of origin of the Earth's biosphere and its evolution" – Program Coordinator Academician E.M. Galimov.

2009–2011 Program of the Presidium of RAS "Chemical aspects of energy" – Program coordinator of Academician I.I. Moiseev.

2004–2007 ISTS project # 2876 "Research and Development of Photo-electrochemical Light Energy Converters Based on Organic Semi-conductors Using the Principles of Photobionics."

2009–2013 ISTS project # 3910 "Modeling of primary stages of photosynthesis on the basis of nano-sized supramolecular systems."

I know Gennady Efremovich Zaikov from the first days of my stay at the Institute of Chemical Physics RAS (since 1967). I was always amazed and is still amazing his lively mind, a fantastic performance, a strong sense of humor. His gait is always swift, handshake is vigorous, at meeting I always see a smiling face. Taking this opportunity, I wish dear Gennady Efremovich good health and new successes, happiness.

KEYWORDS

- artificial photosynthesis
- Becquerel effect
- carbon dioxide
- chlorophyll
- chloroplast
- hydrogen peroxide
- new equation of photosynthesis
- oxygen
- photocells
- photosynthesis
- photovoltaic effect (Becquerel effect)
- phthalocyanine
- quantum yield on a photocurrent
- transpiration
- types of structures of pigment films on electrodes
- water

REFERENCES

1. Komissarov G.G. Photosynthesis (Physicochemical Approach), Moscow: Ed. URSS (2003), 223 p. (in Rus.); Komissarov G.G. Fotosintesis: um enfoque fisicoquimico Ed. URSS, (2005), 258 p. (in Spain)
2. Komissarov G.G. *Abstr. Second Intern. Biophys. Congr.,* Vienna, Austria, (1966*),* 234.
3. Komissarov G.G. *Biophysics,* (1967). 12, #3, 558–561.
4. Becquerel E. C. R. *Academy of Sciences,* Paris (1839), 9, 561.
5. Becquerel E. C.R *Academy of Sciences,* Paris (1874), 79, 185.
6. Chibisov K.V. Comments on history of photography, Art, Moscow (1987), 218 (in Rus.).
7. Komissarov G.G., Shumov Yu.S., The Reports of USSR (1968), 182, 1226–1229.
8. Komissarov G.G., Shumov Yu.S., Borisevich Yu.E., The Reports of USSR, (1969),187, 670–673.
9. Wang J.H. *Proc. Nat. Academy of Sciences. USA,* (1969), 62, 653–660.
10. Ilatovsky V.A., Dmitriev I.B., Komissarov G.G., *Russ. J. Phys. Chem.,* (1978), 52, 66–68.
11. Ilatovsky V.A., Apresyan E.S., Komissarov G.G., *Russ. J. Phys. Chem.,* (1989). 63, 2242–2244.
12. Komissarov G.G., *Russ. J. Phys. Chem.,* (1973), 47, 927–932.
13. Komissarov G.G. *Sov. Sci. Rev.,* (1971*),* 285–290.
14. Ilatovsky V.A., Komissarov G.G. *The Reports of Russian Academy of Sciences* (2008), 420, 66–69.
15. Komissarov G.G., *UPAC Abstr. 5-th Intern. Symp. of Macromoleculare Complexes, Bremen, Germany,* (1993) 420.
16. Komissarov G.G., *UPAC Abstr. XVIth Intern. Symp. of Photochemistry, Helsinky, Finland,* (1996), 332.
17. Ptitsyn G.A., Komissarov G.G., (1994). *Sov. J. Chem.* Phys, (1994), 2137–2147.
18. Das T.N. et al. *J. Indian Chem. Soc.,* (1982), 59, 85–87.
19. Stamm E.V., Purmal A.P., Skurlatov Yu.I., Successes of Chemistry (in Rus.) (1991), 60, 2373–2398.
20. Schumb W.C., Satterfield C.N., Wentworth R.L., Hydrogen Peroxide, Reinold Publishing Corporation, N.Y., (1955), 578p.
21. Mano J., Takahashi M.A., Asada K., *Biochemistry,* (1987), 26, 24995–2497.
22. Komissarov G.G., *Chem. Phys. Rep.* (1995), 14(11), 1723–1732.
23. Komissarov G.G., *Science in Russia* (in Rus.) (1994), № 5, 52–55.
24. Komissarov G.G., *J. Advanc. Chem. Phys.* (2003), 2(1). 28–61.
25. Komissarov G.G., *Optics and Spectroscopy* (1997), 83. 607–610.
26. Komissarov G.G. A new concept of photosynthesis mechanism in Book Problems of ecological security in agriculture Moscow, Sergiev Posad, (2003), 6. 5–25.
27. Vinogradov A.P., Teis R.V. *The Reports of USSR.* (1941), 33, 497–499.
28. Ruben S., et al. *J. Amer. Chem. Soc.,* (1941), 63, 877–879.
29. Vinogradov A.P., Kutyurin V.M. *5-th Int. Biochem. Congr.,* Science (in Rus.), Moscow, (1962), 264–274.

30. Vyurmser Z. *5-th Int. Biochem. Congr.*, Science (in Rus.), Moscow, (1962), 21.
31. Metzner H. *J. Theoret. Biol.* (1975), 51, 201–216.
32. Komissarov G.G. Chemistry and physics of photosynthesis, Knowledge (in Rus.), Moscow, (1980). 63
33. Komissarov G.G. Current Research in Photosynthesis Ed.M.Baltscheffsky Kluwer Academic Publishers. (1990). IV, 107–110.
34. Timashev S.V. Physicochemistry of membrane processes, Chemistry (in Rus.), Moscow, (1988). 237.
35. Rabinovich, E. Photosynthesis, Foreign literature (in Rus.) Moscow, (1959). 3, 936.
36. Kucheva N.S., et al., *Biophysics*, (1997), 42, 628–632.
37. Apasheva L.M., Komissarov G.G, *Biology Bulletin*, (1996), 23, 518–519.
38. Chikov V.I., *Soros Educ.J. Chem*, (1996). 11, 2–6. (in Rus.).
39. 39. Golovko T.K., Breathing of plants (physiological aspects), Science (in Rus.)1999, SPb, 190.
40. Allen J., Nelson M., Space biospheres, Synergetic Press, Arizona, (1989). 108.
41. Nelson M., et al., *Herald of Russian Academy of Sciences, (1993), 63, 1024–1036*
42. Komissarov G.G. Artificial photosynthesis: when? *10 Intern. Conf. Photochem. Conversion and Storage Solar Energy (IPS – 10) Book of Abstracts Editor Gion Calzafferri.* (1994).
43. Lobanov A.V., Kholuiskaya S.N., Komissarov G.G. *Reports of Phys. Chem.* (in Rus.) (2004), 399, Part 1. 266–268.
44. Lobanov A.V., Kholuiskaya S.N., Komissarov G.G. Chemical Physics (in Rus.) 2004, vol. 23, № 5, 44–47.
45. Nevrova O.V., Lobanov A.V., Komissarov G.G. J. Charact. Develop. Novel Mater., (2011), vol. 3, № 3, 172–176.
46. Komissarov G.G., Lobanov A.V., Nevrova O.V. and al. *Reports of Physical Chemistry* (in Rus.) *2013, Vol. 453, Part 2, 275–278.*
47. Houtkooper J.M. and Schulze-Makuch D., Planet. Space Sci. (2009), vol. 57,. 449–453.
48. Bergman P., Parise B., Liseau R. at al. *A&A (2011), 531, L8, 1–4.*
49. Du F., Parise B., Bergman P. A&A (2012), 544, C4, 1–2.
50. Encreannaz T., Greathouse T.K., Lefevre F., Atreya S.K. Planetary Space Sci.(2012) 68, 3-17.
51. Komissarov G.G., Avakyanz G.S., Mazo M.A. Pros. Int. Conf. Nonlinear Word, Astrakhan' (2000), 110.
52. Komissarov G.G, Petrenko U.M., Avakyants G.S., Rubtsova N. A. Abstr.XX Intern. Scien. Conf. "Math. Methods in Technics and Technology " Yaroslavl 2007, Vol. 2, p. 253–256.

CHAPTER 4

ENTROPIC NOMOGRAM*

G. A. KORABLEV,[1] N. G. PETROVA,[3] V. I. KODOLOV,[2]
R. G. KORABLEV,[1] G. E. ZAIKOV,[4] and A. K. OSIPOV[1]

[1]Izhevsk State Agricultural Academy, Russia Federation

[2]Kalashnikov Izhevsk State Technical University, Russia Federation

[3]Ministry of Informatization and Communication of the Udmurt Republic, Russia Federation

[4]Institute of Biochemical Physics, Russian Academy of Science, Russia Federation

CONTENTS

[1]Similarly to the ideas of thermodynamics on the static entropy, the concept of the entropy of spatial-energy interactions is used. The nomogram to assess the entropy of different processes is obtained. The variability of entropy demonstrations is discussed, in biochemical processes and economics, as well.

4.1   INTRODUCTION

The idea of entropy appeared based on the second law of thermodynamics and ideas of the adduced quantity of heat.

In statistic thermodynamics the entropy of the closed and equilibrious system equals the logarithm of the probability of its definite macrostate:

$$S = k \ln W \qquad (1)$$

where W – number of available states of the system or degree of the degradation of microstates; k – Boltzmann's constant.

or:

$$W = e^{s/k} \qquad (2)$$

These correlations are general assertions of macroscopic character, they do not contain any references to the structure elements of the systems considered and they are completely independent from microscopic models [1].

Therefore, the application and consideration of these laws can result in a large number of consequences, which are most fruitfully used in statistic thermodynamics.

The sense of the second law of thermodynamics comes down to the following:

The nature tends from the less probable states to more probable ones. The most probable is the uniform distribution of molecules through the entire volume. From the macrophysical point, these processes consist in equalizing the density, temperature, pressure and chemical potentials, and the main characteristic of the process is the thermodynamic probability W.

In actual processes in the isolated system the entropy growth is inevitable – disorder and chaos increase in the system, the quality of internal energy goes down.

The thermodynamic probability equals the number of microstates corresponding to the given macrostate.

Since the system degradation degree is not connected with the physical features of the systems, the entropy statistic concept can also have other applications and demonstrations (apart from statistic thermodynamics).

"It is clear that out of the two systems completely different by their physical content, the entropy can be the same if their number of possible microstates corresponding to one macroparameter (whatever parameter it is) coincide. Therefore, the idea of entropy can be used in various fields. The increasing self-organization of human society … leads to the increase in entropy and disorder in the environment that is demonstrated, in particular, by a large number of disposal sites all over the earth" [2].

In this research we are trying to apply the concept of entropy to assess the degree of spatial-energy interactions using their graphic dependence, and in other fields.

4.2 ENTROPIC NOMOGRAM OF THE DEGREE OF SPATIAL-ENERGY INTERACTIONS

The idea of spatial-energy parameter (P-parameter) which is the complex characteristic of the most important atomic values responsible for inter-atomic interactions and having the direct bond with the atom electron density is introduced based on the modified Lagrangian equation for the relative motion of two interacting material points [3].

The value of the relative difference of P-parameters of interacting atoms-components – the structural interaction coefficient α is used as the main numerical characteristic of structural interactions in condensed media:

$$\alpha = \frac{P_1 - P_2}{(P_1 + P_2)/2} 100\% \tag{3}$$

Applying the reliable experimental data we obtain the nomogram of structural interaction degree dependence (ρ) on coefficient α, the same for a wide range of structures. This approach gives the possibility to evaluate the degree and direction of the structural interactions of phase formation, isomorphism and solubility processes in multiple systems, including molecular ones.

Such nomogram can be demonstrated [5] as a logarithmic dependence:

$$\alpha = \beta \ln (\rho^{-1}) \tag{4}$$

where coefficient β – the constant value for the given class of structures. β can structurally change mainly within ±5% from the average value. Thus coefficient α is reversely proportional to the logarithm of the degree of structural interactions and therefore can be characterized as the entropy of spatial-energy interactions of atomic-molecular structures.

Actually the more is ρ, the more probable is the formation of stable ordered structures (e.g., the formation of solid solutions), that is, the less is the process entropy. But also the less is coefficient α.

The Eq. (4) does not have the complete analogy with Boltzmann's Eq. (1) as in this case not absolute but only relative values of the corresponding characteristics of the interacting structures are compared which can be expressed in percent. This refers not only to coefficient α but also to the comparative evaluation of structural interaction degree (ρ), for example, the percent of atom content of the given element in the solid solution relatively to the total number of atoms.

Therefore, in Eq. (4) coefficient k = 1. Thus, the relative difference of spatial-energy parameters of the interacting structures can be a quantitative characteristic of the interaction entropy:

4.3   ENTROPIC NOMOGRAM OF SURFACE-DIFFUSIVE PROCESSES

As an example, let us consider the process of carbonization and formation of nanostructures during the interactions in polyvinyl alcohol gels and metal phase in the form of copper oxides or chlorides. At the first stage, small clusters of inorganic phase are formed surrounded by carbon containing phase. In this period, the main character of atomic-molecular interactions needs to be assessed via the relative difference of P-parameters calculated through the radii of copper ions and covalent radii of carbon atoms demonstrated in Table 4.1.

From Table 4.1 it is seen that, in this case, the coefficient $\alpha = 3.50$ corresponding to the complete structural interaction: $\rho = 100\%$. The process takes place only in the gel volume but not on the film surface, which is not formed yet.

In the next main carbonization period the metal phase is formed on the surface of the polymeric structures being formed by the reaction:

$$2CuCl + [-CH = CH-]_n \rightarrow 2Cu + 2HCl + [C_2]_n$$

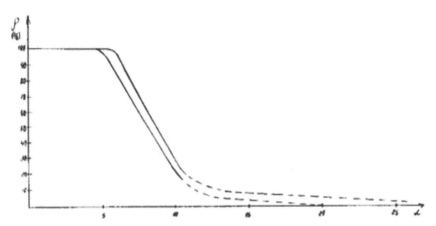

FIGURE 4.1   Nomogram of structural interaction degree dependence (ρ) on coefficient α.

From this point, the binary matrix of the nanosystem C→Cu is being formed. Let us consider the process of building up the film matrix of carbons in copper in the surface diffusion model. In this period of metal phase formation P-parameters calculated via the atom radii are valent-active.

In the liquid, the radius of molecular interaction sphere R≈3r, where r – molecule radius. Liquids are mainly formed by the elements of the system first and second periods. For the second period the following can be written down: R≈3r = (n + 1)r, where n – main quantum number. For both periods (first and second) we obtain R = (<n> + 1)r≈2.5r.

Let us assume that this principle with the definite approximation can be spread to different elements of the other periods, but taking the screening effects into account, introducing the value of the effective main quantum number (n*) instead of n. These values of n* and n* + 1 taken by Slater [4] are given in Table 4.2.

Thus, let us assume that the sphere radius of atomic-molecular interaction during the particle diffusion is defined as:

$$R = (n^* + 1)r,   \tag{5}$$

where r – dimensional characteristic of the atomic structure. The total change of R is from 3r to 5.2r (from the second to sixth period).

TABLE 4.1 Structural Interactions During the Nanofilm Formation in the System C→Cu

Carbon atom				Copper atom			Interaction characteristics					
Orbital	$P_0/(R(n^*+1))$ (eV)	$P_0/(r_u(n^*+1))$ (eV)	K	Orbital	$P_0/(R(n^*+1))$ (eV)	$P_0/(r_u(n^*+1))$ (eV)	α (%)	$1/\alpha$ (%)	ρ (%)	t (hour)	ω (%)	Interaction type
$2P^2 2S^2$	3.1519		1	$4S^2$		3.0436	3.50	0.29	100	0	0	volumetric
$2P^2$	4.3554		1.6	$4S^1 3d^1$	2.2011		21.17	0.05	5–8	0	0	semisurface
$2P^2$	4.3554		1.7	$4S^1 3d^1$	2.2011		15.15	0.07	19–21	0.49	21.5	surface
$2P^2$	4.3554		1.8	$4S^1 3d^1$	2.2011		9.46	0.11	56–58	1.05	63.9	surface
$2P^2$	4.3554		1.9	$4S^1 3d^1$	2.2011		4.06	0.25	~98–100	1.6	95	surface
$2P^2$	4.3554		2.0	$4S^1 3d^1$	2.2011		1.07	0.93	100	2.0	98.3	surface

TABLE 4.2   Effective Quantum Number

n	1	2	3	4	5	6
n*	1	2	3	3.7	4	4.2
n* + 1	2	3	4	4.7	5	5.2

The averaged value of the structural P_c-parameter falling on the radius unit of atomic-molecular interaction is defined by the following equation:

$$P_c = \frac{P_0}{KR} = \frac{P_0}{r(n^* + 1)K} \tag{6}$$

where K – coefficient taking into account the relative number of interacting particles and equal as follows (based on the calculations):

$$K = N_0 / N \tag{7}$$

Here N_0 – number of particles in the sphere volume of the radius R, N – number of particles or realized interactions depending on the process type (internal or surface diffusion).

Inside the liquid, the resultant of molecular interaction forces equals zero below the upper layer 2R thick (Fig. 4.2).

Applying the initial analogy to the internal diffusion, we can accept that such equilibrious state corresponds to the equality $N_0 = N$, then $K = 1$.

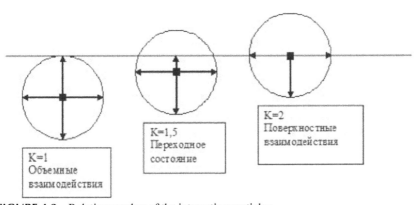

FIGURE 4.2   Relative number of the interacting particles.

On the upper part of the liquid surface layer, the sphere volume of atomic-molecular interaction and number of particles in it is practically twice less in comparison with the internal layers below 2R, that is, $N_o/N \approx 2$ and $K = 2$ for surface diffusion (Fig. 4.2).

Actually, the surface diffusion proceeds with the change of the coefficient K in the range from 1.5 to 2.0, which is taken into account in the calculations. Based on such initial ideas, the values of P-parameter and coefficient $1/\alpha_2$ are calculated by the Eqs. (3), (6) and (7) for carbon and copper atoms (Table 4.1).

The values of the degree of structural interactions from coefficient α are calculated, that is, $\rho_2 = f(1/\alpha_2)$ – curve 2 given in Fig. 4.3. Here, the graphical dependence of the degree of nanofilm formation (ω) on the process time is presented by the data from Ref. [5] – curve 1 and previously obtained nomogram in the form $\rho_1 = f(1/\alpha_1)$ – curve 3.

The analysis of all the graphical dependencies obtained demonstrates the practically complete graphical coincidence of all three graphs: $\omega = f(t)$, $\rho_1 = f(1/\alpha_1)$, $\rho_2 = f(\rho_2 = f(1/\alpha_2)$ with slight deviations in the beginning and end of the process. Thus, the carbonization rate, as well as the functions of many other physical-chemical structural interactions, can be assessed via the values of the calculated coefficient α and entropic nomogram.

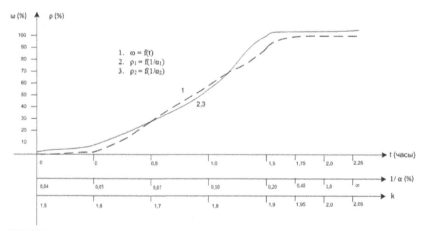

FIGURE 4.3 Dependence of the carbonization rate on the coefficient α.

4.4   NOMOGRAMS OF BIOPHYSICAL PROCESSES

4.4.1   ON THE KINETICS OF FERMENTATIVE PROCESSES

"The formation of ferment-substrate complex is the necessary stage of fermentative catalysis ... At the same time, n substrate molecules can join the ferment molecule" [6].

For ferments with stoichiometric coefficient n not equal one, the type of graphical dependence of the reaction product performance rate (μ) depending on the substrate concentration (c) has [6] a sigmoid character with the specific bending point (Fig. 4.4).

In Fig. 4.4, it is seen that this curve generally repeats the character of the entropic nomogram in Fig. 4.3.

The graph of the dependence of electron transport rate in biostructures on the diffusion time period of ions is similar [6].

In the procedure of assessing fermentative interactions (similarly to the previously applied for surface-diffusive processes) the effective number of interacting molecules over 1 is applied.

In the methodology of P-parameter, a ferment has a limited isomorphic similarity with substrate molecules and does not form a stable compound with them, but, at the same time, such limited reconstruction of chemical bonds which "is tuned" to obtain the final product is possible.

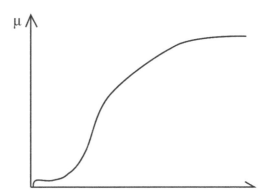

FIGURE 4.4   Dependence of the fermentative reaction rate (μ) on the substrate concentration (c).

4.4.2   DEPENDENCE OF BIOPHYSICAL CRITERIA ON THEIR FREQUENCY CHARACTERISTICS

1. The passing of alternating current through live tissues is character-ized by the dispersive curve of electrical conductivity – this is the graphical dependence of the tissue total resistance (z-impedance) on the alternating current frequency logarithm (log ω). Normally, such curve, on which the impedance is plotted on the coordinate axis, and log ω – on the abscissa axis, formally, completely cor-responds to the entropic nomogram (Fig. 4.1).
2. The fluctuations of biomembrane conductivity (conditioned by random processes) "have the form of Lorentz curve" [7]. In this graph, the fluctuation spectral density (ρ) is plotted on the coor-dinate axis, and the frequency logarithm function (log ω) – on the abscissa axis. The type of such curve also corresponds to the entro-pic nomogram in Fig. 4.1.

4.5   LORENTZ CURVE OF SPATIAL-TIME DEPENDENCE

The intervals between the events in different coordinate systems are deter-mined by Lorentz geometry of space-time. In this geometry, the velocity (β) is not additive by itself, therefore, the concept of the velocity param-eter is introduced (θ). The connection between the velocity β and velocity parameter is simple: $\beta = th\theta$, where means "hyperbolic tangent" and the law of adding two velocities is as follows:

$$th\Theta = th(\Theta_1 + \Theta_2) = \frac{th\Theta_1 + th\Theta_2}{1 + th\Theta_1 th\Theta_2}$$

The dependence between the velocity parameter and velocity itself is demonstrated [8] with Lorentz curve in Fig. 4.5. Both values are used in relative units in respect to the light velocity. The curve type is formally completely corresponds to the entropic nomogram in Fig. 4.3.

Example: "Let the bullet be shot with the velocity $\beta' = 0.75$ from the rocket flying with the velocity $\beta_r = 0.75$. It is necessary to find the bullet speed β relatively to the laboratory system. We know that the velocity parameters are additive, but not the velocities. By the graph, for the point

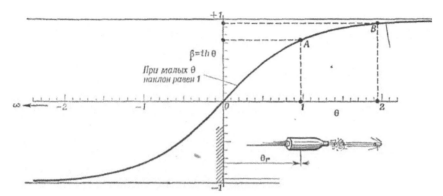

FIGURE 4.5 Connection between the velocity parameter θ and velocity itself β = thθ obtained directly from the addition law.

A we find θ' = θ$_r$ = 0.973. The addition produces θ = θ' + θ$_r$ = 1.946. For this value of the velocity parameter we find the point *B* by the graph and velocity value β = 0.96" [8].

4.6   ENTROPIC CRITERIA IN BUSINESS AND NATURE

The main properties of free market providing its economic advantages are: (i) effective competition, and (ii) maximal personal interest of each worker.

But on different economy concentration levels these ab initio features function and demonstrate themselves differently. Their greatest efficiency corresponds to small business – when the number of company staff is minimal, the personal interest is stronger and competitive struggle for survival is more active. With companies and productions increase the number of staff goes up, the role of each person gradually decreases, the competition slackens as new opportunities for coordinated actions of various business structures appear. The quality of economic relations in business goes down, that is, the entropy increases. Such process is mostly vivid in monostructures at the largest enterprises of large business (syndicates and cartels).

The concept of thermodynamic probability as a number of microstates corresponding to the given macrostate can be modified as applicable to the processes of economic interactions that directly depend on the parameters of business structures.

A separate business structure can be taken as the system macrostate, and as the number of microstates – number of its workers (N) which is the number of the available most probable states of the given business structure. Thus it is supposed that such number of workers of the business structure is the analog of thermodynamic probability as applicable to the processes of economic interactions in business.

Therefore, it can be accepted that the total entropy of business quality consists of two entropies characterizing: (i) decrease in the competition efficiency (S_1), and (ii) decrease in the personal interest of each worker (S_2), that is: $S = S_1 + S_2$. S_1 is proportional to the number of workers in the company: $S \approx N$, and S_2 has a complex dependence not only on the number of workers in the company but also on the efficiency of its management. It is inversely proportional to the personal interest of each worker. Therefore, it can be accepted that $S_2 = 1/\gamma$, where γ – coefficient of personal interest of each worker.

By analogy with Boltzmann's Eq. (1) we have:

$$S = (S_1 + S_2) \sim \left[\ln N + \ln\left(\frac{1}{\gamma}\right)\right] \sim \ln\left(\frac{N}{\gamma}\right)$$

or $S = k\ln\left(\frac{N}{\gamma}\right),$

where k – proportionality coefficient.

Here N shows how many times the given business structure is larger than the reference small business structure, at which N = 1, that is, this value does not have the name.

For nonthermodynamic systems we take k = 1. Therefore:

$$S = \ln\left(\frac{N}{\gamma}\right) \tag{8}$$

The Table 4.3 shows the approximate calculations of business entropy by the Eq. (8) for three main levels of business: small, medium and large. At the same time, it is supposed that number N corresponds to some average value from the most probable values.

When calculating the coefficient of personal interest, it is considered that it can change from 1 (one self-employed worker) to zero (0), if such

TABLE 4.3 Entropy Growth with the Business Increase

Structure parameters	Business		
	Small	Average	Large
$N_1 - N_2$	10–50	100–1000	10,000–100,000
	0.9–0.8	0.6–0.4	0.1–0.01
S	2.408–4.135	5.116–7.824	11.513–16.118
	3.271	6.470	13.816

worker is a deprived slave, and for larger companies it is accepted as $\gamma = 0.1\text{–}0.01$.

Despite of the rather approximate accuracy of such averaged calculations, we can make quite a reliable conclusion on the fact that business entropy, with the aggregation of its structures, sharply increases during the transition from the medium to large business as the quality of business processes decreases. The application of more accurate initial data allows obtaining specific values of business entropy, above which the process of economic relations can reach a critical level.

In live systems the entropy growth is compensated via the negative entropy (negoentropy), which is formed through the interaction with the environment. That is a live system is an open one. And business cannot be an isolated system for a long period without the exchange process and interactions with the environment. The role of the external system diminishing the increase in the business entropy must be fulfilled, for example, by the corresponding state and public structures functionally separated from business. The demonopolization of the largest economic structures carried out from the "top" in the evolution way can be the inevitable process here.

In thermodynamics it is considered that the uncontrollable entropy growth results in the stop of any macrochanges in the systems, that is, to their death. Therefore, the search of methods of increasing the uncontrollable growth of the entropy in large business is topical. At the same time, the entropy critical figures mainly refer to large business. A simple cut-down of the number of its employees cannot give an actual result of entropy decrease. Thus the decrease in the number of workers by 10% results in diminishing their entropy only by 0.6% and this is inevitably followed by the common negative unemployment phenomena.

Therefore for such super-monostructures controlled neither by the state nor by the society the demonopolization without optimization (i.e., without decreasing the total number of employees) is more actual to diminish the business entropy.

Comparing the nomogram (Fig. 4.1) with the data from the Table 3, we can see the additively of business entropy values (S) with the values of the coefficient of spatial-energy interactions (α), that is, S = $\square$.

Therefore, as applicable to business processes, the idea of business quality is similar to the concept of structural interaction degree (ρ). All this allows approximately defining the critical values of these parameters. Thus, at $\rho \approx 10\%$ the value S = $\square \approx 12 - 18\%$, that corresponds to the number of business structures in the range between 10,000 and 100,000 workers (in the average about 55,000).

The optimal criteria of a more qualitative business are defined by the maximal values of their entropies: S = 6 $\square$ 7 (in relative units).

The same values have been obtained earlier and for more complete degree of structural interactions, as continuous solid solutions correspond to the value $\square = 6 - 7$.

It is known that the number of atoms in polymeric chain maximally acceptable for a stable system is about 100 units, which is 10^6 in the cubic volume. Then we again have log $10^6 = 6$.

Now the scientific world is puzzled with the intensification of technological processes based on energy-saving electrical technologies, for instance [9, 10]. The given methodology of P-parameter can be also used in this perspective trend.

KEYWORDS

- business
- carbonization
- diffusion
- entropy
- nomogram
- spatial-energy parameter

REFERENCES

1. F. Reif. Statistic physics. M.: Nauka, 1972, 352 p.
2. L.A. Gribov, N.I. Prokopyeva. Basics of physics. M.: Vysshaya shkola, 1992, 430 p.
3. Korablev G.A. Spatial-Energy Principles of Complex Structures Formation//Brill Academic Publishers and VSP, Netherlands, 2005, 426pp. (Monograph).
4. Batsanov S.S., Zvyagina R.A. Overlap integrals and problems of effective charges. Nauka Publishers, Siberian Branch, Novosibirsk, 1966, 386 p.
5. Kodolov V.I., Khokhriakov N.V., Trineeva V.V., Blagodatskikh I.I. Activity of nano-structures and its manifestation in nanoreactors of polymeric matrixes and active media. Chemical physics and mesoscopy, 2008. V. 10. №4. p. 448–460.
6. Rubin A.B. Biophysics. Book 1. Theoretical biophysics. M.: Vysshaya shkola, 1987, 319 p.
7. Rubin A.B. Biophysics. Book 2. Biophysics of cell processes. M.: Vysshaya shkola, 1987, 303 p.
8. E. Taylor, J. Wheeler. Spacetime physics. Mir Publishers. M., 1971, 320 p.
9. Smirnova A.A., Alexeeva N.A., Pospelova I.G., Vozmischev I.V. New approach in controlling the utilization of recyclable materials in AIC. Russian scientific-practical conference "Innovations in science, engineering and technologies," April 28–30, 2014, Izhevsk, Udmurt University Publishers, p. 249–250.
10. Pospelova I.G. Sublimate drying with combined power drive. Pospelova I.G., Zakha-rova Ya.N., Gabasova F.V. Mechanization and electrification in agriculture. 2009. №6. 30–32.

CHAPTER 5

THE STRUCTURAL ANALYSIS OF PARTICULATE-FILLED POLYMER NANOCOMPOSITES MECHANICAL PROPERTIES

G. E. ZAIKOV,[1] G. V. KOZLOV,[2] and A. K. MIKITAEV[2]

[1]N.M. Emanuel Institute of Biochemical Physics of Russian Academy of Sciences, Kosygin st., 4, Moscow-119334, Russian Federation

[2]Kh.M. Berbekov Kabardino-Balkarian State University, Chernyshevsky st., 173, Nal'chik-360004, Russian Federation

CONTENTS

ABSTRACT

A number of the main mechanical characteristics (yield stress, impact toughness, microhardness) of particulate-filled polymer nanocomposites was described quantitatively within the framework of general conception

– fractal analysis. Such approach allows to study the main specific features of the indicated nanomaterials mechanical behavior. The influence of both nanofiller initial particles size and their aggregation degree on nanocomposites mechanical properties has been shown.

5.1   INTRODUCTION

Mechanical properties represent a very important part of polymer materials characteristics, particularly if the talk is about their application as engineering materials. Nevertheless, even if the indicated materials have another functional assignment, mechanical properties remain always-practical application important factor in this case as well. Particulate-filled polymer nanocomposites mechanical properties have a specific features number, which will be considered below.

As it is well known [1], the yield stress of polymeric materials is an important operating characteristic, restricting the range of their application as engineering materials from above. Therefore, the theoretical treatment of yielding process was always paid special attention, to that resulted in the development of a large number of theoretical models, describing this process [2]. For particulate-filled polymer nanocomposites the specific feature of the dependence of yield stress σ_Y on nanofiller contents is observed [3, 4]: unlike microcomposites of the same class [5], the value σ_Y is not increase to some extent perceptibly at nanofiller contents growth and even can be reduced. It is obvious, that the indicated effect is a negative one from the point of view of these polymeric materials exploitation, since it restricts their using possibilities as engineering materials.

The authors of Refs. [6, 7] found out, that the introduction of particulate nanofiller (calcium carbonate ($CaCO_3$)) into high density polyethylene (HDPE) results in nanocomposites HDPE/$CaCO_3$ impact toughness A_p in comparison with the initial polymer by about 20%. The authors of Refs. [6, 7] performed this effect detailed fractographic analysis and explained the observed A_p increase by nanocomposites HDPE/$CaCO_3$ plastic deformation mechanism change in comparison with the initial HDPE. Without going into details of the indicated analysis, one should note some reasons for doubts in its correctness. In Fig. 5.1, the schematic diagrams load-

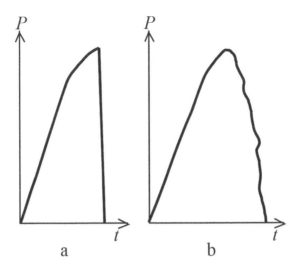

FIGURE 5.1   The schematic diagrams load-time (P-t) in instrumented impact tests. Failure by instable (a) and stable (b) crack.

time (P-t) for two cases of polymeric materials samples failure in impact tests are adduced: by instable (a) and stable (b) cracks. As it is known [8], the value A_p is characterized by the area under P-t diagram, which gives mechanical energy expended on samples failure. The polymeric materials macroscopic fracture process, defined by main crack propagation, begins at the greatest load P. From the schematic diagrams P-t it follows, that fracture process as a matter of fact practically does not influence on A_p value in case of crack instable propagation and influences only partly in case of stable crack. Although the authors [6, 7] performed impact tests on instrumented apparatus, allowing to obtain diagrams P-t, these diagrams were not adduced. Moreover, the structural aspect of fracture process has been considered in Refs. [6, 7] with the usage of secondary structures (crazes, shear zones and so on). Their interconnection with the initial undeformed material structure is purely speculative. It is obvious, that it does not occur possible to obtain quantitative relationships structure properties at such method of analysis.

At present it is known [10–12] that microhardness H_v is the property sensitive to morphological and structural changes in polymeric materials.

For composite materials the existence of the filler, whose microhardness exceeds by far polymer matrix corresponding characteristic, is an additional powerful factor [13]. The introduction of sharpened indentors in the form of a cone or a pyramid in polymeric material a stressed state is localized in small enough microvolume and it is supposed, that in such tests polymeric materials real structure is found [14]. In connection with the fact, that polymer nanocomposites are complex enough [15], the question arises, which structure component reacts on indentor forcing and how far this reaction alters with particulate nanofiller introduction.

The interconnection of microhardness, determined according to the results of the tests in a very localized microvolume, with such macroscopic properties of polymeric materials as elasticity modulus E and yield stress σ_Y is another problem aspect. At present a large enough number of derived theoretically and obtained empirically relationships between H_v, E and σ_Y exists [16].

Proceeding from the said above, the purpose of this chapter is the indicated mechanical properties of particulate-filled polymer nanocomposites treatment within the framework of general structural approach, namely, the fractal analysis.

5.2 EXPERIMENTAL PART

Polypropylene (PP) "Kaplen" of mark 01030 with average weight molecular mass of $\sim (2–3) \times 10^3$ and polydispersity index 4.5 was used as a matrix polymer. Nanodimensional calcium carbonate ($CaCO_3$) in a compound form of mark Nano-Cal P-1014 (production of China) with particles size of 80 nm and mass contents of 1–7 mass % and globular nanocarbon (GNC) (production of corporations group "United Systems," Moscow, Russian Federation) with particles size of 5–6 nm, specific surface of 1400 m^2/g and mass contents of 0.25–3.0 mass % were applied as nanofiller.

Nanocomposites PP/$CaCO_3$ and PP/GNC were prepared by components mixing in melt on twin-screw extruder Thermo Haake, model Reomex RTW 25/42, production of German Federal Republic. Mixing was performed at temperature 463–503 K and screw speed of 50 rpm during 5 min. Testing samples were prepared by casting under pressure method on a casting machine Test Sample Molding Apparate RR/TS MP of firm Ray-Ran (Taiwan) at temperature 483 K and pressure 43 MPa.

Uniaxial tension mechanical tests have been performed on the samples in the shape of a two-sided spade with the sizes according to GOST 112 62–80. The tests have been conducted on a universal testing apparatus Gotech Testing Machine CT-TCS 2000, production of German Federal Republic, at temperature 293 K and strain rate ~ 2×10^{-3} s^{-1}.

The impact tests have been conducted by Sharpy method on samples by sizes of 80×10×4 mm. Samples have V-like notch with length of 0.8 mm. Tests have been performed on pendulum apparatus model Gotech Testing Machine GT-7045-MD, production of Taiwan, with the energy dial of 1 J so that no less than 10% and no more than 80% of energy reserve was consumed on sample failure, with distance between supports (span) of 60 mm. No less than 5 samples were used for each test.

The microhardness H_v measurements by Shore (scale D) were performed according to Gost 24 621–91 on scleroscope HD-3000, model 05–2 of form "Hildebrand," production of German Federal Republic. The samples have cylindrical shape with diameter of 40 mm and height of 3 mm.

5.3   RESULTS AND DISCUSSION

5.3.1   YIELD STRESS

For the dependence of yield stress σ_Y on particulate nanofiller contents theoretical analysis the dispersive theory of strength was used, where nanocomposite yield stress at shear τ_n is determined as follows [17]:

$$\tau_n = \tau'_m + \frac{G_n b_B}{\lambda} \qquad (1)$$

where τ_n is shear yield stress of polymer matrix, G_n is shear modulus of nanocomposite, b_B is Burgers vector, λ is distance between nanofiller initial particles in nanocomposite.

In case of nanofiller particles aggregation the Eq. (1) has the look [17]:

$$\tau_n = \tau'_m + \frac{G_n b_B}{k(r)\lambda} \qquad (2)$$

where $k(r)$ is aggregation parameter.

It is easy to see, that the Eq. (2) describes the initial nanoparticles aggregation influence on nanocomposite yield stress. This effect is important from both theoretical and practical points of view in virtue of well-known nanoparticles tendency to aggregation, which is expressed by the following relationship [15]:

$$k(r) = 7.5 \times 10^{-3} S_u \qquad (3)$$

where S_u is nanofiller specific surface, which is determined as follows [18]:

$$S_u = \frac{6}{\rho_n D_p} \qquad (4)$$

where ρ_n is nanofiller density, D_p is its particles diameter.

From the Eqs. (3) and (4) it follows, that the nanofiller particles size decreasing results in S_u enhancement, that intensifies nanofiller initial particles tendency to aggregation.

Let us consider determination methods of the parameters, included in the Eq. (2). The general relationship between normal stress σ and shear stress τ has the look [19]:

$$\tau = \frac{\sigma}{\sqrt{3}} \qquad (5)$$

The stress τ_n is determined according to the equation [17]:

$$\tau'_m = \tau_m \left(1 - \varphi_n^{2/3}\right) \qquad (6)$$

where τ_m is shear yield stress of matrix polymer, φ_n is nanofiller volume content, determined according to the well-known formula [1]:

$$\varphi_n = \frac{W_n}{\rho_n} \qquad (7)$$

where W_n is nanofiller mass content and the value ρ_n for nanoparticles is determined as follows [15]:

$$\rho_n = 188 \left(D_p\right)^{1/3} \text{ kg/m}^3 \qquad (8)$$

where D_p is given in nm.

The shear modulus G_n is connected with Young's modulus E_n by the following simple relationship [20]:

$$G_n = \frac{E_n}{d_f} \tag{9}$$

where d_f is fractal dimension of nanocomposite structure, which is determined according to the equation [20]:

$$d_f = (d-1)(1+v) \tag{10}$$

where d is dimension of Euclidean space, in which a fractal is considered (it is obvious, that in our case $d = 3$), v is Poisson's ratio, estimated according to the mechanical test results with the aid of the relationship [14]:

$$\frac{\sigma_Y}{E_n} = \frac{1-2v}{6(1+v)} \tag{11}$$

where E_n is nanocomposite elasticity modulus.

The value of Burgers vector b_B for polymeric materials is determined according to the equation [2]:

$$b_B = \left(\frac{60.5}{C_\infty}\right)^{1/2} \text{Å} \tag{12}$$

where C_∞ is characteristic ratio, connected with dimension d_f by the equation [2]:

$$C_\infty = \frac{2d_f}{d(d-1)(d-d_f)} + \frac{4}{3} \tag{13}$$

It is obvious, that for the value τ_n theoretical estimation according to the Eq. (2) an independent method of parameter $k(r)\lambda$ determination is necessary. The following equation gives such method [3]:

$$k(r)\lambda = 2.09 \times 10^{-2} D_p \left(S_u / \varphi_n\right)^{1/2} \tag{14}$$

where D_p is given in nm, S_u – in m²/g.

The value S_u estimation according to the Eq. (4) gave the following results: $S_u = 3280$ and 93 m²/g for GNC and CaCO$_3$, respectively.

In Fig. 5.2, the comparison of the received experimentally σ_Y and calculated according to the described above method σ^r_Y yield stress values

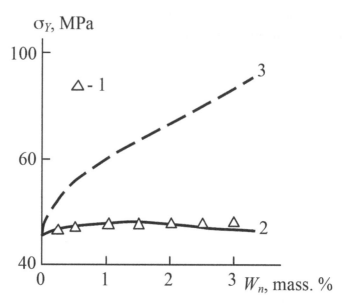

FIGURE 5.2 The dependences of yield stress σ_Y on nanofiller mass contents W_n for nanocomposites PP/GNC. 1 – experimental data; 2, 3 – calculation according to the Eqs. (2) and (1), respectively.

for nanocomposites PP/GNC is adduced. As one can see, the theory and experiment good correspondence is observed (the average discrepancy between σ_Y and σ^r_Y makes up 5.5%). Besides, the value σ_Y for nanocomposites does not differ to some extent significantly from the corresponding parameter for matrix PP: for nanocomposites $\sigma_Y = 36.0$–32.9 MPa, for PP $\sigma_Y = 31.5$ MPa, that is, σ_Y enhancement at nanofiller introduction does not exceed 15%. The causes of such effect can be elucidated by the Eq. (1) using, where the value λ is calculated as follows [17]:

$$\lambda = \left[\left(\frac{4\pi}{3\varphi_n} \right)^{1/3} - 2 \right] \frac{D_p}{2} \tag{15}$$

The Eq. (1) supposes nanofiller initial particles aggregation absence ($k(r) = 1.0$) and the dependence $\sigma^r_Y(W_n)$, calculated according to the indicated equation, is also adduced in Fig. 5.2. The absence of GNC initial particles aggregation results in nanocomposites PP/GNC yield stress strong increasing within the range of $W_n = 0.25–3.0$ mass % – from 44 up to 86 MPa.

In Fig. 5.3, the similar dependences of yield stress on W_n for nanocomposites PP/CaCO$_3$ are adduced. As one can see, σ^r_Y estimation according to the Eq. (2) gives an excellent correspondence to the experiment – the average discrepancy between σ_Y and σ^r_Y makes up 0.7% only. Besides, for nanocomposites PP/CaCO$_3$ σ_Y reduction within the range of $W_n = 1–7$ mass % is observed. And at last, CaCO$_3$ initial nanoparticles aggregation suppression does not give positive effect for these nanocomposites. It is obvious, that the cause of the indicated σ_Y reduction for nonaggregated CaCO$_3$ is a relatively large diameter of its initial nanoparticles, approaching to upper limit of nanoparticles dimensional range, which is equal to ~ 100 nm [21]. Owing to that λ value for nanocomposites PP/CaCO$_3$ varies within the limits of 200–66 nm within the range of $W_n = 1–7$ mass %, whereas for nanocomposites PP/GNC the value λ is essentially smaller:

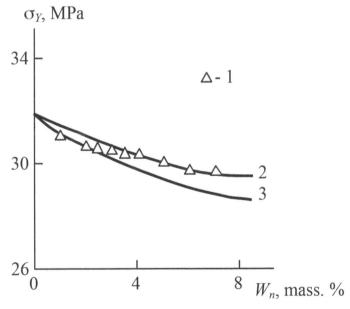

FIGURE 5.3 The dependences of yield stress σ_Y on nanofiller mass contents W_n for nanocomposites PP/CaCO$_3$. 1 – experimental data; 2, 3 – calculation according to the Eqs. (2) and (1), respectively.

17–4 nm within the range of $W_n = 0.25$–3.0 mass %. Thus, in particulate-filled polymer nanocomposites yield stress value definition two competing factors played critical role: nanofiller initial particles size and their aggregation level. It is important to note, that weak dependence of yield stress on nanofiller contents is typical not only for particulate-filled polymer nanocomposites, but also for other classes of these nanomaterials: polymer/organoclay [22] and polymer/carbon nanotubes [23].

Let us consider alternative, specific for nanocomposites with semicrystalline matrix, treatment of yield stress change. The Eqs. (10) and (11) combination at the condition $d = 3$ allows to obtain the following dependence of ratio E_n/σ_Y on the main structural characteristic d_f:

$$\frac{E_n}{\sigma_Y} = \frac{3d_f}{(3 - d_f)} \qquad (16)$$

In Fig. 5.4, the dependence of ratio E_n/σ_Y on dimension d_f is adduced, which demonstrates strong nonlinear growth of the indicated ratio at d_f

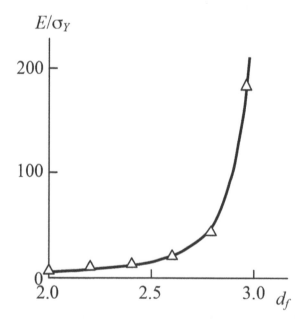

FIGURE 5.4   The dependence of elasticity modulus and yield stress ratio E/σ_Y on structure fractal dimension d_f for polymeric materials.

increasing, specifically at $d_f \geq 2.7$. Thus, the postulated in work [24] σ_Y and E_n proportionality is true in special case only, namely, in case of polymeric material structure invariability, that is, in case d_f = const. This rule is fulfilled for particulate-filled polymer nanocomposites with amorphous matrix: phenylone/β-sialone and phenylone/oxynitride silicium-yttrium [15]. For the indicated nanocomposites d_f = const = 2.416 and then E_n/σ_Y = 12.4, that is confirmed experimentally [15].

Let us consider further yield stress σ_Y behavior as a function of nanofiller mass contents W_n for the considered nanocomposites. The value d_f for them can be estimated by an independent mode, using the following equation [2]:

$$d_f = 3 - 6\left(\frac{\varphi_{cl}}{SC_\infty}\right)^{1/2} \tag{17}$$

where φ_{cl} is local order domains (nanoclusters) relative fraction, S is macromolecule cross-sectional area, which is equal to 27.2 Å^2 for PP [2].

The value φ_{cl} can be estimated according to the following percolation relationship [2]:

$$\varphi_{cl} = 0.03(1 - K)(T_m - T)^{0.55} \tag{18}$$

where K is crystallinity degree, T_m and T are melting and testing temperatures, respectively. For the considered nanocomposites the value K according to DSC data varies within the limits of 0.637–0.694 for PP/GNC and 0.637–0.668 for PP/CaCO$_3$ and the value T_m for PP was accepted equal to 445 K [25].

Since the value φ_{cl} is considered for nanocomposites, where nanoclusters are concentrated in polymer phase only, then one should use its reduced value φ_{cl}^{red}, which is equal to [15]:

$$\varphi_{cl}^{red} = \varphi_{cl}(1 - \varphi_n) \tag{19}$$

In Fig. 5.5, the comparison of the received experimentally and calculated according to the described above method dependences of yield stress σ_Y on nanofiller mass contents W_n for the considered nanocomposites is adduced. As one can see, a good correspondence of theory and experiment is obtained (their mean discrepancy makes up 2.5%). This circum-

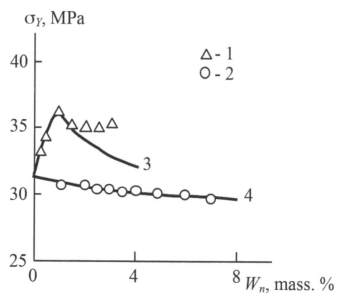

FIGURE 5.5 The dependences of yield stress σ_Y on nanofiller mass contents W_n for nanocomposites PP/GNC (1, 3) and PP/CaCO₃ (2, 4). 1, 2 – the theoretical calculation; 3, 4 – experimental data.

stance allows to explain the cause of insignificant increasing and in case of nanocomposites PP/CaCO₃ even reduction of yield stress at nanofiller contents growth. As it is known [15], the fractal dimension d_f of crystallizing polymeric materials structure depends on their crystallinity degree K as follows:

$$d_f = 2 + K + \varphi_{if} \tag{20}$$

where φ_{if} is a relative fraction of interfacial (crystallizing also) regions.

Therefore, the values d_f for the considered nanocomposites are within the range of ~ 2.75–2.80, that is, within the range, where the ratio E_n/σ_Y strong increase begins (see Fig. 5.4). So, for nanocomposites PP/GNC the ratio E_n/σ_Y value varies within the range of 31.0–41.8 and for PP/CaCO₃ – within the limits of 31.0–37.2. This increase corresponds by absolute value to nanocomposites elasticity modulus enhancement, which makes up ~40% for PP/GNC and ~ 13% for PP/CaCO₃. Hence, the ratio E_n/σ_Y increasing compensates E_n growth owing to nanofiller introduction and in case of nanocomposites PP/CaCO$_3$ E_n small increase results in σ_Y reduction. Let us note, that the proposed model is true for nanocomposites with amorphous matrix as well, having high enough d_f values (e.g., for rubbers) [4].

5.3.2   IMPACT TOUGHNESS

In Fig. 5.6, the dependences of impact toughness A_p on nanofiller volume contents φ_n are adduced for the considered nanocomposites. As it follows from the data of this figure, for both nanocomposites the dependence $A_p(\varphi_n)$ has an extreme character, whose maximum is reached at $\varphi_n \approx 0.03$. A_p increasing for nanocomposites in comparison with the corresponding parameter for matrix polymer can be significant: so, for nanocomposite PP/CaCO$_3$ at $\varphi_n = 0.03$ A_p value exceeds impact toughness for PP in 1.5 times. Nanocomposites PP/GNC and PP/CaCO$_3$ mechanical properties

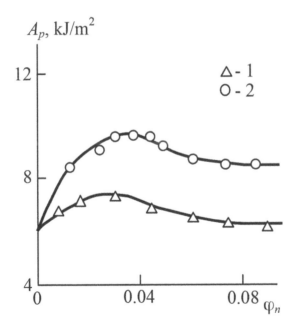

FIGURE 5.6   The dependences of impact toughness A_p on nanofiller volume contents φ_n for nanocomposites PP/GNC (1) and PP/CaCO$_3$ (2).

study has shown that a similar extreme dependence of property on nano-filler contents has yield stress σ_Y only (see Figs. 5.2 and 5.4). As it has been shown above, such dependence $\sigma_Y(\varphi_n)$ shape is due to nanofiller initial particles aggregation, which is intensified at φ_n growth. This interconnection is not accidental: as it has been noted above, A_p value is proportional to area under P-t diagram or curve stress-strain (σ-ε). In its turn, for plastic polymeric materials, which are investigated nanocomposites, a stress is restricted from above by yield stress σ_Y and limiting strain is equal to failure strain ε_f. Therefore, it is to be expected that impact toughness A_p is proportional to product $\sigma_Y\varepsilon_f$. Within the framework of fractal analysis the value ε_f is determined as follows [2]:

$$\varepsilon_f = C_\infty^{D_{ch}-1} - 1 \qquad (21)$$

where D_{ch} is fractal dimension of a polymer chain part between its fixation points (chemical cross-linking nodes, physical entanglements, nanoclusters etc.).

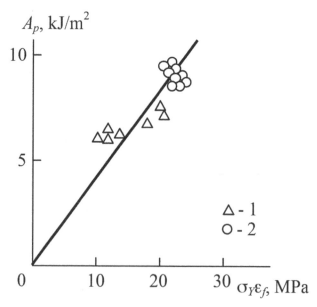

FIGURE 5.7   The dependence of impact toughness A_p on parameter $\sigma_Y\varepsilon_f$ for nanocomposites PP/GNC (1) and PP/CaCO$_3$ (2).

The parameters C_∞ and D_{ch} characterize polymer chain statistical flexibility and molecular mobility level, respectively [2]. The dimension D_{ch} can be determined with the aid of the following equation [2]:

$$\frac{2}{\varphi_{cl}} = C_\infty^{D_{ch}} \tag{22}$$

In Fig. 5.7 the dependence $A_p(\sigma_Y \varepsilon_f)$ for the considered nanocomposites is adduced, which proves to be linear and passing through coordinates origin that allows to describe it analytically by the following empirical equation:

$$A_p = \left(0.4 \times 10^{-3} m\right) \sigma_Y \varepsilon_f \tag{23}$$

The adduced above analysis allows to elucidate the cause of higher values A_p for nanocomposites PP/CaCO$_3$ in comparison with PP/GNC. This cause is higher D_{ch} values: for PP/CaCO$_3$ $D_{ch} = 1.33$–1.34, for PP/GNC D_{ch} $= 1.13$–1.29, that is, higher molecular mobility level for nanocomposites PP/CaCO$_3$, although σ_Y values are somewhat higher for PP/GNC.

5.3.4 MICROHARDNESS

Let us consider the interconnection of microhardness H_v and other mechanical characteristics, among their number yield stress σ_Y for nanocomposites PP/GNC and PP/CaCO$_3$. The following relationship was received by Tabor [26] for metals, which were considered as rigid perfectly plastic solids, between H_v and σ_Y:

$$\frac{H_v}{\sigma_Y} \approx c \tag{24}$$

where c is constant, which is approximately equal to 3.

The relationship (24) implies, that the exerted in microhardness tests pressure under indentor is higher than yield stress in quasistatic tests owing to restriction, imposed by undeformed polymer, surrounding indentor. However, in works [12, 16, 22, 27, 28] it has been shown that the value c can differ essentially from 3 and varied in wide enough limits: ~ 1.5–30. In the work [28] it has been found out, that for the composites HDPE/CaCO$_3$

depending on strain rate $\dot{\varepsilon}$ and type of quasistatic tests, in which the value σ_Y was determined (tensile or compression) c magnitude varies within the limits of 1.80–5.83. To $c = 3$ the ratio H_v/σ_Y approaches only at minimum value $\dot{\varepsilon}$ and at using σ_Y values, received by compression tests. Therefore, in the Ref. [28], the conclusion has been obtained, that the value $c = 3$ can be received only at comparable strain rates in microhardness and quasistatic tests and at interfacial boundaries polymer-filler failure absence.

An elasticity role in indentation process was proposed to consider for the analysis spreading on a wider interval of solids. For the solid, having elasticity modulus E and Poisson's ratio v, Hill has obtained the following equation [16]:

$$H_v = \frac{2}{3}\left[1 + \ln\frac{E}{3(1-v)\sigma_Y}\right]\sigma_Y \qquad (25)$$

and empirical Marsh equation has the look [16]:

$$H_v = \left(0.07 + 0.6\ln\frac{E}{\sigma_Y}\right)\sigma_Y \qquad (26)$$

The Eqs, (25) and (26) allow the microhardness H_v theoretical estimation for particulate-filled polymer nanocomposites at the known E and σ_Y condition and the value v can be calculated according to the Eq. (11).

In Table 5.1, the comparison of experimental H_v and calculated according to the Eq. (26) H_v^T microhardness values for the considered nanocomposites is adduced. The Eq. (26) was chosen according to a simple reason that it gives better correspondence to experiment than the Eq. (25) for all classes of nanocomposites [4, 22, 29]. As it follows from the data of this table, a good enough correspondence of theory and experiment is obtained (the mean discrepancy between H_v and H_v^T makes up ~ 8%). This correspondence indicates, that H_v value for the considered nanocomposites is controlled by their macroscopic mechanical properties to the same extent, as for other materials.

Let us consider the physical nature of the ratio H_v/σ_Y deviation from the constant $c \approx 3$ in the Eq. (24), using for this purpose the Eqs. (25) and (26). The value d_f can be determined according to the Eq. (10) and then the relationships combination allows to obtain the following equations [4, 22]:

$$\frac{H_v}{\sigma_Y} = \frac{2}{3}\left\{1 + \ln\left[\frac{2d_f}{(4-d_f)(3-d_f)}\right]\right\} \qquad (27)$$

TABLE 5.1 The Comparison of Obtained Experimentally H_v and Calculated According to the Eq. (26) H_v^T Microhardness Values for Nanocomposites PP/GNC and PP/CaCO$_3$

Nanocomposite	Nanofiller mass content, mass %	H_v, MPa	H_v^T, MPa	Δ, %
PP/GNC	0	68	66.3	2.5
	0.25	75	71.7	4.4
	0.50	73	75.8	9.4
	1.0	74	78.7	6.4
	1.50	72	76.4	6.1
	2.0	72	75.5	4.9
	2.50	72	75.2	4.4
	3.0	72	75.9	4.8
PP/CaCO$_3$	1.0	72	67.0	10.7
	2.0	75	66.8	10.9
	2.5	76	66.8	12.1
	3.0	75	66.5	10.9
	3.5	75	66.7	11.3
	4.0	75	66.4	11.1
	5.0	75	66.7	11.5
	6.0	75	66.2	11.7
	7.0	75	66.5	11.3

*Δ is relative discrepancy between parameter H_v and H_v^T.

and

$$\frac{H_v}{\sigma_Y} = \left[0.07 + 0.6\ln\left(\frac{3d_f}{3 - d_f}\right) \right]$$ (28)

for case $d = 3$.

From the Eqs. (27) and (28) it follows, that the ratio H_v/σ_Y is defined by structural state of nanocomposite (matrix polymer) only, which is characterized by its fractal dimension d_f. In Fig. 5.8 the dependences of the ratio H_v/σ_Y on d_f is adduced, calculated according to the Eqs. (27) and (28), which found out complete similarity, but absolute values H_v/σ_Y, calculated by the Eq. (27), proved to be on about 15% higher than the analogous magnitudes, calculated according to the Eq. (28). The identical results for extrudates of polymerization-filled compositions on the basis of ultra-high-molecular polyethylene were obtained in Ref. [12] and for polymer nanocomposites of different classes – in Refs. [4, 22, 29]. In Fig. 5.8 the condition $H_\nu/\sigma_Y = c = 3$ according to the Eq. (24) is given by a horizontal stroke line. As it follows from this figure data, the indicated condition is reached at $d_f \approx 2.85$ according to the equation (27) and $d_f \approx 2.93$ according to the Eq. (28). As it is known [20], for real solids the limiting greatest value d_f is equal to 2.95. Thus, these d_f values at $c = 3$ indicate again, that the empirical Marsh equation (the Eq. (28)) gives more precise estimation of ratio H_ν/σ_Y, than more strictly derived Hill relationship (the Eq. (27)). Hence, the adduced above results show, that the ratio H_ν/σ_Y value is defined by polymer nanocomposites structural state only and Tabor criterion $c = 3$ is realized for Euclidean solids only.

In Fig. 5.8, the dependence of the obtained experimentally values H_ν/σ_Y on nanocomposites structure fractal dimension d_f is shown by points. As one can see, the obtained experimentally dependence $H_\nu/\sigma_Y(d_f)$ corresponds well to the theoretical curve, calculated according to the Eq. (28)

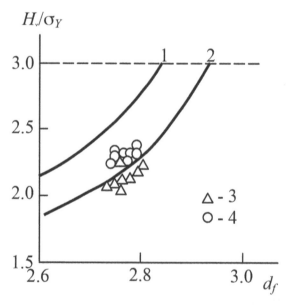

FIGURE 5.8 The dependences of ratio H_ν/σ_Y on structure fractal dimension d_f. 1, 2 – calculation according to the equations (27) (1) and (28) (2); 3, 4 – experimental data for nanocomposites PP/GNC (3) and PP/CaCO$_3$ (4). The horizontal stroked line indicates Tabor criterion $H_\nu/\sigma_Y = c = 3$.

(the mean discrepancy between theory and experiment makes up 8.5%), whereas calculation according to the Eq. (27) gives overstated absolute values of the indicated ratio.

5.4 CONCLUSIONS

The yield stress value of particulate-filled polymer nanocomposites is controlled by two competing factors: diameter of nanofiller initial particles and their aggregation degree. The cause of weak increasing (and even reduction) of the indicated nanocomposites yield stress at nanofiller contents growth is the initial nanoparticles strong aggregation. The application of nanoparticles disaggregation artificial methods might be worthwhile only for nanocomposites with a small diameter of the initial nanoparticles. High enough values of nanocomposites structure fractal dimension are an additional factor, influencing on this effect.

The impact toughness of particulate-filled polymer nanocomposites is defined by a number of factors on various structural levels: molecular, topological and suprasegmental ones. The indicated levels characteristics are interconnected and changed at nanofiller introduction. The molecular mobility level is the main parameter, defining the considered nanocomposites impact toughness.

The ratio of microhardness and yield stress for particulate-filled polymer nanocomposites is defined by these polymer nanomaterials structural state only. Tabor criterion is correct for Euclidean (or close to them) solids only.

KEYWORDS

- fractal analysis
- high density polyethylene
- mechanical properties
- nanocomposite
- structure
- ultrafine particles

REFERENCES

1. Narisawa I. Strength of Polymeric Materials. Moscow, Chemistry, 1987, 400 p.
2. Kozlov G.V., Zaikov G.E. Structure of the Polymer Amorphous State. Utrecht, Boston, Brill Academic Publishers, 2004, 465 p.
3. Kozlov G.V., Zaikov G.E. Structure and Properties of Particulate-Filled Polymer Nanocomposites. Saarbrücken, Lambert Academic Publishing, 2012, 112 p.
4. Kozlov G.V., Yanovskii Yu.G., Zaikov G.E. Particulate-Filled Polymer Nanocomposites. Structure, Properties, Perspectives. New York, Nova Science Publishers, Inc. 2014, 273 p.
5. Kozlov G.V., Yanovskii Yu.G., Zaikov G.E. Structure and Properties of Particulate-Filled Polymer Composites: The Farctal Analysis. New York, Nova Science Publishers, Inc. 2010, 282 p.
6. Tanniru M., Misra R.D.K. Mater. Sci. Engng., 2005, v. A405, № 1, p. 178–193.
7. Deshmane C., Yuan Q., Misra R.D.K. Mater. Sci. Engng., 2007, v. A452–453, № 3, p. 592–601.
8. Bucknall C.B. Toughened Plastics. London, Applied Science, 1977, 315 p.
9. Bartenev G.M., Frenkel S.Ya. Physics of Polymers. Leningrad, Chemistry, 1990, 432 p.
10. Balta-Calleja F.J., Kilian H.G. Colloid Polymer Sci., 1988, v. 266, № 1, p. 29–34.
11. Balta-Calleja F.J., Santa Cruz C., Bayer R.K., Kilian H.G. Colloid Polymer Sci., 1990, v. 268, № 5, p. 440–446.
12. Aloev V.Z., Kozlov G.V. The Physics of Orientational Effects in Polymeric Materials. Nal'chik, Polygraphservice and T, 2002, 288 p.
13. Perry A.J., Rowcliffe D.J. J. Mater. Sci. Lett., 1973, v. 8, № 6, p. 904–907.
14. Kozlov G.V., Sanditov D.S. Anharmonic Effects and Physical-Mechanical Properties of Polymers. Novosibirsk, Science, 1994, 261 p.
15. Mikitaev A.K., Kozlov G.V., Zaikov G.E. Polymer Nanocomposites: Variety of Structural Forms and Applications. New York, Nova Science Publishers, Inc. 2008, 319 p.
16. Kohlstedt D.L. J. Mater. Sci., 1973, v. 8, № 6, p. 777–786.
17. Sumita M., Tsukumo Y., Miyasaka K., Ishikawa K. J. Mater. Sci., 1983, v. 18, № 5, p. 1758–1764.
18. Bobryshev A.N., Kozomazov V.N., Babin L.O., Solomatov V.I. Synergetics of Composite Materials. Lipetsk, NPO ORIUS, 1994, 154 p.
19. Honeycombe R.W.K. The Plastic Deformation of Metals. Cambridge, Edwards Arnold Publishers, 1968, 402 p.
20. Balankin A.S. Synergetics of Deformable Body. Moscow, Publishers of Ministry Defence SSSR, 1991, 404 p.
21. Buchachenko A.L. Achievements of Chemistry, 2003, v. 72, № 5, p. 419–437.
22. Kozlov G.V., Mikitaev A.K. Structure and Properties of Nanocomposites Polymer/Organoclay. Saarbrücken, LAP LAMBERT Academic Publishing GmbH and Comp., 2013, 318 p.
23. Yanovsky Yu.G., Kozlov G.V., Zhirikova Z.M., Aloev V.Z., Karnet Yu.N. Intern. J. Nanomechanics Science and Technology, 2012, v. 3, № 2, p. 99–124.

24. Brown N. Mater. Sci. Engng., 1971, v. 8, № 1, p. 69–73.
25. Kalinchev E.L., Sakovtseva M.B. Properties and Processing of Thermoplactics. Leningrad, Chemistry, 1983, 288 p.
26. Tabor D. The Hardness of Metals. New York, Oxford University Press, 1951, 329 p.
27. Aphashagova Z.Kh., Kozlov G.V., Burya A.I., Zaikov G.E. Theoretical Fundamentals of Chemical Technology, 2007, v. 41, № 6, p. 699–702.
28. Suwanprateeb J. Composites, Part A, 2000, v. 31, № 3, p. 353–359.
29. Zhirikova Z.M., Kozlov G.V., Aloev V.Z. Main Problems of Modern Materials Science, 2012, v. 9, № 1, p. 82–85.

CHAPTER 6

THE REINFORCEMENT OF PARTICULATE-FILLED POLYMER NANOCOMPOSITES BY NANOPARTICLES AGGREGATES

G. E. ZAIKOV,[1] G. V. KOZLOV,[2] and A. K. MIKITAEV[2]

[1]N.M. Emanuel Institute of Biochemical Physics of Russian Academy of Sciences, Kosygin st., 4, Moscow-119334, Russian Federation

[2]Kh.M. Berbekov Kabardino-Balkarian State University, Chernyshevsky st., 173, Nal'chik-360004, Russian Federation

CONTENTS

ABSTRACT

The applicability of irreversible aggregation model for theoretical description of nanofiller particles aggregation process in polymer nanocomposites has been shown. The main factors, influencing on nanoparticles aggregation process, were revealed. It has been shown that strongly expressed particulate nanofiller particles aggregation results in sharp (in about four times) formed fractal aggregates real elasticity modulus reduction. Nanofiller particles aggregation is realized by cluster-cluster mechanism and results in the formed fractal aggregates density essential reduction, that is the cause of their elasticity modulus decreasing. As distinct from microcomposites nanocomposites require consideration of interfacial effects for elasticity modulus correct description in virtue of a well-known large fraction of phases division surfaces for them.

6.1 INTRODUCTION

In the course of technological process of particulate-filled polymer composites in general [1] and nanocomposites [2–4] in particular preparation of the initial filler powder particles aggregation in more or less large particles aggregates always occurs. The aggregation process exercises essential influence on composites (nanocomposites) macroscopic properties [1–5]. For nanocomposites the aggregation process gains special significance, since its intensity can be such, that nanofiller particles aggregates size exceeds 100 nm – the value, which is assumed (although conditionally enough [6]) as upper dimensional limit for a nanoparticle. In other words, the aggregation process can be resulted in the situation, when initially supposed nanocomposite ceases to be as such. Therefore, at present a methods number exists, allowing to suppress nanoparticles aggregation process [4, 7].

Analytically this process is treated as follows. The authors [5] obtain the relationship:

$$k(r) = 6.5 \times 10^{-3} S_u \qquad (1)$$

where $k(r)$ is aggregation parameter, S_u is specific surface of nanofiller initial particles, which is given in m²/g.

In its turn, the value S_u is determined as follows [8]:

$$S_u = \frac{6}{\rho_n D_p} \qquad (2)$$

where ρ_n is nanofiller density, D_p is diameter of its initial particles.

From the Eqs. (1) and (2) it follows, that D_p reduction results in S_u growth, that in its turn reflects in the aggregation intensification, characterized by the parameter $k(r)$ increasing. Therefore, in polymer nanocomposites strengthening (reinforcing) element are not nanofiller initial particles themselves, but their aggregates [9]. This results in essential changes of nanofiller elasticity modulus, the value of which is determined with the aid of the equation [9]:

$$E_{agr} = E_{nan}\left(\frac{a}{R_{agr}}\right)^{3+d_l} \tag{3}$$

where E_{agr} is nanofiller particles aggregate elasticity modulus, E_{nan} is elasticity modulus of material, from which the nanofiller was obtained, a is an initial nanoparticles size, R_{agr} is a nanoparticles aggregate radius, d_l is chemical dimension of the indicated aggregate, which is equal to ~ 1.1 [9].

As it follows from the Eq. (3), the initial nanoparticles aggregation degree enhancement, expressed by R_{agr} growth, results in E_{agr} decrease (the rest of parameters in the Eq. (3) are constant) and, as consequence, in nanocomposite elasticity modulus reduction.

Very often the elasticity modulus (or reinforcement degree) of polymer composites (nanocomposites) is described within the frameworks of numerous micromechanical models, which proceed from elasticity modulus of matrix polymer and filler (nanofiller) and the latter volume contents [10]. Additionally it is supposed, that the indicated above characteristics of a filler are approximately equal to the corresponding parameters of compact material, from which a filler is prepared. This practice is inapplicable absolutely in case of polymer nanocomposites with fine-grained nanofiller, since in this case a polymer is reinforced by nanofiller fractal aggregates, whose elasticity modulus and density differ essentially from compact material characteristics (see the Eq. (3)) [5, 9]. Therefore, the microcomposite models application, as a rule, gives a large error at polymer composites elasticity modulus evaluation, that in its turn results in the appearance of an indicated models modifications large number [10].

Proceeding from the said above, the present work purpose is the theoretical treatment of particulate nanofiller aggregation process and elasticity modulus (reinforcement degree) particulate-filled polymer nanocomposites with due regard for the indicated effect within the framework of irreversible aggregation models and fractal analysis.

6.2 EXPERIMENTAL

Polypropylene (PP) "Kaplen" of mark 01030 with average weight molecular mass of ~ $(2-3)\times10^3$ and polydispersity index 4.5 was used as matrix polymer. Nanodimensional calcium carbonate ($CaCO_3$) in compound form of mark Nano-Cal P-1014 (production of China) with particles size of 80 nm and mass contents of 1–7 mass % and globular nanocarbon (GNC) (production of corporations group "United Systems," Moscow, Russian Federation) with particles size of 5–6 nm, specific surface of 1400 m^2/g and mass contents of 0.25–3.0 mass % were applied as nanofiller.

Nanocomposites PP/$CaCO_3$ and PP/GNC were prepared by components mixing in melt on a twin-screw extruder Thermo Haake, model Reomex RTW 25/42, production of German Federal Republic. Mixing was performed at temperature 463–503 K and screw speed of 50 rpm during 5 min. Testing samples were prepared by casting under pressure method on a casting machine Test Sample Molding Apparatus RR/TS MP of firm Ray-Ran (Taiwan) at temperature 483 K and pressure 43 MPa.

The nanocomposites melt viscosity was characterized by a melt flow index (MFI). MFI measurements were performed on an extrusion-type plastometer Noselab ATS A-MeP (production of Italy) with capillary diameter of 2.095±0.005 mm at temperature 513 K and load of 2.16 kg. The sample was maintained at the indicated temperature during 4.5±0.5 min.

Uniaxial tension mechanical tests have been performed on the samples in the shape of a two-sided spade with sizes according to GOST 112 62–80. The tests have been conducted on a universal testing apparatus Gotech Testing Machine CT-TCS 2000, production of German Federal Republic, at temperature 293 K and strain rate ~ 2×10^{-3} s^{-1}.

6.3 RESULTS AND DISCUSSION

The particulate nanofiller aggregation degree can be evaluated and aggregates diameter D_{agr} quantitative estimation can be performed within the framework of strength dispersive theory [11], where shear yield stress of nanocomposite τ_n is determined as follows:

$$\tau_n = \tau'_m + \frac{G_n b_B}{\lambda} \tag{4}$$

where τ_m is shear yield stress of polymer matrix, b_B is Burgers vector, G_n is nanocomposite shear modulus, λ is distance between nanofiller particles.

In case of nanofiller particles aggregation the Eq. (4) has the look [11]:

$$\tau_n = \tau'_m + \frac{G_n b_B}{k(r)\lambda} \tag{5}$$

where $k(r)$ is aggregation parameter.

The parameters, included in the Eqs. (4) and (5) are determined as follows. The general relationship between normal stress σ and shear stress τ has the look [12]:

$$\tau = \frac{\sigma}{\sqrt{3}} \tag{6}$$

The intercommunication of matrix polymer τ_m and nanocomposite polymer matrix τ'_m shear yield stresses is given as follows [5]:

$$\tau'_m = \tau_m \left(1 - \varphi_n^{2/3}\right) \tag{7}$$

where φ_n is nanofiller volume content, which can be determined according to the well-known formula [5]:

$$\varphi_n = \frac{W_n}{\rho_n} \tag{8}$$

where W_n is nanofiller mass contents, ρ_n is its density, which for nanoparticles is determined according to the equation [5]:

$$\rho_n = 188\left(D_p\right)^{1/3} \text{ kg/m}^3 \tag{9}$$

where D_p is given in nm.

The value of Burgers vector b_B for polymeric materials is determined as follows [13]:

$$b_B = \left(\frac{60.5}{C_\infty}\right)^{1/2} \text{ Å} \tag{10}$$

where C_∞ is characteristic ratio, connected with nanocomposite structure dimension d_f by the equation [13]:

$$C_\infty = \frac{2d_f}{d(d-1)(d-d_f)} + \frac{4}{3} \tag{11}$$

where d is dimension of Euclidean space, in which a fractal is considered (it is obvious, that in our case $d = 3$).

The value d_f can be calculated according to the equation [14]:

$$d_f = (d-1)(1+v) \qquad (12)$$

where v is Poisson's ratio, estimated according to the mechanical tests results with the aid of the relationship [15]:

$$\frac{\sigma_Y}{E_n} = \frac{1-2v}{6(1+v)} \qquad (13)$$

where σ_Y and E_n are yield stress and elasticity modulus of nanocomposite, respectively.

Nanocomposite moduli E_n and G_n are connected between themselves by the relationship [14]:

$$G_n = \frac{E_n}{d_f} \qquad (14)$$

And at last, the distance λ between nanofiller nonaggregated particles is determined according to the equation [11]:

$$\lambda = \left[\left(\frac{4\pi}{3\varphi_n} \right)^{1/3} - 2 \right] \frac{D_p}{2} \qquad (15)$$

From the Eqs. (5) and (15), $k(r)$ growth from 5.65 up to 43.70 within the range of $W_n = 0.25\text{–}3.0$ mass % for nanocomposites PP/GNC and from 1.0 up to 2.87 within the range of $W_n = 1\text{–}7$ mass % for nanocomposites PP/CaCO$_3$ follows. Let us note, that the indicated variation $k(r)$ for the considered nanocomposites corresponds completely to the Eqs. (1) and (2). Let us consider, how such $k(r)$ growth is reflected on nanofiller particles aggregates diameter D_{agr}. The Eqs. (8), (9) and (15) combination gives the following relationship:

$$k(r)\lambda = \left[\left(\frac{0.251\pi D_{agr}^{1/3}}{W_n} \right)^{1/3} - 2 \right] \frac{D_{agr}}{2} \qquad (16)$$

allowing at D_p replacement on D_{agr} to determine real, that is, with accounting of nanofiller particles aggregation, nanoparticles aggregates diameter of the used nanofiller. Calculation according to the Eq. (16) shows D_{agr} increasing (corresponding to $k(r)$ growth) from 25 up to 125 nm within the range of $W_n = 0.25$–3.0 mass % for GNC and from 80 up to 190 nm within the range of 1–7 mass % for $CaCO_3$. Further nanofiller particles aggregates density can be calculated according to the Eq. (9) at the condition of D_p replacement by D_{agr}.

Within the framework of irreversible aggregation model D_{agr} value is given by the following relationship [16]:

$$D_{agr} \sim \left(\frac{4c_0 kT}{3\eta m_0} \right)^{1/d_f^{agr}} t^{1/d_f^{agr}} \tag{17}$$

where c_0 is nanoparticles initial concentration, k is Boltzmann constant, T is temperature, η is medium viscosity, m_0 is mass of initial nanoparticle, d_f^{agr} is fractal dimension of particles aggregate, t is aggregation process duration.

Let us consider estimation methods of the parameters, included in the Eq. (17). In the simplest case it can be accepted that all particles of nanofiller initial powder have the same size and mass. In this case $c_0 \approx \varphi_n$, where φ_n value is determined according to the Eq. (8) with using nanofiller particles aggregates diameter D_{agr}. η value is accepted equal to reciprocal of MFI value and m_0 magnitude was calculated as follows. In supposition of nanofiller initial particles spherical shape the nanoparticle volume was calculated according to the known values of their diameter D_p and then, using ρ_n value, calculated according to the Eq. (8), their mass m_0 can be estimated. T value is accepted as constant and equal to nanocomposites processing duration, that is, 300 s.

The fractal dimension of nanofiller particles aggregates structure d_f^{agr} was calculated with the aid of the equation [17]:

$$\rho_n = \rho_{dens} \left(\frac{D_{agr}}{2a} \right)^{d_f^{agr} - d} \tag{18}$$

where ρ_{dens} is density of compact material of nanofiller particles, a is self-similarity (fractality) lower scale of nanofiller particles aggregates.

ρ_{dens} value for carbon is accepted equal to 2700 kg/m^2, for $CaCO_3$ – 2000 kg/m^2 [5] and a value is accepted equal to the initial GNC particle radius, that is, 2.5 nm. d_f^{agr} values, calculated according to the Eq. (18), are equal to 2.09–2.67 and 2.47–2.75 for GNC and $CaCO_3$ nanoparticles aggregates, respectively.

In Fig. 6.1 the dependences $D_{agr}(W_n)$, plotted according to the Eqs. (16) and (17), comparison is adduced. As one can see, the good enough correspondence of estimations according to both indicated methods was obtained (the average discrepancy of D_{agr} values, calculated with the usage of these relationships, makes up ~ 16%). This circumstance indicates, that irreversible aggregation models can be used for the theoretical description of particulate nanofiller particles aggregation processes. Besides, the Eq. (17) analysis demonstrates various factors influence on nanofiller particles aggregates size (or their aggregation degree). So, c_0, T and t increasing results in aggregation processes intensification and η, m_0 and d_f^{agr} enhancement – to their weakening.

Let us note in conclusion, that proportionality coefficient in the Eq. (17) for GNC and $CaCO_3$ (c_{GNC} and c_{CaCO_3}, respectively) can be approximated by the following relationship:

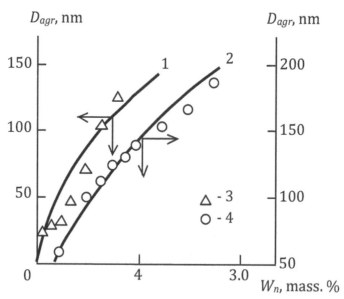

FIGURE 6.1   The dependences of nanofiller particles aggregates diameter D_{agr} on nanofiller mass contents W_n for nanocomposites PP/GNC (1, 3) and PP/CaCO$_3$ (2, 4). 1, 2 – calculation according to the Eq. (16); 3, 4 – calculation according to the relationship (17).

$$\frac{c_{CaCO_3}}{c_{GNC}} = \left(\frac{m_0^{CaCO_3}}{m_0^{GNC}}\right)^{1/d_f^{av}}$$

(19)

where $m_0^{CaCO_3}$ and m_0^{GNC} are masses of the initial particles of $CaCO_3$ and GNC, respectively, d_f^{av} is average fractal dimension of the indicated nanoparticles aggregates.

Further elasticity modulus E_{agr} of nanofiller particles aggregates according to the Eq. (3) can be determined. Let us consider the concrete conditions of this equation usage in reference to nanocomposites PP/GNC. Two possible variants exist at parameter a choice in the indicated equation. The first from them supposes, that the value a is equal to GNC initial particles diameter [9], that is, 5.5 nm. Such supposition means, that GNC nanoparticles aggregates are formed by particle-cluster (P-Cl) mechanism, that is, by separate particles GNC joining to a growing aggregate [18]. However, such supposition gives unreal high E_{agr} values of order of 5×10^5 GPa. The other variant assumes, that nanofiller aggregation is realized by a cluster-cluster (Cl-Cl) mechanism, that is, small clusters association in larger ones [18]. In such model aggregate radius R_{agr}^{i-1} on the previous $(i-1)$th aggregation stage is accepted as a and then the Eq. (3) can be rewritten as follows:

$$E_n = E_{agr}\left(\frac{R_{agr}^{i-1}}{R_{agr}^i}\right)^{4.1}$$

(20)

The elasticity modulus E_{agr} real values within the range of 21.3–5.0 GPa were obtained at such calculation method. Further the simplest microcomposite models can be used for nanocomposite elasticity modulus E_n estimation. For the case of uniform strain in nanocomposite phases the theoretical value E_n (E_n^T) is given by a parallel model [10]:

$$E_n^T = E_{agr}\varphi_n + E_m(1-\varphi_n)$$

(21)

where E_m is elasticity modulus of matrix polymer.

For the case of uniform stress in nanocomposite phases the lower theoretical boundary E_n^T is determined according to the serial model [10]:

$$E_n^T = \frac{E_{agr}E_m}{E_{agr}(1-\varphi_n)+E_m\varphi_n}$$

(22)

In Fig. 6.2, the comparison of the received experimentally E_n and calculated according to the Eqs. (21) and (22) E_n^T elasticity modulus values of the considered nanocomposites PP/GNC is adduced. As one can see, the experimental data correspond better to the determined according to the Eq. (21) E_n^T upper boundary (in this case average discrepancy of E_n and E_n^T makes up ~ 8 %). The indicated discrepancy is due to objective causes. As it is known [10], at the Eqs. (21) and (22) derivation the equality of Poisson's ratio for nanocomposite both phases was supposed. In practice this condition nonfulfillment defines discrepancy between experimental and theoretical data.

In Fig. 6.2, the dependence $E_n^T(W_n)$, calculated according to the equation (21) in supposition E_{agr} = const = 21.3 GPa, is also adduced. As one can see, in this case the theoretical values of elasticity modulus E_n^T exceed essentially experimentally received ones E_n. Hence, the good correspondence of experiment and calculation according to the Eq. (21) is due to real values E_{agr} usage only.

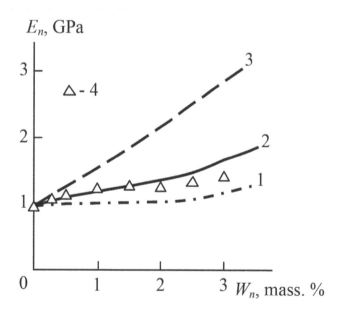

FIGURE 6.2   The dependences of elasticity modulus E_n on nanofiller mass contents W_n for nanocomposites PP/GNC. 1 – calculation according to the Eq. (22); 2 – according to the Eq. (21) at E_{agr} = variant; 3 – according to the equation (21) at E_{agr} = const = 21.3 GPa; 4 – experimental data.

It is obvious, that nanoparticles aggregates elasticity modulus reduction is due to the indicated aggregates diameter growth and, as consequence, their density ρ_n reduction, which can be calculated according to the Eq. (18). In Fig. 6.3 the dependence $E_{agr}(\rho_n)$ is adduced, which, as was expected, proves to be linear, passing through coordinates origin and is described analytically by the following empirical equation:

$$E_{agr.} = 12.6 \times 10^{-3}\rho_n \text{ GPa} \tag{23}$$

where ρ_n is given in kg/m³.

The limiting magnitude $\rho_n = \rho_{dens}$ allows to obtain the greatest value $E_{agr} \approx 34$ GPa for GNC aggregates, that is the real value of this parameter [1].

The authors [19] proposed to use for nanocomposites elasticity modulus E_n determination a modified mixtures rule, which in original variant gives upper limiting value of composites elasticity modulus [10]:

$$E_n = E_m(1 - \varphi_n) + bE_{nan}\varphi_n \tag{24}$$

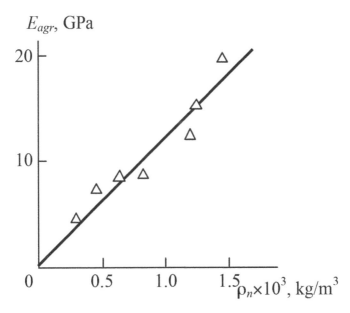

FIGURE 6.3 The dependence of GNC nanoparticles fractal aggregates elasticity modulus E_{agr} on their density ρ_n for nanocomposites PP/GNC.

where $b<1$ is coefficient, reflected nanofiller properties realization degree in polymer nanocomposite. In the present work context the parameter bE_{nan} as a matter of fact presents nanofiller effective modulus or, more precisely, its aggregates modulus E_{agr} (compare with the Eq. (21)).

In Fig. 6.4, the dependence of parameter b in the Eq. (24) on nanofiller particles aggregates diameter D_{agr}, calculated according to the Eq. (16), for the studied nanocomposites is adduced. As one can see, this dependence disintegrates on two linear parts: at small D_{agr} fast decay of b at D_{agr} growth is observed and at large enough D_{agr} the value $b{\approx}const{\approx}0.175$. Let us note, that dimensional interval of the indicated transition, showed in Fig. 6.4 by a shaded area, makes up $D_{agr}{\approx}70$–100 nm, that is, it coincides approximately with upper dimensional boundary of nanoparticles interval (although and conditional enough [6]), which is equal to about 100 nm. As a matter of fact, the indicated dimensional interval defines the transition from nanocomposites to microcomposites, the dependence $b(D_{agr})$ for which differs actually qualitatively. The adduced in Fig. 6.4 dependence $b(D_{agr})$ can be described analytically by the following integrated equation:

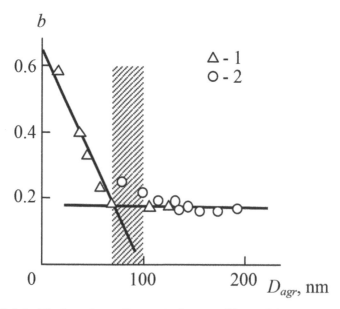

FIGURE 6.4 The dependence of parameter b on nanofiller particles aggregates diameter D_{agr} for nanocomposites PP/GNC (1) and PP/CaCO$_3$ (2). The shaded region indicates transition of nanofiller particles aggregates from nano- to microbehavior.

$$b = 0.67 - 6.7 \times 10^{-3} D_{agr}, \text{ for } D_{agr} \leq 70 \text{ nm}$$

$$b = \text{const} = 0.175, \text{ for } D_{agr} > 70 \text{ nm} \qquad (25)$$

In Fig. 6.5, the comparison of experimentally obtained and calculated according to the Eq. (25) dependences $E_n(\varphi_n)$ is adduced for the studied nanocomposites. In this case the parameter b value was estimated according to the Eq. (25) and values E_{nan} were accepted equal to 30 GPa for GNC and 15 GPa for $CaCO_3$. As one can see, the good correspondence of theory and experiment is obtained (their mean discrepancy makes up 3%, that approximately equal to the experimental error of E_n determination). Higher values E_n for nanocomposites PP/GNC in comparison with PP/$CaCO_3$ even at $D_{agr} > 100$ nm are due to two factors: the initial nanoparticles smaller size, that gives higher values φ_n at the same W_n values (see the Eqs. (8) and (9)) and higher value E_{nan}. It is important to note close

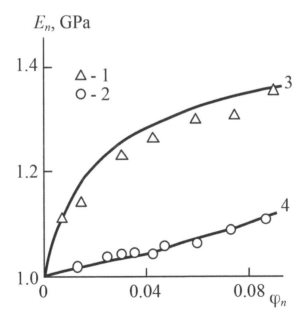

FIGURE 6.5   The comparison of experimentally received (1, 2) and calculated according to the Eqs. (24) and (25) (3, 4) dependences of elasticity modulus E_n on nanofiller volume contents φ_n for nanocomposites PP/GNC (1, 3) and PP/$CaCO_3$ (2, 4).

values E_{agr} for nanocomposites PP/GNC, determined according to the Eq. (20) and as bE_{nan}.

The authors [8] proposed the following percolation relationship for polymer microcomposites reinforcement degree E_c/E_m description:

$$\frac{E_c}{E_m} = 1 + 11(\varphi_n)^{1.7} \tag{26}$$

where E_c is elasticity modulus of microcomposite.

Later the Eq. (26) was modified in reference to the polymer nanocomposites case [20]:

$$\frac{E_n}{E_m} = 1 + 11(\varphi_n + \varphi_{if})^{1.7} \tag{27}$$

where φ_{if} is relative fraction of interfacial regions.

It is easy to see, that the modified relationship (27) takes into consideration a factor of sharp increase of division surfaces polymer matrix-nanofiller [21]. In Fig. 6.6, the comparison of experimentally obtained and

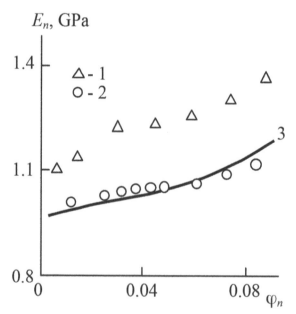

FIGURE 6.6   The comparison of experimentally received (1, 2) and calculated according to the Eq. (26) (3) dependences of elasticity modulus E_n on nanofiller volume contents φ_n for nanocomposites PP/GNC (1, 3) and PP/CaCO$_3$ (2, 4).

calculated according to the Eq. (26) dependences $E_n(\varphi_n)$ for the considered nanocomposites is adduced. As it follows from this figure data, the Eq. (26) describes well the experimental data for nanocomposites PP/CaCO$_3$, but the corresponding data for nanocomposites PP/GNC set essentially higher than theoretical curve. This discrepancy cause is obvious from the Eqs. (26) and (27) comparison – for nanocomposites PP/GNC interfacial effects accounting is necessary, that is, parameter φ_{if} accounting. Hence, in the considered case only compositions PP/GNC are true nanocomposites.

6.4 CONCLUSIONS

The applicability of irreversible aggregation models for theoretical description of particulate nanofiller particles aggregation processes in polymer nanocomposites has been shown. Analysis within the framework of the indicated models allows to reveal either factors influence on aggregation degree.

Strongly expressed aggregation of particulate nanofiller particles results in sharp (in about 4 times) formed fractal aggregates real elasticity modulus reduction. In its turn, this process defines nanocomposites as the whole elasticity modulus reduction. Nanofiller particles aggregation is realized by a cluster-cluster mechanism and results in the formed fractal aggregates density essential reduction, that is the cause of their elasticity modulus decreasing.

A nanofiller elastic properties realization degree is defined by the aggregation of its initial particles level. Unlike microcomposites nanocomposites require interfacial effects accounting for elasticity modulus correct description in virtue of well-known large fraction of phases division surfaces for them.

KEYWORDS

- **Astrakhan region**
- **Cyanobacteria**
- **Cyano-bacterial community**
- **Phyto- and growth-stimulate activity**
- **Soil algae**

REFERENCES

1. Kozlov G.V., Yanovskii Yu.G., Zaikov G.E. Structure and Properties of Particulate-Filled Polymer Composites: The Farctal Analysis. New York, Nova Science Publishers, Inc. 2010, 282 p.
2. Kozlov G.V., Zaikov G.E. Structure and Properties of Particulate-Filled Polymer Nanocomposites. Saarbrücken, Lambert Academic Publishing, 2012, 112 p.
3. Kozlov G.V., Yanovskii Yu.G., Zaikov G.E. Particulate-Filled Polymer Nanocomposites. Structure, Properties, Perspectives. New York, Nova Science Publishers, Inc. 2014, 273 p.
4. Edwards D.C. J. Mater. Sci., 1990, v. 25, № 12, p. 4175–4185.
5. Mikitaev A.K., Kozlov G.V., Zaikov G.E. Polymer Nanocomposites: Variety of Structural Forms and Applications. New York, Nova Science Publishers, Inc. 2008, 319 p.
6. Buchachenko A.L. Achievements of Chemistry, 2003, v. 72, № 5, p. 419–437.
7. Kozlov G.V., Yanovsky Yu.G., Zaikov G.E. Synergetic and Fractal Analysis of Polymer Composites Filled with Short Fibers. New York, Nova Science Publishers, Inc. 2011, 223 p.
8. Bobryshev A.N., Kozomazov V.N., Babin L.O., Solomatov V.I. Synergetics of Composite Materials. Lipetsk, NPO ORIUS, 1994, 154 p.
9. Witten T.A., Rubinstein M., Colby R.H. J. Phys. II France, 1993, v. 3, № 3, p. 367–383.
10. Ahmed S., Jones F.R. J. Mater. Sci., 1990, v. 25, № 12, p. 4933–4942.
11. Sumita M., Tsukumo Y., Miyasaka K., Ishikawa K. J. Mater. Sci., 1983, v. 18, № 5, p. 1758–1764.
12. Honeycombe R.W.K. The Plastic Deformation of Metals. Cambridge, Edwards Arnold Publishers, 1968, 402 p.
13. Kozlov G.V., Zaikov G.E. Structure of the Polymer Amorphous State. Utrecht, Boston, Brill Academic Publishers, 2004, 465 p.
14. Balankin A.S. Synergetics of Deformable Body. Moscow, Publishers of Ministry Defence SSSR, 1991, 404 p.
15. Kozlov G.V., Sanditov D.S. Anharmonic Effects and Physical-Mechanical Properties of Polymers. Novosibirsk, Science, 1994, 261 p.
16. Weitz D.A., Huang J.S., Lin M.Y., Sung J. Phys. Rev. Lett., 1984, v. 53, № 17, p. 1657–1660.
17. Brady L.M., Ball R.C. Nature, 1984, v. 309, № 5965, p. 225–229.
18. Shogenov V.N., Kozlov G.V. Fractal Clusters in Physics-Chemistry of Polymers. Nal'chik, Polygraphservice and T, 2002, 268 p.
19. Komarov B.A., Dzhavadyan E.A., Irzhak V.I., Ryabenko A.G., Lesnichaya V.A., Zvereva G.I., Krestinin A.V. Polymer Science, Series A, 2001, v. 53, № 6, p. 897–905.
20. Malamatov A.Kh., Kozlov G.V., Mikitaev M.A. Reinforcement Mechanisms of Polymer Nanocomposites. Moscow, Publishers of D.I. Mendeleev RKhTU, 2006, 240 p.
21. Andrievsky R.A. Russian Chemical Journal, 2002, v. 46, № 5, p. 50–56.

CHAPTER 7

THE FRACTAL MODEL OF POLYMER PAIRS ADHESION

G. E. ZAIKOV,[1] KH. SH. YAKH'YAEVA,[2] G. M. MAGOMEDOV,[3]
G. V. KOZLOV,[4] B. A. HOWELL,[5] and R. YA. DEBERDEEV[6]

[1]N.M. Emanuel Institute of Biochemical Physics, of Russian Academy of Sciences, Kosygin st., 4, Moscow-119334, Russian Federation

[2]FSBEI HPE "M.M. Dzhambulatov Daghestan State Agrarian University," M. Gadzhiev st, 180, Makhachkala-367032, Russian Federation

[3]FSBEI HPE "Daghestan State Pedagogical University," M. Yaragskii st., 57, Makhachkala-367003, Russian Federation

[4]FSBEI HPE "Kh.M. Berbekov Kabardino-Balkarian State University," Chernyshevsky st., 173, Nal'chik-360004, Russian Federation

[5]Central Michigan University, Mount Pleasent, Michigan, USA

[6]Kazan National Research Technological University, Kazan, Tatarstan, Russia

CONTENTS

ABSTRACT

The generalized fractal model allowing quantitative description shear strength of polymer pairs adhesional bonding on various factors (temperature, contact duration and pressure at the indicated bonding formation) was proposed. This model also allows to take into consideration molecular and structural characteristics of polymers, forming adhesional bonding. The indicated model applicability for interfacial interactions level in polymer-polymer composites description has been shown.

7.1 INTRODUCTION

As it is known [1, 2], at two polymer samples contact a macromolecular coils interdiffusion through a division boundary occurs, owing to that macromolecular entanglements are formed, ensuring certain finite adhesional contact strength. The increase of temperature, duration and pressure in such contact zone results in its mechanical properties improvement in case of compatible polymers [2]. This observation has been explained by two effects realization in adhesional contact formation process: wetting of one polymer surface by another one and macromolecular segments interdiffusion through division boundary [2]. As a rule [2–4], interdiffusion through the indicated boundary is described within the framework of polymer chains reptation model [5], since the adhesional contact shear strength value is proportional to its formation duration to the fourth power. On the whole such models have qualitative [1, 3, 4], or at best semiquantitative [2] character. The present work purpose is elaboration of a quantitative structural model of adhesional contact strength development as a function of the indicated above factors with the fractal analysis notions attraction [6].

7.2 EXPERIMENTAL

Amorphous polystyrene (PS, $M_w = 23 \times 10^4$, $M_w/M_n = 2.84$) and poly(2,6-dimethyl-1,4-phenylene oxide) (PPO, $M_w = 44 \times 10^3$, $M_w/M_n = 1.91$), obtained from Dow Chemical and General Electric, respectively, were used [2]. Polymer films of about 100 μm in thickness were prepared by extrusion method. The glass transition temperature T_g was measured by using Perkin-Elmer DSC-4 differential scanning calorimeter, at a heating rate of 20 K/min (T_g = 376 K for PS and 489 K for PPO [2]). For adhesional bonding's formation PS and PPO samples, having width of 5 mm, were driven in contact in a lap-shear joint geometry on the area of 5×5 mm² in a laboratory press carver at constant temperature and pressure of 0.8 MPa. Division boundaries PS-PPO are healed during 60–88,700s within the temperatures range of 343–386 K. Besides, PS-PS contacts were obtained at temperature 353 K within the pressures range of 0.02–80 MPa with their formation duration of 30 min.

Mechanical tests of the formed contacts have been performed at temperature 293 K on an Instron tensile tester, model 1130, at the crosshead speed of 3×10⁻² m/s with shear strength determination in contact zone (or on division boundary).

7.3   RESULTS AND DISCUSSION

7.3.1   THE DEPENDENCE OF ADHESIONAL CONTACT STRENGTH ON ITS FORMATION TEMPERATURE

As it is known [6], within the framework of fractal analysis shear strength τ_k of adhesional bonding is described by the following general equation:

$$\ln \tau_k = A \ln N_k - B, \tag{1}$$

where A and B are constants, which can be varied depending on polymer nature, temperature and specific tests conditions, N_k is a macromolecular coils contacts number, which is determined as follows [7]:

$$N_k \approx R_g^{Df_1 + Df_2 - d} \tag{2}$$

where R_g is macromolecular coil gyration radius, D_{f1} and D_{f2} are fractal dimensions of coils structure, forming adhesional bonding, d is dimension of Euclidean space, in which a fractal is considered (it is obvious, that in our case $d = 3$).

For the autohesion case $D_{f1} = D_{f2} = D_f$ and $d = 3$ the Eq. (2) is simplified up to [6]:

$$N_k \approx R_g^{2D_f - 3} \tag{3}$$

Let us consider the estimation methods of the parameters, included in the Eq. (3), that is, D_f and R_g. For D_f estimation the approximated method will be used, which consists in the following [8]. As it is known [9], between D_f and structure dimension d_f of linear polymers in solid-phase state the following relationship exists:

$$D_f = (d_f/1.5) \tag{4}$$

where d_f estimation can be performed according to the formula [8]:

$$d_f = 3 - 6 \, (\varphi_{cl}/SC_\infty)^{1/2}, \tag{5}$$

where φ_{cl} is a relative fraction of local order domains (nanoclusters), S is cross-sectional area of a macromolecule, C_∞ is characteristic ratio, which is an indicator of polymer chain statistical flexibility [10].

The value φ_{cl} was estimated according to the following percolation relationship [8]:

$$\varphi_{cl} = 0.03 \, (T_g - T)^{0.55}, \tag{6}$$

where T_g and T are temperatures of glass transition and adhesional contact formation, respectively.

For PS $C_\infty = 9.8$ [11], $S = 54.8$ Å^2 [12], for PPO $C_\infty = 3.8$ [11], $S = 27.9$ Å^2 [12]. Further the value of macromolecular coil gyration radius R_g was calculated as follows [1]:

$$R_g = l_0 \, (C_\infty M_w/6m_0)^{1/2} \tag{7}$$

where l_0 is the length of the main chain skeletal bond, which is equal to 0.154 nm for PS and 0.541 nm for PPO [11], m_0 is molar mass of the main chain skeletal bond ($m_0 = 52$ for PS and $m_0 = 25$ PPO [1]).

Let us note an important methodological aspect. At R_g value calculation according to the Eq. (7) the value C_∞ has been accepted as a variable one and calculated according to the following equation [8]:

$$C_\infty = \frac{2d_f}{d(d-1)(d-d_f)} + \frac{4}{3}$$

(8)

The constants A and B in the Eq. (1) were chosen empirically and accepted equal to $A = 2.15$ and $B = 6.0$ for both considered polymers.

In Fig. 7.1, the comparison of the obtained experimentally τ_k and calculated according to the Eq. (1) at the indicated above conditions τ_k^T values of adhesional bonding PS-PS and PPO-PPO shear strength is adduced. As one can see, the good enough correspondence of theory and experiment was obtained. Let us pay attention to a very important aspect of the indicated theoretical estimations. From the Eq. (1) it follows, that the value τ_k depends only on a macromolecular coils contacts number N_k for both considered polymers and the stated above N_k estimation methods assumes, that this parameter depends only on polymers molecular characteristics,

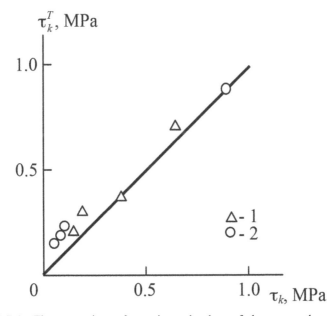

FIGURE 7.1   The comparison of experimental values of shear strength τ_k and those calculated according to the Eq. (1) τ_k^T of autohesional bonding for PS (1) and PPO (2). A straight line corresponds to a 1:1 ratio.

namely M_w, $C_□$, m_0, l_0 and D_f. In other words, this method does not assume the dependence of N_k and, hence, τ_k on any parameters, characterizing intermolecular interaction, the dependence of τ_k on which is postulated in the work of Ref. [13]. Besides, the shear strength τ_k of adhesional bonding change is described by one common equation for so strongly differing polymers as PS and PPO.

As it is known [1], at $M_w > 12 M_e$ (where M_e is molecular weight of a chain part between macromolecular entanglements) the boundary layer thickness α_i in the autohesion case can be determined according to the formula [1]:

$$\alpha_i = l_0 \left(\frac{12 C_\infty M_e}{6 m_0} \right)^{1/2} \tag{9}$$

In the considered polymers case $M_e = 18000$ for PS [1] and $M_e = 3620$ for PPO [14], that is, the indicated above condition $M_w > 12 M_e$ is fulfilled and the Eq. (9) can be used for value α_i estimation. In Fig. 7.2, the comparison of macromolecular coil gyration radius R_g, calculated according to the Eq. (7), and boundary layer thickness α_i, calculated according to the Eq. (9), is adduced. As one can see, at contact formation duration $t = 10$ min α_i value reaches R_g magnitude for both considered polymers. Let us also note, that the Eq. (9) does not suppose the dependence of α_i on contact formation duration t, although the dependence $\tau_k(t)$ exists [13]. This circumstance assumes boundary layer structure change for adhesional bonding at $t > 10$ min.

Within the framework of thermodynamical approach the segments of PS and PP interaction level in boundary layer can be estimated with the aid of Flory-Huggins interaction parameter χ_{AB}, which is determined with the help of the following equation [1]:

$$\alpha_i = \frac{2b}{\left(\chi_{AB} \cdot c \right)^{1/2}} \tag{10}$$

where b is Kuhn segment length, c is constant, which is equal to 6 in the limit $\alpha_i << R_g$ and to 9 in the limit $\alpha_i >> R_g$ [1]. Since in the considered case $\alpha_i \approx R_g$ (see Fig. 7.2), then further $c = 7.5$ was accepted.

In Fig. 7.3, the dependence of χ_{AB} on $\Delta T = T_g - T$ for the considered polymers is adduced. As one can see, for both PS and PPO χ_{AB} reduction at adhesional bonding formation temperature T growth is observed, that is

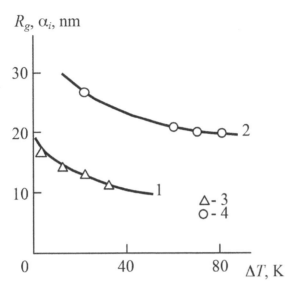

FIGURE 7.2 The dependences of the macromolecular coil gyration radius R_g (1, 2) and the boundary layer thickness α_i (3, 4) on difference of the glass transition and autohesional contact formation temperatures $\Delta T = T_g - T$ for PS (1, 3) and PPO (2, 4).

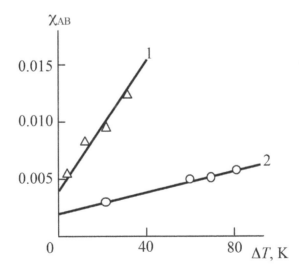

FIGURE 7.3 The dependences of Flory Huggins interaction parameter χ_{AB} on difference of the glass transition and autohesional contact formation temperatures $\Delta T = T_g - T$ for PS (1) and PPO (2).

a general effect [1]. Let us note, that the values χ_{AB} for PS are essentially higher than this parameter values for PPO (at higher values τ_k for the latter), which also does not correspond to the treatment, proposed in Ref. [13].

7.3.2 THE DEPENDENCE OF ADHESIONAL BONDING STRENGTH ON ITS FORMATION DURATION

The influence of adhesional contact formation duration t, which changes boundary layer structure owing to macromolecular coils inter-diffusion for the value τ_k, can be taken into account within the framework of anomalous (strange) diffusion conception [15]. The main equation of this conception can be written as follows [16]:

$$\left\langle r^2(t) \right\rangle^{1/2} = D_{gen} t^\beta \tag{11}$$

where $\langle r^2(t) \rangle^{1/2}$ is mean-square displacement diffusible particle (size of the region, visited by this particle), D_{gen} is generalized diffusivity, β is diffusion exponent. For classical (by Fick) diffusion $\beta = 1/2$, for slow diffusion $\beta < 1/2$, for the fast one – $\beta > 1/2$. The condition $\beta \neq 1/2$ is a definition of anomalous (strange) diffusion.

The boundary structural condition for slow and fast diffusion processes is the value $d_f = 2.5$ [15]. Since for the considered polymers (PS and PPO) $d_f \geq 2.603$, then all diffusion processes occurring in them are slow ones. The value β is connected with the main parameter in fractional derivatives theory (the fractional exponent α) by the following relationship [15]:

$$\beta = (1 - \alpha)/2. \tag{12}$$

In its turn, the fractional exponent α is determined according to the equation [15]:

$$\alpha = d_f - (d - 1) \tag{13}$$

The Eqs. (12) and (13) combination allows to obtain the direct interconnection between rate (intensity) of interdiffusion processes in the adhesional

bonding boundary layer, characterized by exponent β, and this layer structure, characterized by dimension d_f [15]:

$$\beta = \frac{d - d_f}{2} \tag{14}$$

It is obvious, that in the adhesion case the value $<r^2(t)>^{1/2}$ will be equal
to the boundary layer thickness α_i, determined according to the equation
(9). As the value α_i for compatible PS and PPO contact the mean value of
this parameter α_i^m for the indicated polymers ($\alpha_i^m = 15.9$ nm) was accepted.
Let us consider further two variants of dimension D_f application in the Eq.
(2) for the studied polymers pair PS-PPO: macromolecular coils of both
contacting polymers preserve their individuality in a boundary layer and
then the exponent for R_g in the indicated equation is equal to $D_{f1} = D_{f2} - d$
or PS and PPO macromolecular coils form secondary structure with the
dimension D_f, determined as follows [17]:

$$D_f = \frac{d\left(2D_{f_1} - D_{f_2}\right)}{d + 2\left(D_{f_1} - D_{f_2}\right)} \tag{15}$$

at the condition $D_{f1} \geq D_{f2}$.

In Fig. 7.4, the comparison of the dependences of theoretical values
τ_k^T (where the exponent for R_g in the Eq. (2) is equal to $2D_f - d$ and $D_{f1} = D_{f2} - d$, respectively) and experimental value τ_k of shear strength adhesional
contact on the indicated contact formation duration t to the ¼-th power.
The calculation of τ_k^T was performed according to the equation (1) at $A =$
2.15 and $B = 6.0$. As it follows from the data of Fig. 7.4, the variant, where
as an exponent in the Eq. (2) value $D_{f1} = D_{f1} - d$ was used, gives a better correspondence to experiment, that is, PS and PPO macromolecular coils preserve their individuality. The mean discrepancy between τ_k and τ_k^T in this
case makes up 15.7%, that is comparable with standard deviation ±15% at
τ_k experimental determination [2]. Let us note the following methodological closer definition. The experimental data of Fig. 7.4 were obtained at
contact formation temperature $T = 386$ K, that is, above T_g for PS. At $T>T_g$
polymer structure has the greatest possible dimension $d_f = 2.95$ [18] for
real solids and then according to the Eq. (4) $D_f = 1.967$. In this case the
value C_∞, which is necessary for macromolecular coil gyration radius R_g
determination according to the Eq. (7), was calculated according to the Eq.
(8), that for the indicated temperature T gives the mean value $R_g = 18.7$ nm.

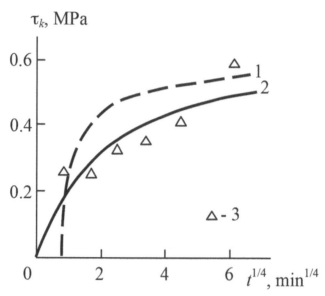

FIGURE 7.4   The dependences of autohesional contact PS-PPO shear strength τ_k on its formation duration t at $T = 386$ K. 1, 2 – calculation according to the Eq. (1) at using in the Eq. (2) of exponents $2D_f$-3 (1) and $D_{f1} + D_{f2}-d$ (2); 3 – experimental data [4].

The adhesional bonding shear strength τ_k calculation at the indicated above constant A and B values according to the Eq. (1) does not give a good correspondence to experiment at $T = 343$–373 K, that is, at $T<T_g$. Therefore, the constant A values in the indicated equation were chosen empirically, so as to obtain the best correspondence to the experiment at the condition $B = $ const. It turns out, that the value A decreases systematically at T reduction, that is shown in diagram form in Fig. 7.5. The dependence $A(T)$, adduced in Fig. 7.5, can be expressed analytically by the following empirical equation:

$$A = 1.88 \times 10^{-2} (T - 293).   \tag{16}$$

From the Eqs. (1) and (16) it follows, that at $T = 293$ K ln $\tau_k = $ -6.0 and $\tau_k = 0.0025$ MPa, that is smaller by order of magnitude than τ_k value at $T = 343$ K [2] and by two order of magnitude smaller than τ_k, obtained at $T = $

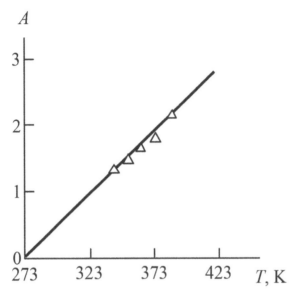

FIGURE 7.5 The dependence of constant A in the equation (1) on autohesional contact PS-PPO formation temperature T.

386 K [4]. These estimations show, that adhesional contact formation will occur the temperature smaller than the room one as well, but very slowly. With the Eq. (16) accounting the Eq. (1) can be rewritten as follows:

$$\ln \tau_k = 1.88 \times 10^{-2}(T - 293)\ln N_k - 6.0 \qquad (17)$$

The comparison of the obtained experimentally $\tau_k(t^{1/4})$ and calculated according to the Eq. (17) $\tau_k^T(t^{1/4})$ dependences of adhesional bonding shear strength on its formation duration t for PS-PPO pair within the range of $T = 343–373$ K is adduced in Fig. 7.6. As one can see, the Eq. (17) gives the results, corresponding to the experimental data (the mean discrepancy between τ_k and τ_k^T makes up 15.6%).

Let us consider in conclusion two features of the linear dependences $\tau_k(t^{1/4})$, obtained in Refs. [2–4]: these dependences extrapolation to finite values τ_k at $t = 0$ and the indicated plots linearity actually. Strictly speaking, the condition $t = 0$ means adhesional contact absence and contact finite strength at its absence is a kind of physical nonsense. This

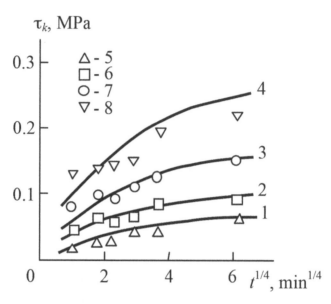

FIGURE 7.6   The dependences of autohesional contact PS-PPO shear strength τ_k on it formation duration t at formation temperatures T: 343 (1, 5), 353 (2, 6), 363 (3, 7) and 373 K (4, 8). 1–4 – calculation according to the Eq. (17); 5–6 – experimental data.

effect is explained by polymer surfaces wetting and macromolecular coils fast interdiffusion on a contact initial stage. Nevertheless, wetting at $t = 0$ should also be zero and values β, calculated according to the Eq. (14), are within the range of 0.025–0.198, that is, as it was noted above, they characterize slow, but not fast, anomalous diffusion. Therefore, the theoretical dependence $\tau_k^T(t^{1/4})$ extrapolation to $\tau_k^T = 0$ at $t = 0$ presents itself physically justified. As to the plots $\tau_k(t^{1/4})$ linearity, it should be noted, that τ_k increasing at t growth is due to macromolecular entanglements density enhancement in boundary layer, which is limited from above by its dependence on polymer molecular characteristics [8, 14]. The linear dependence $\tau_k(t^{1/4})$ does not assume such restrictions. Therefore, the dependences $\tau_k^T(t^{1/4})$ of reaching plateau tendency, supposed by the Eqs. (1) and (17) (see Figs. 7.4 and 7.6), presents itself as a well-founded one from the point of view of the very general physical postulates. An adhesional contact strength restriction from above was also shown experimentally [1].

7.3.3   THE DEPENDENCE OF ADHESIONAL BONDING STRENGTH ON PRESSURE AT ITS FORMATION

As it is known [2], pressure P influence at amorphous polymers adhesional contact formation has specific character. At small pressures (£1 MPa) its application does not change the indicated contact strength and at P enhancement above the indicated limit adhesional bonding strength increasing is observed. The last effect is connected with interdiffusion intensification of macromolecular coils in adhesional bonding boundary layer, induced by the applied from outside pressure [2]. However, such explanation has qualitative character and based on adhesional bonding shear strength dependence on its formation duration t to the ¼-th power, that in the reptation theory is typical for the processes, controlled by diffusion [5]. The adhesional bonding shear strength τ_k dependences on the applied at its formation pressure P theoretical description of two variants will be considered below. The first variant is a matter of fact an empirical one and uses the Eq. (1). The constant A empirical choice at $B = \text{const} = 6.0$ and the condition of theory [the Eq. (1)] and experiment the best correspondence has been shown, that at the adhesional bonding formation pressure $P£1.0$ MPa the value A is independent on P and is equal to ~ 1.90, that was to be expected [2]. At $P>1.0$ MPa linear growth A at P increasing is observed (see Fig. 7.7), described by the following empirical equation:

$$A = 1 + 0.70P^{1/6} \tag{18}$$

from which weak enough dependence of A on P (to the 1/6-th power) follows.

In Fig. 7.8, the comparison of the obtained experimentally $\tau_k(P)$ and calculated according to the Eqs. (1) and (18) $\tau_k^T(P)$ dependences of adhesional bonding PS-PS shear strength on its formation pressure P is adduced. As one can see, the proposed empirical approach gives a good correspondence to experiment (the mean discrepancy between τ_k and τ_k^T makes up 11.6%).

The second variant is more strict in physical significance. It uses the following equation of fluctuational free volume kinetic theory:

$$f_g = \exp\left(-\frac{E_h + PV_h}{kT}\right) \tag{19}$$

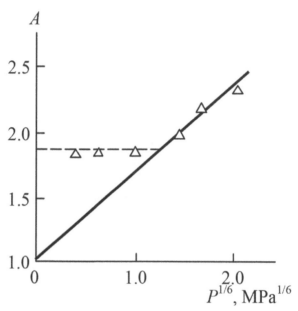

FIGURE 7.7   The dependence of constant A on autohesional contact formation pressure P for PS-PS.

where f_g is relative fluctuational free volume, E_h is activation energy of free volume formation, V_h is free volume microvoid volume, k is Boltzmann constant, T is temperature of adhesional bonding formation.

The value E_h within the framework of the indicated theory is determined as follows [19]:

$$E_h = kT_g \ln(1/f_g)$$

(20)

where T_g is glass transition temperature, equal to 376 K for PS and the value f_g at the used in the present work temperature $T = 353$ K is equal to 0.07 [15].

In its turn, the value V_h is determined according to the equation [19]:

$$V_h = \frac{3(1-v)kT_g}{f_g E}$$

(21)

where v is Poisson's ratio, E is elasticity modulus, which is equal to 0.5 GPa [8].

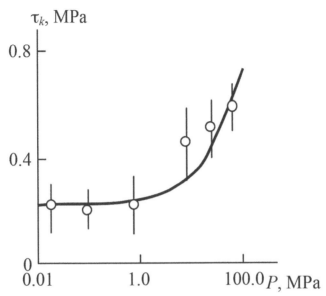

FIGURE 7.8   The experimental (points) and calculated according to the Eq. (1) (solid curve) dependences of autohesional contact PS-PS shear strength τ_k on its formation pressure P. The standard error of τ_k experimental determination is indicated.

Poisson's ratio v is connected with dimension d_f by the following relationship [18]:

$$d_f = (d-1)(1+v) \qquad (22)$$

To change d_f estimation (and, hence, D_f, see the Eq. (4)) the formula was used [20]:

$$f_g = 2.6 \times 10^{-3}\left(\frac{d_f}{d-d_f}\right) \qquad (23)$$

And at last, the Eq. (1) following variant was used for adhesional bonding shear strength τ_k^T theoretical estimation [21]:

$$\ln \tau_k = N_k - 16.6D_f + 20.5 \qquad (24)$$

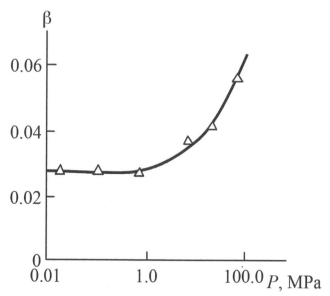

FIGURE 7.9   The experimental (points) and calculated according to the Eq. (24) (solid curve) dependences of autohesional contact PS-PS shear strength τ_k on its formation pressure P.

The comparison of experimental $\tau_k(P)$ and calculated according to the Eq. (24) $\tau_k^T(P)$ dependences of adhesional bonding shear strength on pressure at its formation is shown in Fig. 7.9. As one can see, the second variant of the dependence $\tau_k(P)$ calculation also gives a good correspondence to experiment (the mean discrepancy between τ_k and τ_k^T makes up 15%).

The Eq. (24) demonstrates clearly the physical significance of pressure P influence on adhesional bonding strength. At P growth within the range of 0.02–80 MPa D_f reduction from 1.962 up to 1.922 occurs, that is, macromolecular coil compactization, that was to be expected. D_f reduction results in coils interdiffusion processes intensification, that is expressed by the second member in the Eq. (24) right-hand part absolute value decreasing and, as consequence, by τ_k increasing. At the same time the good correspondence of τ_k to τ_k^T can be obtained only at the condition $N_k = $ const, although d_f decreasing at P growth results in characteristic ratio C_∞ reduction according to the Eq. (8) and, as consequence, to R_g reduction according to the Eq. (7), that is, to macromolecular coil compactization again. The decreasing of both R_g and D_f results in the long run in N_k reduction according to the Eq. (2). The stated above considerations suppose, that PS macromolecular coils forced interdiffusion, which is due to the applied pressure, compensates in abundance their compactization negative effect.

This supposition is confirmed by anomalous diffusion exponent β enhancement at P growth (Fig. 7.10). The dependence $\beta(P)$ by its shape is similar to the dependence $\tau_k(P)$ (see Figs. 7.8 and 7.9), that allows to obtain the following empirical equation:

$$\tau_k = 10\ \beta\ \text{MPa.} \tag{25}$$

In Fig. 7.11, the comparison of the obtained experimentally $\tau_k(P)$ and calculated according to the Eq. (25) dependences of shear strength of adhesional bonding on its formation pressure is adduced. The good correspondence of theory and experiment is obtained again (the mean discrepancy between τ_k and τ_k^T makes up 13%). The data of Fig. 7.11 confirm correctness of the made above supposition in respect to the defining influence of forced interdiffusion on adhesional bonding strength.

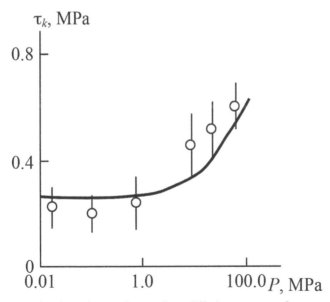

FIGURE 7.10 The dependence of anomalous diffusion exponent β on autohesional contact PS-PS formation pressure P.

7.3.4   AUTOHESION IN POLYMER-POLYMERIC COMPOSITES

The problems of autohesion (surfaces of identical materials coupling) acquire particular importance in virtue of polymer-polymeric composites development [22]. For composites of low-density polyethylene/ultrahigh molecular polyethylene (LDPE/UHMPE) it has been shown that these materials elasticity modulus corresponds to its upper boundary (mixtures rule). The interfacial autohesion LDPE/UHMPE level estimation can be performed with the aid of the equation [2]:

$$\frac{E_c}{E_m} = 1 + 11\left(\varphi_f b_\alpha\right)^{1.7} \tag{26}$$

where E_c and E_m are elasticity moduli of composite and matrix polymer (LDPE), respectively, φ_f is volume contents of filler (UHMPE), b_α is dimensionless parameter, characterizing autohesion LDPE-UHMPE level.

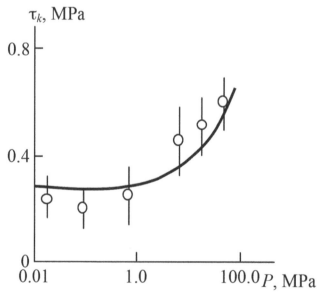

FIGURE 7.11   The experimental (points) and calculated according to the Eq. (25) (solid curve) dependences of autohesional contact PS-PS shear strength τ_k on its formation pressure P.

Estimations according to the Eq. (26) have shown that the value $b_\alpha \approx 1.0$, that is, it corresponds to perfect adhesion by Kerner [22]. This means, that autohesional bonding LDPE-UHMPE strength is close to polymer matrix (LDPE) strength. Let us consider this effect treatment within the framework of the stated above conception.

As it is known, LDPE is a branched polymer and UHMPE is a linear one. Within the framework of fractal analysis polymer chain branching degree is characterized by its spectral dimension d_s, which is varied from 1.0 for linear chain up to 1.33 for strongly branched or cross-linked macromolecule [23]. In the present paper it is accepted that $d_s = 1.0$ for UHMPE and $d_s = 1.30$ – for LDPE. Further macromolecular coil fractal dimension D_f can be determined according to the equation [7]:

$$D_f = \frac{d_s(d+2)}{d_s + 2} \tag{27}$$

According to the Eq. (27) the value $D_f = 1.667$ for UHMPE and $D_f = 1.970$ for LDPE. Since on the boundary between the indicated polyethylene's their macromolecular coils mixture is formed owing to self-diffusion process, then the value D_f^m for such mixture can be determined according to the Eq. (15), where $D_f^m = 1.891$.

For the considered polyethylenes molecular weight is equal to: $M_w = 3 \times 10^5$ for LDPE and 2×10^6 for UHMPE. Accepting the average $M_w^{av} = 1.15 \times 10^6$ and monomer link molecular weight $M_{mon} = 28$, let us obtain mean polymerization degree N_{pol}:

$$N_{pol} = \frac{M_w^{av}}{M_{mon}} \tag{28}$$

which is equal to ~4.1×10^4. Further the macromolecular coil gyration radius R_g can be determined [23]:

$$R_g = N_{pol}^{1/D_f^m} \tag{29}$$

which is equal to ~ 276 relative units.

Then the number of macromolecular coils contacts N_k on the division boundary LDPE-UHMPE was calculated according to the Eq. (3), which proved to be equal: $N_k \approx 81$.

And at last, autohesional bonding shear strength τ_k can be determined according to the equation, similar to the Eq. (1):

$$\ln \tau_k = 0.10 N_k - 6.0, \text{ MPa}. \tag{30}$$

The value $\tau_k = 8.17$ MPa according to the relationship (30). In its turn, polymer matrix (LDPE) shear strength $\tau_m = 8.67$ MPa. In other words, the obtained comparable values τ_k and τ_m (their discrepancy makes up less than 6%) confirm correctness of the condition $b_\alpha \approx 1.0$, estimated according to the Eq. (26).

The indicated condition correctness theoretical calculation E_c according to the Eq. (26) allows to confirm that at $E_m = 150$ MPa, $b_\alpha = 1.0$ and φ_f value is accepted equal to UHMPE mass contents, since polyethylenes density is close to 1000 kg/m³. As it follows from the comparison, adduced in Fig. 7.12, the good correspondence of theory and experiment is obtained

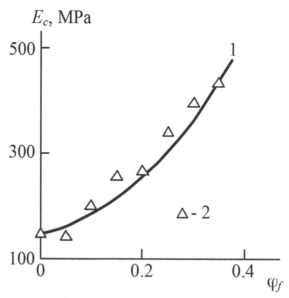

FIGURE 7.12 The dependences of elasticity modulus E_c on UHMPE volume contents φ_f for composite LDPE/UHMPE. 1 – calculation according to the Eq. (26); 2 – experimental data.

(their mean discrepancy makes up ~9%), that confirms again the condition $b_\alpha \approx 1.0$ correctness.

7.4 CONCLUSIONS

The generalized fractal model was proposed, allowing quantitative description of polymer pairs adhesional bonding shear strength dependence on different kinds of factors (temperature, contact duration and pressure at the indicated bonding formation). This model also allows to take into account molecular and structural characteristics of polymers, forming adhesional bonding. The indicated model applicability for the description of interfacial interactions level in polymer-polymeric composites has been shown.

KEYWORDS

- **adhesion**
- **fractal analysis**
- **polymer pair**
- **shear strength**
- **structure**

REFERENCES

1. Schnell R., Stamm M., Creton C. Macromolecules, 1998, v. 31, № 7, p. 2284–2292.
2. Boiko Yu.M., Prud'homme R.E. Macromolecules, 1998, v. 31, № 19, p. 6620–6626.
3. Boiko Yu.M., Prud'homme R.E. Macromolecules, 1997, v. 30, № 12, p. 3708–3710.
4. Boiko Yu.M., Prud'homme R.E. J. Polymer Sci.: Part B: Polymer Phys., 1998, v. 36, № 4, p. 567–572.
5. Doi M., Edwards S.F. The Theory of Polymer Dynamics. Oxford, Clarendon Press, 1986, 432 p.
6. Yakh'yaeva Kh.Sh., Kozlov G.V., Magomedov G.M. Polymer Science D. Glues and Sealing Materials, 2013, v. 6, № 2, p. 93–95.
7. Vilgis T.A. Physica A, 1988, v. 153, № 2, p. 341–354.
8. Kozlov G.V., Zaikov G.E. Structure of the Polymer Amorphous State. Utrecht, Boston, Brill Academic Publishers, 2004, 465 p.

9. Kozlov G.V., Mikitaev A.K., Zaikov G.E. The Fractal Physics of Polymer Synthesis. Toronto, New Jersey, Apple Academic Press, 2014, 359 p.
10. Budtov V.P. Physical Chemistry of Polymer Solutions. Sankt-Peterburg, Chemistry, 1992, 384 p.
11. Aharoni S.M. Macromolecules, 1983, v. 16, № 9, p. 1722–1728.
12. Aharoni S.M. Macromolecules, 1985, v. 18, № 12, p. 2624–2630.
13. Boiko Yu.M. Mechanics of Composite Materials, 2000, v. 36, № 1, p. 127–134.
14. Wu S. J. Polymer Sci.: Part B: Polymer Phys., 1989, v. 27, № 4, p. 723–741.
15. Kozlov G.V., Zaikov G.E., Mikitaev A.K. The Fractal Analysis of Gas Transport in Polymers. The Theory and Practical Applications. New York, Nova Science Publishers, Inc., 2009, 238 p.
16. Zelenyi L.M., Milovanov A.V. Achievements of Physical Sciences, 2004, v. 174, № 8, p. 809–852.
17. Hentschel H.G.E., Deutch J.M. Phys. Rev. A, 1984, v. 29, № 12, p. 1609–1611.
18. Balankin A.S. Synergetics of Deformable Body. Moscow, Publishers of Ministry Defence SSSR, 1991, 404 p.
19. Kozlov G.V., Sanditov D.S., Ovcharenko E.N., Mikitaev A.K. Physics and Chemistry of Glass, 1997, v. 23, № 4, p. 369–373.
20. Kozlov G.V., Novikov V.U. Achievements of Physical Sciences, 2001, v. 171, № 7, p. 717–764.
21. Kozlov G.V., Naphadzokova L.Kh., Zaikov G.E., Yarullin A.F. Herold of Kazan Technological University, 2012, v. 15, № 9, p. 97–100.
22. Mikitaev A.K., Kozlov G.V., Zaikov G.E. Polymer Nanocomposites: Variety of Structural Forms and Applications. New York, Nova Science Publishers, Inc. 2008, 319 p.
23. Kozlov G.V., Dolbin I.V., Zaikov G.E. The Fractal Physical Chemistry of Polymer Solutions and Melts. Toronto, New Yersey, Apple Academic Press, 2014, 316 p.

CHAPTER 8

A REVIEW ON HYALURONAN BIOPOLYMER: PROPERTIES AND PHARMACEUTICAL APPLICATIONS

A. M. OMER,[1-3] T. M. TAMER,[1] and M. S. MOHYELDIN[1]

[1]Polymeric Materials Department, Advanced Technologies and New Materials Research Institute, City of Scientific Research and Technological Applications, Alexandria, Egypt

[2]Laboratory of Bioorganic Chemistry of Drugs, Institute of Experimental Pharmacology and Toxicology, Slovak Academy of Sciences, Bratislava, Slovakia

[3]Materials Delivery Group, Polymeric Materials Department, Advanced Technologies and New Materials Research Institute (ATNMRI), City of Scientific Research and Technological Applications (SRTA- City), New Borg El-Arab City 21934, Alexandria, Egypt. E-Mail: Ahmedomer_81@yahoo.com.com

CONTENTS

ABSTRACT

Hyaluronan (HA) is a high-molecular weight, naturally occurring linear polysaccharide and found in all tissues and body fluids of higher animals. The excellent properties of HA such as biodegradability, biocompatibility, safety, excellent mucoadhesive capacity and high water retaining ability make it well-qualified for using in various bio-medical applications. In addition; HA is nontoxic, noninflammatory and nonimmunogenic. Because of all these advantages, HA has received much attention as a matrix for drug delivery system. This review will summarize our present knowledge about HA, properties and its development in some pharmaceutical applications.

8.1 HISTORICAL PERSPECTIVE OF HYALURONAN

Hyaluronan is one of the most interesting and useful natural biopolymer macromolecules and considered as a member of a similar polysaccharides group, and also known as mucopolysaccharides, connective tissue polysaccharides, or glycosaminogylcans [56, 64, 92]. The popular name of hyaluronic acid (HA) is derived from "hyalos," which is the Greek word for glass + uronic acid, and it was discovered and investigated in 1934 by Karl Meyer and his colleague John Palmer [66]. Firstly, they isolated a previously unknown chemical substance from the vitreous body of cows' eyes as an acid form but it behaved like a salt in physiological conditions (sodium hyaluronate) [27, 28, 71], they solved the chemical structure of HA and found that its composed from two sugar molecules (D-glucuronic acid (known as uronic acid) and D-Nacetyl glucosamine) and they named the molecule "hyaluronic acid" because of the hyaloid appearance of the substance when swollen in water and the probable presence of hexuronic acid as one of the components. Hyaluronan (HA) is the currently used name; hence it represents a combination of "hyaluronic acid" and "hyaluronate," in order to indicate the different charged states of this polysaccharide [3].

In 1942, HA was applied for the first time as a substitute for egg white in bakery products [71], and shortly afterward, in 1950s HA was isolated from umbilical cord and then from rooster combs [27], and finally it was isolated from other sources. HA is present in synovial fluid (SF) with final physiological concentration about 2–3 mg/ml, and the largest amounts of HA are found in the extracellular matrix (ECM) of soft connective tissues [49, 57] and so its widely distributed in vertebrate connective tissues, particularly in umbilical cord, vitreous humor, dermis, cartilage, and intervertebral disc [52, 94]. Also, it was reported that HA is present in the capsules of some bacteria (e.g., strains of Streptococci) but it's absent completely in fungi, plants, and insects [50].

8.2 PHYSICOCHEMICAL PROPERTIES OF HYALURONAN

8.2.1 CHEMICAL STRUCTURE

HA is an un-branched nonsulfated glycosaminoglycan (GAG) composed of repeating disaccharides and present in the acid form [51, 54, 105], and composed of repeating units from D-glucuronic acid and N-acetyl-D- glucosamine linked by a glucuronidic β(1–3) bond [44] as shown in Fig. 8.1. Also HA forms specific stable tertiary structures in aqueous solution.

Both sugars are spatially related to glucose which in the β-configuration allows all of its bulky groups (the hydroxyls, the carboxylate moiety, and the anomeric carbon on the adjacent sugar) to be in sterically favorable equatorial positions while all of the small hydrogen atoms occupy the less

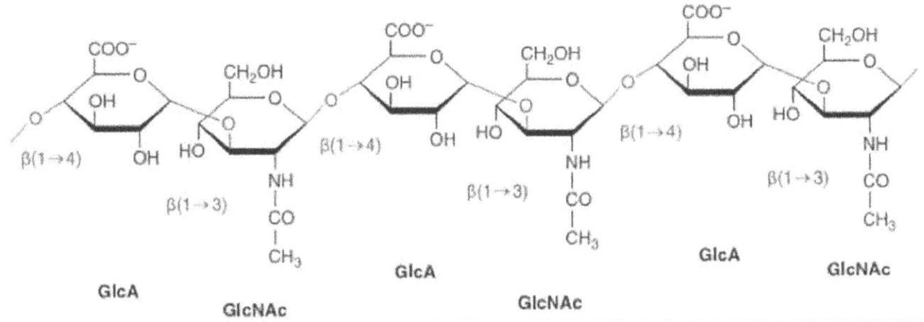

FIGURE 8.1   Hyaluronan is composed of repeating polymeric disaccharides D-glucuronic acid (GlcA) and *N*-acetyl-D-glucosamine (GlcNAc) linked by a glucuronidic (1–3) bond. Three disaccharide GlcA-GlcNAc are shown [44].

sterically favorable axial positions. Thus, the structure of the disaccharide is energetically very stable [96]. Several thousand sugar molecules can be included in the backbone of HA. The structure of HA called a coiled structure, and this can attributed to that the equatorial side chains form a more polar face (hydrophilic), while the axial hydrogen atoms form a nonpolar face (relatively hydrophobic), and this, led to a twisted ribbon structure for HA (i.e., a coiled structure) [71].

8.2.2   SOLUBILITY AND VISCOSITY

According to hygroscopic and homeostatic properties of HA; the molecules of HA can be readily soluble in water and this property prompt the proteoglycans for hydration producing a gel like a lubricant [74]. HA also exhibit a strong water retention property and this advantage can be explained by the fact that HA is a natural hydrophilic polymer, (i.e., water soluble polymer), where its contain carboxylic group and also high number of hydroxyl groups which impart hydrophilicity to the molecule, and so increase affinity of water molecules to penetrate in to the HA network and swells the macromolecular chains consequently. The water retention ability of HA can also attributed to the strong anionic nature of HA, where the structure of the HA chains acts to trap water between the coiled chains and giving it a high ability to uptake and retain water molecules. It was stated that stated that HA molecules can retain water up to 1,000 times from own weight [64]. The water holding capacity of HA increases with increasing relative humidity [67], therefore, the hydration parameters are independent of the molecular weight of the HA [20].

On the other hand, the viscosity is one of the most important properties of HA gel, in which several factors affecting the viscosity of this molecule such as the length of the chain, molecular weight, cross-linking, pH and chemical modification [45]. The rotational viscometry is considered one of the successful and simplest instruments, which used for identification of the dynamic viscosity and the 'macroscopic' properties of HA solutions. It was indicated that the viscosity is strongly dependent on the applied shear-stress. At concentrations less than 1 mg/mL HA start to entangle.

Morris and his co-workers identified the entanglement point by measuring the viscosity, they confirmed that the viscosity increases rapidly and exponentially with concentration (~c) beyond the entanglement point, also, the viscosity of a solution with concentration 10 g/L at low shear probably equal to 10^6 times the viscosity of the solvent [69]. While, at high shear the viscosity may drop as much as ~10 times [29]. However; in the synovial fluid (SF), unassociated high molar mass HA confers its unique viscoelastic properties, which required for maintaining proper functioning of the synovial joints [110].

8.2.3   VISCOELASTICITY

Viscoelasticity is another characteristic of HA resulting from the entanglement and self-association of HA random coils in solution [28]. Viscoelasticity of HA can be related to the molecular interactions which are also dependent on the concentration and molecular weight of HA. The higher the molecular weight and concentration of HA, the higher the viscoelasticity the solutions possess. In addition, with increasing molecular weight, concentration or shear rate, HA in aqueous solution is undergo a transition from Newtonian to non-Newtonian characteristics [30]. The dynamic viscoelasticity of HA gels was increased relative to HA–HA networks when the network proteoglycan–HA aggregates shift the Newtonian region to lower shear rates [11]. In addition to the previous properties of HA, the shape and viscoelasticity of HA molecule in aqueous solution like a polyanion is undergo the pH sensitivity (i.e., pH dependent) and effected by the ionic strength. Indeed, HA has a pKa value of about 3.0 and therefore, the extent of ionization of the HA chains was affected by the change in pH. The intermolecular interactions between the HA molecules may be affected by the shift in ionization, which its rheological properties changes consequently [12].

8.3   DEGRADATION OF HA

In principle there are many ways of HA degradation depends on biological (enzymatic) or physical and chemical (non enzymatic) methods.

8.3.1　BIOLOGICAL METHODS

In the biological methods, the degradation of HA can take place using enzymes. It was reported that there are three types of enzymes which are present in various forms, in the intercellular space and in serum (hyaluronidase, β-D-glucuronidase, and β–N-acetyl-hexosaminidase) are involved in enzymatic degradation [26]. Hyaluronidase (HYAL) is considered as a most powerful degradation enzyme for hyaluronan [23].Volpi et al. [114] reported that Hyaluronidase cleaves high molecular weight HA into smaller fragments of varying size via hydrolyzing the hexosaminidic β (1–4) linkages between N-acetyl-D-glucosamine and D-glucuronic acid residues in HA [72], while the degradation of fragments via removal of nonreducing terminal sugars can be done by the other two enzymes. However, it was found that the HYAL enzymes are present with very low concentrations and the measuring of its activity, characterization and purification of it are difficult, in addition, measuring their activity, which is high but unstable, so that this family has received little attention until recently.

8.3.2　PHYSICAL METHODS

By physical methods, the degradation and depolymerization of HA can be performed by different techniques, it was reported that HA can be degraded using ultrasonication in a nonrandom fashion and the obtained results shows that high molecular weight HA chains degrade slower than low molecular weight HA chains. However, it was noted that the degradation of HA into monomers is not fully completed when using different HA samples under applying different ultrasound energies, and the increasing of absorbance at 232 nm after sonication is not observed. Heat is another type of the physical methods used for HA degradation, in which with increasing temperature the degradation increased consequently and the viscosity strongly decreased [26]. In case of thermal degradation method, it was reported that the treatment of different HA samples at temperatures from 60 to 90°C for 1 h results in only moderate degradation and a small increase of polydispersity. Bottner and his co-workers have proved that that thermal degradation of HA occurs in agreement with the random-scission mechanism during the study of two high-molar-mass HA samples that were extensively degraded at 128°C in an autoclave [9]. HA can also degrade by other physical methods like irradiation.

8.3.3 CHEMICAL METHODS

HA like other polysaccharides can be degraded by acid and alkaline hydrolysis or by a deleterious action of free radicals.

Stern et al. [90] reported the degradation of HA by acid and alkaline conditions occurs in a random fashion often resulting in disaccharide fragment production. Where, the glucuronic acid moiety of HA degraded via acidic hydrolysis, while the alkaline hydrolysis occurs on N-acetylglucosamine units and giving rise to furan containing species. Also, the oxidation processes can degrade HA via reactive oxygen species (ROS) such as hydroxyl radicals and superoxide anions which generated from cells as a consequence of aerobic respiration. It was found that acceleration of degradation of high-molecular-weight HA occurring under oxidative stress produces an impairment and loss of its viscoelastic properties. Figure 8.2. describes the fragmentation mechanism of HA under free radical stress.

The ROS are involved in the degradation of essential tissue or related components such as synovial fluid (SF) of the joint which contains high-molar-mass HA. It is well known that most of rheumatic diseases are resulting from reduction of HA molar mass in the synovial fluid of patients. Numerous studies have been reported to study the effect of various ROS on HA molar mass. Soltes and his team focused their research on the hydroxyl radicals resulting from the reaction mechanism of (H_2O_2 + transitional metal cation H_2O_2 in the presence of ascorbic acid as a reducing agent) under aerobic conditions and studied its effect on the degradation of HA molar mass, in which the system of acerbates and metal cation as copper(II) ions educes hydrogen peroxide (H_2O_2) to turn into OH radicals by a Fenton-like reaction [37] and this system is called Weissberger's oxidative system. They observed a decrease of the dynamic viscosity value of the HA solution, and this indicate the degradation of the HA by the system containing Cu (II) cations. Therefore, agents that could delay the free-radical-catalyzed degradation of HA may be useful in maintaining the

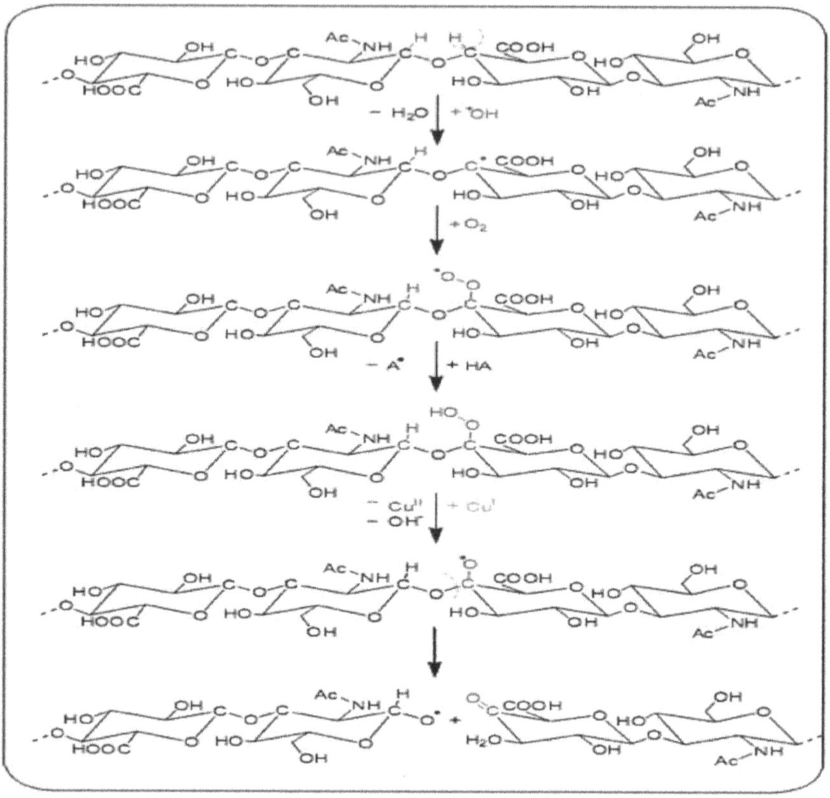

FIGURE 8.2   Schematic degradation of HA under free radical stress [38].

integrity of dermal HA in addition to its moisturizing properties. It should be noted that the concentrations of ascorbate and Cu(II) were comparable to those that may occur during an early stage of the acute phase of joint inflammation.

8.4   APPLICATIONS OF HA

8.4.1   PHARMACEUTICAL APPLICATIONS

Indeed, HA and its modified forms have been extensively investigated and widely used for various pharmaceutical applications. In the current review, we presented in brief some pharmaceutical applications of HA biopolymer.

8.4.1.1   HA in Drug Delivery Systems

It is well known that macromolecular drug forms are composed basically of three components: (i) the carrier; (ii) the drug; and (iii) a link between them.

It was reported that polysaccharide-based microgels are considered as one class of promising protein carriers due to their large surface area, high water absorption, drug loading ability, injectability, nontoxicity, inherent biodegradability, low cost and biocompactibility. Among various polysaccharides, HA, has been recently most concerned [62].

The physicochemical and biological properties of hyaluronan qualify this macromolecule as a prospective carrier of drugs. This natural anionic polysaccharide has an excellent mucoadhesive capacity and many important applications in formulation of bioadhesive drug delivery systems. It was found that this biopolymer may enhance the absorption of drugs and proteins via mucosal tissues. In addition, it is immunologically inert, safely degraded in lysosomes of many cells and could be an ideal biomaterial for drug and gene delivery. Therefore, HA biopolymer has become the topic of interest for developing sustained drug delivery devices of peptide and protein drugs in subcutaneous formulations. The recent studies also suggested that HA molecules may be used as gel preparations for nasal and ocular drug delivery. Also, HA has been used for targeting specific intracellular delivery of genes or anticancer drugs.

The applications of HA in the above mentioned drug delivery systems and its advantages in formulations for various administration routes delivery were summarized in Table 8.1.

8.4.1.2   Nasal Delivery

Over the last few decades nasal route has been explored as an alternative for drug delivery systems (Nonparenteral). This is due to the large surface area and relatively high blood flow of the nasal cavity and so the rapid

TABLE 8.1   Summary of Some Drug Delivery Applications of HA and Its Advantages

Administration route	Advantages	References
Intravenous	• Enhances drug solubility and stability	[10, 21]
	• Promotes tumor targeting via active (CD44 and other cell surface receptors) and passive (EPR) mechanisms	
	• Can decrease clearance, increase AUC and increase circulating half-life	
Dermal	• Surface hydration and film formation enhance the permeability of the skin to topical drugs	[11, 97]
	• Promotes drug retention and localization in the epidermis	
	• Exerts an anti-inflammatory action	
Subcutaneous	• Sustained/controlled release from site of injection	[25, 75]
	• Maintenance of plasma concentrations and more favorable pharmacokinetics	
	• Decreases injection frequency	
Intra-articular	• Retention of drug within the joint	[65, 81]
	• Beneficial biological activities include the anti-inflammatory, analgesic and chondroprotective properties of HA	
Ocular	• The shear-thinning properties of HA hydrogels mean minimal effects on visual acuity and minimal resistance to blinking	[85]
	• Mucoadhesion and prolonged retention time increase drug bioavailability to ocular tissues	
Nasal	• Mucoadhesion, prolonged retention time, and increased permeability of mucosal epithelium increase bioavailability	[60, 100]
Oral	• Protects the drug from degradation in the GIT	[71]
	• Promotes oral bioavailability	
Gene	• Dissolution rate modification and protection	[43, 117]

absorption is possible. It was reported that viscous solutions of polymer have been shown to increase the residence time of the drug at the nasal mucosa and thereby promote bioavailability. The mucoadhesive properties of HA could promote the drugs and proteins absorption through mucosal tissues. The mucoadhesive property of HA can be increased by conjugating it with other bioadhesive polymers such as Chitosan and polyethylene glycol. Lim and his team prepared biodegradable microparticles using chitosan (CA) and HA by the solvent evaporation method, they used gentamicin used as a model drug for intranasal studies in rats and sheep. The results showed that the release of gentamicin is prolonged when formulated in HA, CH and HA/CH and that the resultant microparticles are mucoadhesive in nature. In addition, much attention has received for delivery of drugs to the brain via the olfactory region through nasal route, that is, nose-to-brain transport. Horvat group developed a formulation containing sodium hyaluronate in combination with a nonionic surfactant to enhance the delivery of hydrophilic compounds to the brain via the olfactory route, the results proved that HA, a nontoxic biomolecule used as a excellent mucoadhesive polymer in a nasal formulation, increased the brain penetration of a hydrophilic compound, the size of a peptide, via the nasal route.

8.4.1.3 Ocular Delivery

The current goals in the design of new drug delivery systems in ophthalmology are to achieve directly: (a) precorneal contact time lengthening; (b) an increase in drug permeability; and (c) a reduction in the rate of drug elimination. The excellent water-holding capacity of HA makes it capable of retaining moisture in eyes [80]. Also, the viscosity and pseudoplastic behavior of HA providing mucoadhesive property can increase the ocular residence time. Nancy and her group work reported that HA solutions have tremendous ocular compatibility both internally (when used during ophthalmic surgery) and externally, at concentrations of up to 10 mg/ml (1%). Also, topical HA solutions (0.1–0.2%) have been shown to be effective therapy for dry eye syndrome. It was noted also that HA may interact with the corneal surface and tear film to stabilize the tear film and provide effective wetting, lubrication and relief from pain caused by exposed and often damaged corneal epithelium. The ability to interact with and to stabilize the natural tear film is a property unique to hyaluronan. Pilocarpine – HA vehicle is considered the most commonly studied HA delivery system. Camber and his group proved that 1% pilocarpine solution dissolved in HA increased the 2-fold absorption of drug, improving the bioavailability, and miotic response while extending the duration of action. In another study gentamicin bioavailability was also reported to be increased when formulated with a 0.25% HA solution. It was found that HA in the form of Healon can be used in artificial tears for the treatment of dry eye syndrome, and its efficacy for the treatment was evaluated. On the other hand, A few studies have reported on the use of HA with contact lenses in different applications. Pustorino and his group were conducted a study to determine whether HA could be used to inhibit bacterial adhesion on the surface of contact lenses. He showed that HA did not act as an inhibitor or a promoter of bacterial adhesion on the contact lens surfaces [76].

Also, in another application, Van Beek and others evaluated the use of HA containing hydrogel contact lenses to determine the effect on protein adsorption. Protein deposition on the contact lens surface can result in reduced vision, reduced lens wettability, inflammatory complications, and reduced comfort. They incorporated releasable and chemically cross-linked HA of different molecular weights as a wetting agent in soft contact lenses. The results showed that the addition of HA had no effect on the modulus or tensile strength of the lens regardless of molecular weight and no effect on the optical transparency of the lens. While, the protein adsorption on the lens did not affected by the releasable HA at either molecular weight [113].

8.4.1.4 Protein Sustained Delivery

Indeed, during the past few decades HA has been shown to be useful for sustained release (SR) formulations of protein and peptide drugs via parenteral delivery [24]. Because of the hydrophilic nature of HA, hydrogels can provide an aqueous environment preventing proteins from denaturation. The swelling properties of hydrogel were shown to be affecting the protein diffusion; hence the diffusion of protein was influenced by the crosslink structure itself. In addition, the sustained delivery of proteins without denaturation is realized by tailoring the crosslink network of HA microgels. Luo and his group studied the sustained delivery of bovine serum albumin (BSA) protein from HA microgels by tailoring the crosslink network. He prepared a series of HA microgels with different crosslink network using an inverse microemulsion method, and studied the effect of different crosslink network in HA microgels on the loading capacity and sustained delivery profile of BSA as a model protein.

The date showed that the BSA loading had no obvious influence on the surface morphology of HA microgels but seemed to induce their aggregation. Increase of crosslink density slowed down the degradation of HA microgels by hyaluronidase and reduced the BSA loading capacity as well, but prolonged the sustained delivery of BSA. However, physically cross-linked hydrogel behave very soft and easily disintegrated, thus, an initial burst and rapid protein release resulted hybrid hyaluronan hydrogel encapsulating nanogel was developed to overcome the above mentioned problems. The nanogels were physically entrapped and well dispersed in a three-dimensional network of chemically cross-linked HA (HA gel).

8.4.1.5   Anticancer Drug Delivery

To date, the potentialities of HA in drug delivery have been investigated as carrier of antitumoral and anti-inflammatory drugs. HA is considered one of the major components of the extracellular matrix (ECM), also it is the main ligand for CD44 and RHAMM, which are overexpressed in a variety of tumor cell surfaces including human breast epithelial cells, colon cancer, lung cancer and acute leukemia cells. In fact, it's essential in treatment and prevention of cancer cell metastasis that the localization of drug not only to the cancerous cells, but also to the surrounding lymph. HA is known as a bioadhesive compound capable of binding with high affinity to both cell-surface and intracellular receptors, to the extracellular matrix (ECM) components and to itself. HA can bind to receptors in cancer cells, and this is involved in tumor growth and spreading. CD44 regulates cancer cells proliferation and metastatic processes. In addition, disruption of HA–CD44 binding was shown to reduce tumor progression. Also, administration of exogenous HA resulted in arrest of tumor spreading.

Therefore, anticancer drug solubilization, stabilization, localization and controlled release could be enhanced via coupling with HA. Yang and his team work reported that the degradation of HA by intratumoral administration of hyaluronidases (HYAL) resulted in improved tumor penetration of conventional chemotherapeutic drugs. Also, they stated that high HA level has been detected at the invasive front of growing breast tumors, 3.3-fold higher than in central locations within the tumor. In addition, HA over production is associated with poor prognosis of breast cancer. In women <50 years, breast tumor HA level could predict cancer relapse. It was reported that HA conjugates containing anticancer drugs include sodium butyrate, cisplatin, doxorubicin and mitomycin C. Therefore, depending on the degree of substitution of HA with drugs, these exhibited enhanced targeting ability to the tumor and higher therapeutic efficacy compared to free-anticancer drugs. It is well known that Cisplatin (cis-diaminedi-chloroplatinum or CDDP) is an extensively employed chemotherapeutic agent for the treatment of a wide spectrum of solid tumors. Xiea and others presented a successful drainage of hyaluronan–cisplatin (HA–Pt) conjugates into the axillary lymph nodes with reduced systemic toxicities after local injection in a breast cancer xenograft model in rodents. They also observed that the pulmonary delivery of the HA–Pt conjugate to the lungs may be useful in the treatment of lung cancer by reducing systemic toxicities and increasing CDDP deposition and retention within lung tumors, surrounding lung tissues, and the mediastinal lymph.

8.4.1.6   Gene Delivery

HA could be an ideal biomaterial for gene delivery. Since it has been introduced as a nanocarrier for gene delivery since 2003 [115]. Yang and his team work demonstrated the utility of HA microspheres for DNA gene delivery, they showed that show that DNA can be easy incorporated before derivatization. Once the HA-DNA microspheres degrade, the released DNA is structurally intact and able to transfect cells in culture and in vivo using the rat hind limb model. In addition, they found that the release of the encapsulated plasmid DNA can be sustained for months and is capable of transfection in vitro or in vivo and concluded that the native HA can be used to delivers DNA at a controlled rate and adaptable for site-specific targeting [115]. Another study for the same group, the DNA-HA matrix crosslinking with adipic dihydrazide (ADH) was able to sustain gene release while protecting the DNA from enzymatic degradation. It has been reported that HA combined with polyethyleneimine (PEI), poly (L-lysine) (PLL) and poly (L-arginine) (PLR). Therefore, the biocompatibility of the anionic HA is achieved by shielding the positive surface of gene/polycation complexes and by inhibiting nonspecific binding to serum proteins, thereby reducing the cytotoxicity of cationic polymers. Figure 8.3 shows the binding of anionic HA (negatively charged) to the surface of DNA/PEI complexes, where PEI is polycation (positively charged) through electrostatic interactions between them to form ternary complex.

8.4.2   HA FOR EVALUATION THE ACTIVITY OF ANTIOXIDANTS

It is well known that the fast HA turnover in SF of the joints of healthy individuals can be attributed to the oxidative/degradative action of the reactive oxygen species (ROS), which generated among others by the catalytic effect of transition metal ions on the autoxidation of ascorbate. It has been

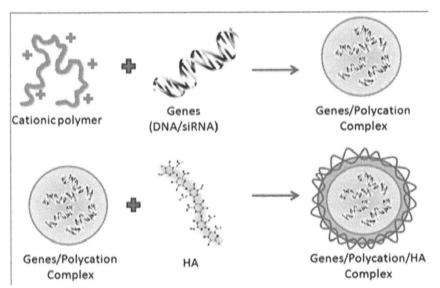

FIGURE 8.3   Schematic representation for the formation of electrostatic complex between negatively charged HA and positively charged polycation to produce gene/polycation (PEI)/HA ternary complex (adapted from Ref. [45]).

reported that among the ROS, hydroxyl radical (OH) represents the most active substance in terms of degradation of HA. Therefore,chondrocytes are able to protect against free radical damage by means of endogenous antioxidants such as catalase and the glutathione peroxidase/reductase system, which are both involved in the removal of H_2O_2. The degradation behavior of HA via ROS has been studied on applying several in vitro models. The team of laboratory of bioorganic chemistry of drugs at the institute of experimental pharmacology and toxicology, Slovak academy of science, focused their attention on studying the effect of various antioxidants on the degradation behavior of HA which used as a model for evaluation the extent of the activity of these antioxidants. As it well known that among various thiol compounds, D-penicillamine and L-glutathione are considered the best antioxidants compounds. Valachová and her work group studied monitored the pro and antioxidative effects of an anti-rheumatoid drug (D-penicillamine (D-PN)), on the degradation kinetics of high-molar-mass HA using the method of rotational viscometry. They observed that the addition of d-PN dose dependently prolonged the period of complete inhibition of the degradation of HA macromolecules, and the initial antioxidative action of D-PN is followed by induction of prooxidative conditions due to the generation of reactive free radicals. It has been reported that L-glutathione (GSH) composed of L-glutamate (glu), L-cysteine (cys) and glycine (gly) moieties and is commonly named the "mother" of all antioxidants.

In addition, GSH is endogenic antioxidant, belongs among the most efficient substances protecting the cells against reactive oxygen species (ROS) escaping from mitochondria and maintains the intracellular reduction oxidation (redox) balance and regulates signaling pathways during oxidative stress/conditions. Valachová et al. studied the antioxidative effect of L-glutathione (GSH) using HA high-molar mass in an oxidative system composed of Cu(II) plus ascorbic acid by using rotational viscometry. Results showed GSH added to the oxidative system in sufficient amount resulted in total inhibition of HA degradation. By the same way, Šoltés et al. used HA solutions as in vitro model for studying the scavenging effect of ibuprofen isomers using H_2O_2 and Cu^{2+} to prove that ibuprofen can be used as an anti-inflammatory drug.

In another study, HA was used for evaluation the activity of stobadine as an antioxidant drug. The protective effect of stobadine·2HCl on ascorbate plus Cu (II)-induced HA degradation was published by Rapta et al. [77]. Surovciková et al. [94] studied the antioxidative effects of two HHPI derivatives, namely SM1dM9dM10•2HCl and SME1i-ProC2•HCl, and compared those effects with that of stobadine·2HCl. From data it was observed that the most effective scavengers of •OH and peroxy-type radicals were recorded to be stobadine•2HCl and SME1i-ProC2•HCl, respectively. On the other hand, the most effective scavenger, determined by applying the ABTS assay, was stobadine•2HCl.

8.5   ACKNOWLEDGEMENTS

The corresponding author would like to thank the Institute of Experimental Pharmacology & Toxicology for having invited him and oriented him in the field of medical research. He would also like to thank Slovak Academic Information Agency (SAIA) for funding him during his work in the Institute.

KEYWORDS

- antioxidant
- drug delivery system
- glycosaminoglycan
- hyaluronan
- hydrogel

REFERENCES

1. Baker M, Feigan J, Lowther D. (1988). Chondrocyte antioxidant defenses: the roles of catalase and glutathione peroxidase in protection against H_2O2 dependent inhibition of PG biosynthesis. J Rheumatol, 15:670–677.
2. Balazs E, Denlinger J. (1993). Viscosupplementation: a new concept in the treatment of osteoarthritis. Journal of rheumatology Supplement, 20:3–9.

3. Balazs EA, Laurent TC, Jeanloz RW. (1986). Nomenclature of hyaluronic acid. Biochemical Journal 235: 903.

4. Baňasová M, Valachová K, Hrabárová E, Priesolová E, Nagy M, Juránek I, Šoltés L. (2011). Early stage of the acute phase of joint inflammation. In vitro testing of bucillamine and its oxidized metabolite SA981 in the function of antioxidants. 16th Interdisciplinary Czech-Slovak Toxicological Conference in Prague. Interdiscip Toxicol 4(2): 22.

5. Baňasová M, Valachová K, Rychly J, Priesolová E, Nagy M, Juránek I, Šoltés L. (2011a). Scavenging and chain breaking activity of bucillamine on free-radical mediated degradation of high molar mass hyaluronan. ChemZi, 7: 205– 206.

6. BeMiller JN, Whistler RL. (1962). Alkaline degradation of amino sugars. J Org Chem,7:1161–4.

7. Bertrand P, Girard N, Delpech B, Duval C, d'Anjou J, Dauce J.(1992). Hyaluronan (hyaluronic acid) and hyaluronectin in the extracellular matrix of human breast carcinomas: comparison between invasive and noninvasive areas. Int J Cancer,52(1):1–6.

8. Bothner H, Wik O. (1987). Rheology of hyaluronate. Acta Otolaryngol Suppl, 442: 25–30.

9. Bottner H, Waaler T, Wik O. (1988). Limiting viscosity number and weight average molecular weight of hyaluronate samples produced by heat degradation. Int J Biol Macromol, 10:287–91.

10. Bourguignon L, Zhu H, Shao L, Chen Y. (2000). CD44 interaction with tiam1 promotes Rac1 signaling and hyaluronic acid-mediated breast tumor cell migration. Journal of Biological Chemistry, 275 (3): 1829–38.

11. Brown M, Jones S. (2005). Hyaluronic acid: a unique topical vehicle for the localized delivery of drugs to the skin. Journal of the European Academy of Dermatology and Venereology, 19:308–318.

12. Brown MB, Jones SA. (2005). Hyaluronic acid: a unique topical vehicle for the localized delivery of drugs to the skin. JEADV,19: 308–318.

13. Camber O, Edman P, Gurny R. (1987). Influence of sodium hyaluronate on the meiotic effect of pilocarpine in rabbits. Current eye research, 6: 779–784.

14. Casalini P, Carcangiu ML, Tammi R, Auvinen P, Kosma VM, Valagussa P.(2008). Two distinct local relapse subtypes in invasive breast cancer: effect on their prognostic impact. Clin Cancer Res,14(1):25–31.

15. Luo C, Zhao J, Tu M, Zeng R, Rong J. (2013). Hyaluronan microgel as a potential carrier for protein sustained delivery by tailoring the crosslink network, Materials Science and Engineering C, 36: 301–308

16. Cortivo R, Brun P, Cardarelli L, O'Regan M, Radice M, Abatangelo G.(1996). Antioxidant Effects of Hyaluronan and Its a-Methyl-Prednisolone Derivative in Chondrocyte and Cartilage Cultures. Seminars in Arthritis and Rheumatism, 26 (1): 492–501.

17. Cuixia Yang, Yiwen Liu, Yiqing He, Yan Du, Wenjuan Wang, Xiaoxing Shi, Feng Gao. (2013). The use of HA oligosaccharide-loaded nanoparticles to breach the endogenous hyaluronan glycocalyx for breast cancer therapy, Biomaterials, 34:6829–6838

18. Naor D, Wallach-Dayan SB, Zahalka MA, Sionov RV. (2008). Involvement of CD44, a molecule with a thousand faces, in cancer dissemination, Semin. Cancer Biol, 18: 260–267.

19. Dalit Landesman-Milo, Meir Goldsmith, Shani Leviatan Ben-Arye, Bruria Witenberg, Emily Brown, Sigalit Leibovitch, Shalhevet Azriel, Sarit Tabak, Vered Morad, Dan Peer. (2013). Hyaluronan grafted lipid-based nanoparticles as RNAi carriers for cancer cells, Cancer Letters, 334:221–227

20. Davies A, Gormally J, Wyn-Jones E. (1982). A study of hydration of sodium hyaluronate from compressibility and high precision densitometric measurements. Int J Biol Macromol, 4:436.

21. Dollo G, MalinovskyJ, Peron A, ChevanneF, Pinaud M, Verge R, Corre P.(2004). Prolongation of epidural bupivacaine effects with hyaluronic acid in rabbits. International Journal of Pharmaceutics, 272:109–119.

22. Dráfi F, Valachová K, Hrabárová E, Juránek I, Bauerová K, Šoltés L. (2010). Study of methotrexate and β-alanyl-L-histidine in comparison with L-glutathione on high-molar-mass hyaluronan degradation induced by ascorbate plus Cu (II) ions via rotational viscometry. 60th Pharmacological Days in Hradec Králové. Acta Medica, 53(3): 170.

23. Drobnik J. (1991). Hyaluronan in drug delivery. Adv Drug Dev Rev, 7: 295–308.

24. Elbert D, Pratt A, Lutolf M, Halstenberg S, HubbellJ. (2001). Protein delivery from materials formed by self-selective conjugate addition reactions. Journal of Controlled Release, 76: 11- 25.

25. Esposito E, Menegatti E, Cortesi R. (2005). Hyaluronan-based microspheres as tools for drug delivery: a comparative study. International Journal of Pharmaceutics, 288: 35–49.

26. Fakhari A. (2011). Biomedical Application of Hyaluronic Acid Nanoparticles. PhD. Faculty of the University of Kansas

27. Falcone S, Palmeri D. (2006). Berg R, editors. Biomedical applications of hyaluronic acid, ACS Publications.

28. Garg G, Hales A. (2004). Chemistry and biology of hyaluronan. Elsevier Science.

29. Gibbs DA, Merrill EW, Smith KA, Balazs EA.(1968). Rheology of hyaluronic acid. Biopolymers,6: 777–91.

30. Gribbon P, Heng BC, Hardingham TE.(2000).The analysis of intermolecular interactions in concentrated hyaluronan solutions suggests no evidence for chain–chain association. Biochem J, 350: 329–335.

31. Gurny R, Ibrahim H, Aebi A, Buri P, Wilson CG, Washington N, Edman P, Camber O. (1987). Design and evaluation of controlled release systems for the eye, J. Controlled Release, 6: 367 373.

32. Hirakura T, Yasugi K, Nemoto T, Sato M, Shimoboji T, Aso Y, Morimoto N, Akiyoshi K. (2009). Hybrid hyaluronan hydrogel encapsulating nanogel as a protein nanocarrier: New system for sustained delivery of protein with a chaperone-like function. Journal of Controlled Release.

33. Horvát S, Fehér A, Wolburg H, Sipos P, Veszelka S, Tóth A, Kis L, Kurunczi A, Balogh G, Kürti L, Eros I, Szabó-Révész P, Deli M. (2009). Sodium hyaluronate as a mucoadhesive component in nasal formulation enhances delivery of molecules to brain tissue. European Journal of Pharmaceutics and Biopharmaceutics, 72: 252–259.

34. Hrabárová E, Gemeiner P, Šoltés L. (2007). Peroxynitrite: In vivo and in vitro synthesis and oxidant degradative action on biological systems regarding biomolecular injury and infl ammatory processes. Chem Pap, 61: 417–437.
35. Hrabárová E, Valachová K, Juránek I, Šoltés L.(2012). Free-radical degradation of high-molar-mass hyaluronan induced by ascorbate plus cupric ions: evaluation of antioxidative eff ect of cysteine-derived compounds. Chemistry & Biodiversity, 9: 309–317.
36. Hrabárová E, Valachová K, Rapta P, Šoltés L. (2010). An alternative standard for trolox-equivalent antioxidant-capacity estimation based on thiol antioxidants. Comparative 2,2′-azinobis[3-ethylbenzothiazoline-6-sulfonic acid] decolorization and rotational viscometry study regarding hyaluronan degradation. Chemistry & Biodiversity, 7(9): 2191–2200.
37. Hrabárová E, Valachová K, Rychly J, Rapta P, Sasinková V, Maliková M, Šoltés L. (2009). High-molar-mass hyaluronan degradation by Weissberger's system: Pro- and antioxidative effects of some thiol compounds. Polymer Degradation and Stability, 94: 1867–1875.
38. Hrabárová E, Valachova K, Juránek I, Soltés L. (2012). Free-radical degradation of high-molar-mass hyaluronan induced by ascorbate plus cupric ions: evaluation of antioxidative effect of cysteine-derived compounds. Chemistry & Biodiversity, 9: 309–317.
39. Illum L, Farraj NF, Critchley H, Davis SS. (1988). Nasal administration of gentamicin using a novel microsphere delivery system. Int. J. Pharm., 46: 261–265.
40. Illum L. (2000).Transport of drugs from the nasal cavity to the central nervous system, Eur. J. Pharm. Sci,11: 1–18.
41. Ito T, Iidatanaka N, Niidome T, Kawano T, Kubo K, Yoshikawa K, Sato T, Yang Z, Koyama Y. (2006). Hyaluronic acid and its derivative as a multifunctional gene expression enhancer: Protection from nonspecific interactions, adhesion to targeted cells, and transcriptional activation. Journal of Controlled Release,112: 382–388.
42. Calles JA, Tártara LI, Lopez-García A, Diebold Y, Palma SD, Vallés EM. (2013). Novel bioadhesive hyaluronan–itaconic acid crosslinked films forocular therapy, International Journal of Pharmaceutics, 455: 48– 56
43. Jeong B, Bae Y, Kim S. (2000). Drug release from biodegradable injectable thermosensitive hydrogel of PEG-PLGA-PEG triblock copolymers. Journal of Controlled Release, 63:155–163.
44. Jiang D, Liang J, Noble P. (2011). Hyaluronan as an Immune Regulator in Human Diseases. Physiol Rev, 91: 221–264.
45. Jin Y, Ubonvan T, Kim D. (2010). Hyaluronic Acid in Drug Delivery Systems. Journal of Pharmaceutical Investigation, 40: 33–43.
46. Kalal J, Drobnik J. Rypacek F. (1982). Affinity chromatography and affinity therapy. In:T.C.J. Gribnau, J. Visser and R.J.F. Nivard (Eds.), Affinity Chromatography and Related Techniques, Elsevier, Amsterdam.
47. Kim A, Checkla DM, Chen W. (2003). Characterization of DNA hyaluronan matrix for sustained gene transfer. J Control Release, 90:81–95.
48. Kobayashi Y, Okamoto A, Nishinari K. (1994) Viscoelasticity of hyaluronic-acid with different molecular-weights. Biorheology, 31: 235–244.

49. Kogan G, Soltés L, Stern R, Gemeiner P. (2007a). Hyaluronic acid: A natural bio-polymer with a broad range of biomedical and industrial applications. Biotechnol Lett, 29: 17–25.

50. Kogan G, Soltés L, Stern R, Mendichi R. (2007). Hyaluronic acid: A biopolymer with versatile physicochemical and biological properties. Chapter 31 – in: Handbook of Polymer Research: Monomers, Oligomers, Polymers and Composites. Pethrick R. A, Ballada A, Zaikov G. F. (eds.), Nova Science Publishers, New York, pp. 393–439.

51. Kogan G. (2010). Hyaluronan – A High Molar mass messenger reporting on the status of synovial joints: part 1. Physiological status In: New Steps in Chemical and Biochemical Physics. ISBN: 97 8-1-61668–923 –0. pp. 121–133.

52. Kongtawelert P, Ghosh P. (1989): An enzyme-linked immunosorbent-inhibition as-say for quantitation of hyaluronan (hyaluronic acid) in biological fluids. Anal. Bio-chem, 178: 367–372.

53. Kreil G. (1995). Hyaluronidases–a group of neglected enzymes. Protein Sci 4:1666–9.

54. Kurisawa M, Chung J, Yang Y, Gao S, Uyama H. (2005). Injectable biodegradable hydrogels composed of hyaluronic acid–tyramine conjugates for drug delivery and tissue engineering. Chemical communications.,(34):4312–4.

55. Langer R. (2003). Biomaterials in drug delivery and tissue engineering: one labora-tory's experience. Acc Chem Res, 33:94–101.

56. Lapcík L, De-Smedt S, Demeester J, Chabrecek P, Lapcík L Jr. (1998).Hyaluronan: Preparation, structure, properties, and applications. Chem Rev, 98:2663–84.

57. Laurent TC, Fraser JRE. (1992). Hyaluronan. FASEB J, 6: 2397–2404.

58. Laurent TC, Ryan M, Pictruszkiewicz A. (1960). Fractionation of hyaluronic acid. The polydispersity of hyaluronic acid from the vitreous body. Biochim Biophys Acta, 42: 476–85.

59. Le Bourlais C, Acar L, Zia H, Sado P, Needham T, Leverge R. (1998). Ophthal-mic drug delivery systems–recent advances. Progress in retinal and eye research, 17: 33–58.

60. Lim S, Martin G, Berry D, Brown M. (2000). Preparation and evaluation of the in vitro drug release properties and mucoadhesion of novel microspheres of hyaluronic acid and chitosan. Journal of Controlled Release, 66: 281–292.

61. Lim T, Forbes B, Berry J, Martin G, Brown M. (2002). In vivo evaluation of novel hyaluronan/Chitosan microparticulate delivery systems for the nasal delivery of gen-tamicin in rabbits. International Journal of Pharmaceutics, 231: 73–82.

62. Luo C, Zhao J, Tua M, Zenga R, Rong J. (2014). Hyaluronan microgel as a potential carrier for protein sustained delivery by tailoring the crosslink network. Materials Science and Engineering C, 36: 301–308

63. Maeda H, Seymour L, Miyamoto Y. (1992). Conjugates of anticancer agents and polymers: advantages of macromolecular therapeutics in vivo. Bioconjugate chemis-try, 3: 351–362.

64. Marjorie J. (2011). A review of hyaluronan and its ophthalmic applications, Optom-etry, 82: 38–43.

65. Marshall K. (2000). Intra-articular hyaluronan therapy. Current opinion in rheuma-tology, 12: 468–474.

66. Meyer K, Palmer JW. (1934).The polysaccharide of the vitreous humor. Journal of Biology and Chemistry, 107: 629–634.
67. Milas M, Rinaudo M. (2005). Characterization and properties of hyaluronic acid (hyaluronan). In: Dumitriu S, ed. Polysaccharides Structural Diversity and Functional Versatility. New York, NY: Marcel Dekker 535–49.
68. Morimoto K, Morisaka K, Kamada A. (1985). Enhancement of nasal absorption of insulin and calcitonin using polyacrylic acid gel. J. Pharm. Pharmacol., 37: 134–136.
69. Morris ER, Rees DA, Welsh EJ. (1980). Conformation and dynamic interactions in hyaluronate solutions. J Mol Biol, 138: 383–400.
70. Nancy E, Larsen, Endre A. Balazs.(1991). Drug delivery systems using hyaluronan and its derivatives, Advanced Drug Delivery Reviews, 7:279–293
71. Necas J, Bartosikova L, Brauner P, Kolar J. (2008). Hyaluronic acid (hyaluronan): a review, Veterinarni Medicina, 53(8):397–411.
72. Papakonstantinou E, Roth M, Karakiulakis G. (2012). Hyaluronic acid: A key molecule in skin aging. Dermato-Endocrinology, 4:3, 1–6.
73. Pethrick P, Petkov A, Zlatarov GE, Zaikov S K, Rakovsk. Nova Science Publishers, N.Y, Chapter 7, pp. 113–126.
74. Price R, Berry M, Navsaria H. (2007). Hyaluronic acid: the scientific and clinical evidence. Journal of Plastic, Reconstructive & Aesthetic Surgery, 60: 1110–1119.
75. Prisell P, Camber O, Hiselius J, Norstedt G. (1992). Evaluation of hyaluronan as a vehicle for peptide growth factors. International Journal of Pharmaceutics, 85: 51–56.
76. Pustorino R, Nicosia R, Sessa R. (1996). Effect of bovine serum, Hyaluronic acid and netilmicine on the in vitro adhesion of bacteria isolated from human-worn disposable soft contact lenses. Ann Ig,8:469–75.
77. Rapta P, Valachová K, Zalibera M, Šnirc V, Šoltés L. (2010). Hyaluronan degradation by reactive oxygen species: scavenging eggect of the hexapyridoindole stobadine and two of its derivatives. In Monomers, Oligomers, Polymers, Composites, and Nanocomposites, Ed: R. A.
78. Rapta P, Valachova' K, Gemeiner P, Šoltés L. (2009). High-molar-mass hyaluronan behavior during testing its radical scavenging capacity in organic and aqueous media: Eff ects of the presence of Manganese (II) ions. Chem Biodivers, 6: 162–169.
79. Reháková M, Bakoš D, Soldán M,Vizárová K. (1994).Depolymerization reactions of hyaluronic acid in solution. Int J Biol Macromol, 16:121–4.
80. Robert L, Robert AM, Renard G. (2010). Biological effects of hyaluro-nan in connective tissues, eye, skin, venous wall. Role Aging Pathol.Biol. 58 (3): 187–198,
81. Rydell N, Balazs E. (1971). Effect of intraarticular injection of hyaluronic acid on the clinical symptoms of osteoarthritis and on granulation tissue formation. Clinical Orthopaedics and Related Research, 80: 25–32.
82. Šoltés L, Brezová V, Stankovská M, Kogan G, Gemeiner P.(2006a) Degradation of high molecular-weight hyaluronan by hydrogen peroxide in the presence of cupric ions. Carbohydr Res, 341:639–44.
83. Šoltés L, Lath D, Mendichi R, Bystrický P.(2001). Radical degradation of high molecular weight hyaluronan: Inhibition of the reaction by ibuprofen enantiomers. Meth Find Exp Clin Pharmacol, 23:65–71.

84. Šoltés L, Mendichi R, Kogan G, Schiller J, Stankovska M, Arnhold J. (2006). Degradative action of reactive oxygen species on hyaluronan, Biomacromolecules, 7(3):659–68.

85. Šoltés L, Mislovičová D, Sebille B. (1996) Insight into the distribution of molecular weights and higher-order structure of hyaluronans and some β-(1→3)-glucans by size exclusion chromatography. Biomed Chromatogr, 10:53–9.

86. Šoltés L, Stankovská M, Brezová V, Schiller J, Arnhold J, Kogan G, Gemeiner P.(2006b). Hyaluronan degradation by copper (II) chloride and ascorbate: rotational viscometric, EPR spin-trapping, and MALDI-TOF mass spectrometric investigation. Carbohydr Res, 341:2826–34.

87. Šoltés L, Stankovská M, Kogan G, Gemeiner P, Stern R. (2005). Contribution of oxidative-reductive reactions to high-molecular-weight hyaluronan catabolism. Chem Biodivers, 2:1242–5.

88. Šoltés L, Valachova' K, Mendichi R, Kogan G, Arnhold J, Gemeiner P. (2007). Solution properties of high-molar-mass hyaluronans: the biopolymer degradation by ascorbate. Carbohydr Res, 342: 1071–1077.

89. Stankovská M, Šoltés L, Vikartovská A, Mendichi r, Lath D, Molnarová M, Gemeiner P. (2004). Study of hyaluronan degradation by means of rotational Viscometry: Contribution of the material of viscometer. Chem Pap, 58: 348–352.

90. Stern R, Kogan G, Jedrzejas MJ, ˇSoltés L. (2007). The many ways to cleave hyaluronan. Biotechnology advances, 25 (6):537–57.

91. Stern R, Maibach HI. (2008). Hyaluronan in skin: aspects of aging and its pharmacologic modulation. Clin Dermatol, 26:106–22.

92. Stuart C, Linn G. (1985). Dilute sodium hyaluronate (Healon) in the treatment of ocular surface disorders. Ann Ophthalmol, 17:190–2.

93. Stuart JC, Linn JG.(1985). Dilute sodium hyaluronate (Healon) in the treatment of ocular surface disorders. Ann Ophthalmol,17:190–2.

94. Surovcikova L, Valachová K, Baňasová M, Snirc V, Priesolová E, Nagy M, Juránek I, Šoltés L. (2012). Free-radical degradation of high-molar-mass hyaluronan induced by ascorbate plus cupric ions: Testing of stobadine and its two derivatives in function as antioxidants. General Physiol Biophys, 31: 57–64.

95. Takei Y, Maruyama A, Ferdous A, Nishimura Y, Kawano S, Ikejima K, Okumura S, Asayama S, Nogawa M, Hashimoto M.(2004). Targeted gene delivery to sinusoidal endothelial cells: DNA nanoassociate bearing hyaluronanglycocalyx. The FASAB Journal, 18: 699–701

96. Tamer TM. (2013). Hyaluronan and synovial joint function, distribution and healing. Interdiscip Toxicol Vol. 6(3): 101–115.

97. Tammi R, Ripellino J, Margolis R, Tammi M. (1988). Localizationof epidermal hyaluronic acid using the hyaluronate binding region of cartilage proteoglycan as a specific probe. Journal of Investigative Dermatology, 90: 412–414.

98. Topolska D, Valachova K, Hrabárová E, Rapta P, Banasova M, Juránek I, Soltés L. (2014). Determination of protective properties of Bardejovske Kupele spa curative waters by rotational viscometry and ABTS assay. Balneo Research Journal, 5 (1): 3–15.

99. Turker S, Onur E, Ozer Y. (2004). Nasal route and drug delivery systems. Pharmacy World & Science, 26: 137–142.

100. Ugwoke M, Agu R, Verbeke N, Kinget R. (2005). Nasal mucoadhesive drug delivery: Background, applications, trends and future perspectives. Advanced drug delivery reviews, 57: 1640–1665.

101. Valachova K, Banasova M, Machova L, Juranek I, Bezek S, Soltes L. (2013a). Antioxidant activity of various hexahydropyridoindoles. Journal of Information Intelligence and Knowledge, 5: 15–32.

102. Valachová K, Hrabárová E, Dráfi F, Juránek I, Bauerová K, Priesolová E, Nagy M, Šoltés L. (2010). Ascorbate and Cu(II) induced oxidative degradation of high-molar-mass hyaluronan. Pro- and antioxidative eff ects of some thiols. Neuroendocrinol Lett, 31(2): 101–104.

103. Valachová K, Hrabárová E, Gemeiner P, Šoltés L. (2008). Study of pro and antioxidative properties of d-penicillamine in a system comprizing highmolar-mass hyaluronan, ascorbate, and cupric ions. Neuroendocrinol Lett, 29 (5): 697–701.

104. Valachová K, Hrabárová E, Juránek I, Šoltés L. (2011b). Radical degradation of high-molar-mass hyaluronan induced by Weissberger oxidative system. Testing of thiol compounds in the function of antioxidants. 16th Interdisciplinary Slovak-Czech Toxicological Conference in Prague. Interdiscip Toxicol, 4(2): 65.

105. Valachová K, Hrabárová E, Priesolová E, Nagy M, Baňasová M, Juránek I, Šoltés L. (2011). Free-radical degradation of high-molecular-weight hyaluronan induced by ascorbate plus cupric ions. Testing of bucillamine and its SA981-metabolite as antioxidants. J Pharma & Biomedical Analysis, 56: 664–670.

106. Valachová K, Kogan G, Gemeiner P, Šoltés L. (2008a). Hyaluronan degradation by ascorbate: Protective effects of manganese (II). Cellulose Chem. Technol, 42(9–10): 473−483.

107. Valachová K, Kogan G, Gemeiner P, Šoltés L. (2009a). Hyaluronan degradation by ascorbate: protective effects of manganese (II) chloride. In: Progress in Chemistry and Biochemistry. Kinetics, Thermodynamics, Synthesis, Properties and Application, Nova Science Publishers, N.Y, Chapter 20, pp. 201–215.

108. Valachová K, Mendichi R, Šoltés L. (2010b). Effect of L-glutathione on high-molar-mass hyaluronan degradation by oxidative system Cu(II) plus ascorbate. In: Monomers, Oligomers, Polymers, Composites, and Nanocomposites, Ed: R. A. Pethrick P. Petkov, A. Zlatarov G. E. Zaikov, S. K. Rakovsky, Nova Science Publishers, N.Y, Chapter 6, pp. 101–111.

109. Valachová K, Rapta P, Kogan G, Hrabárová E, Gemeiner P, Šoltés L. (2009). Degradation of high-molar-mass hyaluronan by ascorbate plus cupric ions: eff ects of D-penicillamine addition. Chem Biodivers 6: 389–395.

110. Valachová K, Rapta P, Slováková M, Priesolová E, Nagy M, Mislovičová D, Dráfi F, Bauerová K, Šoltés L. (2013). Radical degradation of high-molar-mass hyaluronan induced by ascorbate plus cupric ions. Testing of arbutin in the function of antioxidant. In: Advances in Kinetics and Mechanism of Chemical Reactions, G. E. Zaikov, A. J. M. Valente, A. L. Iordanskii (eds), Apple Academic Press, Waretown, NJ, USA, pp. 1–19.

111. Valachová K, Šoltés L. (2010a). Effects of biogenic transition metal ions Zn(II) and Mn(II) on hyaluronan degradation by action of ascorbate plus Cu(II) ions. In: New Steps in Chemical and Biochemical Physics. Pure and Applied Science, Nova Sci-

ence Publishers, Ed: E. M. Pearce, G. Kirshenbaum, G.E. Zaikov, Nova Science Publishers, N.Y, Chapter 10, pp. 153–160

112. Valachová K, Vargová A, Rapta P, Hrabárová E, Drafi F, Bauerová K, Juránek I, Šoltés L. (2011a). Aurothiomalate as preventive and chain-breaking antioxidant in radical degradation of high-molar-mass hyaluronan. Chemistry & Biodiversity 8: 1274–1283.

113. Van Beek M, Jones L, Sheardown H. (2008) Hyaluronic acid containing hydrogels for the reduction of protein adsorption. Biomaterials, 29:780–9.

114. Volpi N, Schiller J, Stern R, Soltés L. (2009). Role, metabolism, chemical modifications and applications of hyaluronan. Current medicinal chemistry, 16(14):1718–45.

115. Yang H. Yuna, Douglas J. Goetzb, Paige Yellena, Weiliam Chen, (2004) Hyaluronan microspheres for sustained gene delivery and site-specific targeting, Biomaterials, 25: 147–157

116. Yumei Xie, Kristin L Aillon, Shuang Cai, Jason M. Christian, Neal M. Davies, Cory J. Berkland, M. Laird Forrest.(2010). Pulmonary delivery of cisplatin–hyaluronan conjugates via endotracheal instillation for the treatment of lung cancer, International Journal of Pharmaceutics, 392:156–163

117. Yun YH, Goetz DJ, Yellen P, Chen W. (2004). Hyaluronan microspheres for sustained gene delivery and site-specific targeting. Biomaterials, 25: 147–157.

CHAPTER 9

OIL NEUTRALIZATION IN THE TURBULENT APPARATUS CONFUSER-DIFFUSER DESIGN OF CRUDE OIL MANUFACTURE

F. B. SHEVLYAKOV,[1] T. G. UMERGALIN,[1] V. P. ZAKHAROV,[2]
O. A. BAULIN,[1] D. A. CHUVASHOV,[2] N. I. VATIN,[3]
and G. E. ZAIKOV[4]

[1]*Ufa State Petroleum Technical University, 1 Kosmonavtov St. Ufa, 45006, Russia*

[2]*Bashkir State University, 32, Validy Str., Ufa, 450076, Ufa, Russia; E-mail: ZaharovVP@mail.ru*

[3]*Civil Engineering Institute Saint-Petersburg State Polytechnical University, 29 Polytechnicheskaya street, Saint-Petersburg 195251 Russia; E-mail: vatin@mail.ru*

[4]*Institute of Biochemical Physics, Russian Academy of Sciences, 4 Kosygin str., Moscow 119334 Russia; E-mail: Chembio@sky.chph.ras.ru*

CONTENTS

ABSTRACT

The regularities of crude oil preparation, particularly its neutralization by an alkali, is been considered in current study. In order to neutralize the crude oil at electric desaltation stage, it is proposed to use the high-performance compact tubular turbulent apparatus of confusor-diffuser design.

9.1 INTRODUCTION

Hydrolysis of inorganic salts which runs on electric desalting plant (EDP) is accompanied by the formation of an acidic environment. It is used to conduct the neutralization of such crude oil by alkalis as well as by organic amines. Because of the difference of the density and viscosity of the oil and alkaline solutions, the neutralization occurs not equimolarly under diffusion control and is accompanied by increase consumption rates of the neutralizing agent. The excess amount of the lye supplied to neutralize is reflected on the thermal oil-processing, reducing the activity of cracking catalysts. Reducing the impact of excess alkali can be achieved by intensification of mixing alkali point of application to the flow of oil.

It is known way of oil neutralization by aqueous alkaline solution, which for the dispersion of aqueous alkali proposes, in particular, carry out their partial premixing with oil (1%) [1], or use a different nozzle design [2]. Experience has shown that the effectiveness of both methods is low. In addition, they are characterized by disadvantages associated with the need to service the mixers and control when the flow of oil and/or the lye is changeable.

Another known on the electric desalting plants method [3] uses mixing valves, which process is working at high pressures drops across the valve, which is associated with significant energy costs to provide the required performance EDP on oil, moreover, the hydraulic losses will increase significantly during the works with heavy and viscous oils.

A more effective way is to neutralize the oil on electric desalting installation with prior addition of de-emulsifier, which produces a mixture of oil with a soda-alkaline solution in static mixer type like Sulzer SMV Hemiteh [4, 5]. Disadvantage of this method is the complexity of device design, large quantity of metal usage, high pressures drops of at high flows of oil.

The purpose of this study was developing a compact mixing reactor design, which can reduce the working pressure drop during the neutralization and amount of the alkaline agent required.

In order to solve the problem indicated, the oil neutralization was carried out in a turbulent reactor confusor-diffuser design [6]. In this case, the oil prepared and preheated to 110–120°C enters the inlet of the first section of a turbulent tubular reactor confusor-diffuser design with a flow rate of 680–750 m³/h, where the dispersion of the two-phase system is occurring. Later, alkaline solution 1–2% by wt. is coaxially inserted in the first section of the reactor-mixer through the frontal atomizers. This allows improve the efficiency of oil neutralization due to a significant reduction of fresh and spent volumes of lye. The advantages of the device are absence of mixing devices, low pressure drop, and low quantity of metal usage.

Due to possibility to increase the quality of alkali and oil mixing via fine dispersion and uniform distribution of the alkali in the entire volume of oil, the conditions for the creation of a homogeneous emulsion-phase model of system "liquid–liquid" in tubular turbulent apparatus were studied [7, 8].

Droplets distribution of the dispersed phase in size to the formation of fine homogeneous systems in the confusor-diffuser channels is narrowed by increasing speed of immiscible fluid streams. Increase in volumetric flow velocity ω and the number of diffuser confused N_C sections 1–4 leads to reduction of the volume-surface diameter of droplets of the dispersed phase and, consequently, to increase in the specific surface of the interface, which in the case of fast chemical reactions intensify the total process. Inadvisability of using the apparatus with the number of diffuser sections N_C confused over 5 ± 1, making these devices simple and inexpensive to manufacture and operate as well as compact, for example, length does not exceed 8–10 caliber (L/d_D).

There is a range of volume velocity of two-phase flow, which corresponds to the cone-channel confused with optimal diameter of the diffuser to confuser (further indicated as d_D/d_C). The distance is limited from bottom by seating stratified two-phase flow, and is limited from top by energy costs arising from the increased pressure on the ends of the device ($\Delta p \sim w^2$). In particular, the ratio $d_D/d_C = 3$ corresponds to the interval 44 < w < 80 cm³/s, and $d_D/d_C = 1.6$ corresponds to the interval 80 < w < 180 cm³/s, and further increase in the velocity of the dispersed system (w > 180 cm³/s) determines the need to further reduce the ratio d_D/d_C until $d_D/d_C = 1$, that is, small units cylindrical structure are effective enough in this case. Thus, the flow, in which the dispersed particles are uniformly dispersed in the unit of confusor-diffuser design in comparison with the cylindrical channel, is formed at the lower velocities of the dispersed system, and the higher the ratio d_D/d_C, the lower the required value W (due to changing the value of the Reynolds number Re according to the ratio Re ≈ d_D/d_C).

Thus, the change in the rate of fluid flow in tube W devices and relation d_D/d_C is almost the only, but very effective way to affect the nature of the dispersion and the quality of the emulsion. These patterns of relationships allow under optimal conditions and without nontechnical or technical problems create thin homogeneous dispersion systems "liquid-liquid" with a minimum residence time of the reactants in the mixing zone, and use simple apparatus to design small confusor-diffuser design.

Another important quantity that characterizes the quality of the emulsion is the polydispersity coefficient k. Ratio L_c/d_D almost no effects on the polydispersity of emulsions obtained. The increase in the spread of the dispersed phase in size is observed during increasing of the ratio d_D/d_C, and quite homogeneous emulsion is formed in the diffuser confused channels tubular device with $d_D/d_C = 1.6$. In particular, the value of k in $d_D/d_C = 1.6$ for $L_c/d_D = 2$–3 equals 0.72–0.75, whereas the k is reduced to 0.63 and 0.41 when the ratio d_D/d_C is 2 and 3, respectively.

Creation of intensive longitudinal mixing in a two-phase system in tubular turbulent apparatus with the ability to increase the surface of contact between the phases allows intensify the flow of fast chemical reactions at the interface.

The dependences obtained allow predict the dispersion of droplets of alkali in oil, which makes it possible to design a mixer for use in a wide range of flow rates of mixed liquids.

The process at low differential pressure is necessary to carry out for effective mixing of oil-base, which is directly related to the energy needed to provide the performance ELOU required. However, the hydraulic losses increase significantly when working with heavy and viscous oil.

The pressure drop is expressed by the relationship

$$\Delta P = (\lambda(L/d) + \zeta))(\rho w/2)$$

where ζ is a coefficient of local resistance, λ is a friction coefficient, L is a length, d is diameter, ρ is a density, w is a speed.

The coefficient of local resistance to the unit area with sudden expansion is calculated (in the calculation of the velocity head speed in a smaller cross section) by the formula $(1 - (S_1/S_2))^2$ (Fig. 9.1), and cylindrical portion of the apparatus for $\zeta = 1$, while coefficient of local resistance to the plot device of a sudden contraction (in the calculation of the velocity head speed in a smaller section) $\zeta = 0.38$. The values of friction coefficient for turbulent flow can be calculated by the Blasius formula: $\lambda = (0.316/Re^{0.25})$.

The pressure drop in the section is the sum of the pressure drop in a smooth tube, expansion (diffuser) and narrowing (confuser) (Fig. 9.2)

$$\Delta P = (P_1 - P_2) + (P_2 - P_3) + (P_3 - P_4)$$

The total pressure drop is the sum of pressure drops in each section. Calculation by these formulas was done according to experimental data

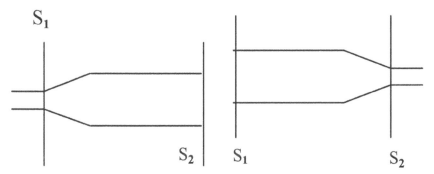

FIGURE 9.1   Scheme for calculation of coefficient of local resistance.

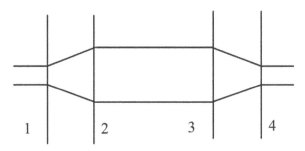

FIGURE 9.2   Scheme for calculation of pressure drop in tube.

of measuring the pressure at the ends of tubular turbulent apparatus consisting of 20 sections with a water flow. Comparison of calculated data obtained with respect to the model system shows correlations with the experimental data for the pressure in the apparatus: $\Delta P_{practical} = 0.955$ atm, $\Delta P_{theoretical} = 1.062$ atm. Calculation of diameter of the narrow section (confuser) section on the proposed formulas, based on the requirements for the pressure drop in the apparatus $\Delta P \leq 0.6$ atm, was done (Table 9.1).

The pressure drop at the ends of the device with a diameter of confuser $d_C = 0.2$ m is $\Delta P_{5unit} \approx 0.52$ atm, which is optimal for steel neutralization of oil.

Sharp rise in temperature is observing while using of concentrated solutions of the reagents during the neutralization of acidic environments.

TABLE 9.1   Calculation of the Diameter of the Diffuser at D_C for Oil in the Apparatus of Confusor-Diffuser Design

$\Delta P_{5\ section}$	0.6	atm
$\Delta P_{section}$	0.118	atm
ΔP_{3-4}	6643	kgF/m²
ΔP_{2-3}	51.1	KgF/m²
ΔP_{1-2}	7875	KgF/m²
ζ_C	0.38	
ζ	0	
ζ_D	0.46	
ρ	762	Kg/m³
L	0.875	m
λ_C	0.0087	
λ_D	0.0115	
d_D	0.35	m
Re	177×10^4	
Re_D	577×10^3	
w_C	6.62	m/s
W	2.163	m/s
S_D	0.096	m²
S_C	0.031	m²
d_C	0.198	m

In this case, the small tubular turbulent reactors confusor diffuser designs define the ability to effectively regulate the temperature field in the reaction zone in several variants: the radius of the apparatus and the speed of the flow of the reactants, the use of the band model of a rapid chemical process and the use of shell and tube apparatus with a bundle of small-radius intensification of convective heat transfer at profiling apparatus.

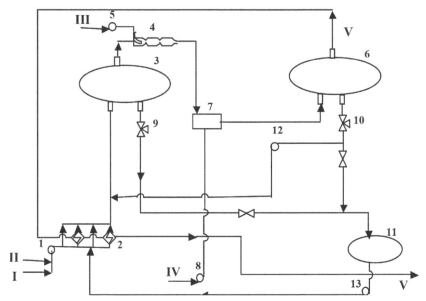

FIGURE 9.3 Scheme of the site electric desalting oil. 1, 5, 8, 12, 13 – pumps; 2 – heat exchanger; 3, 6 – electric dehydrators, 4 – turbulent tubular reactor, 7 – diaphragm mixer, 9, 10 – valve automatically reset the salt water, 11 – the sump.

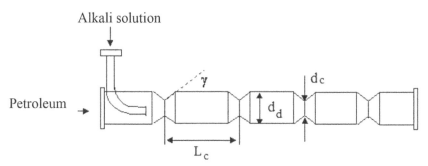

FIGURE 9.4 General view of the tubular turbulent apparatus for neutralization of the petroleum with alkali.

The process of neutralizing the oil in accordance with the proposed method is following (Fig. 9.3). The main flow of commercial oil from the pipeline (I) is mixed with the de-emulsifier (II), the pump (1) is directed to the heat exchanger (2), where it is heated to 110–120°C. The oil with de-emulsifier comes to the first stage of separation in electric dehydrators E1 (3). The oil from electric dehydrators (3) from top comes with the flow rate 680- 750 m³/h at the inlet of the first section of a turbulent tubular reactor (4). Dispersion occurs in the five sections of the tubular turbulent reactor (Fig. 9.4), which is less than 4 meters with a pressure drop at the ends of the device to 0.52 atm. Aqueous alkaline solution (III) by pump (5) is sent to the coaxial connector of the first section of a turbulent tubular reactor (4) confusor-diffuser design with end nozzles (Fig. 9.5). Pipe is perforated

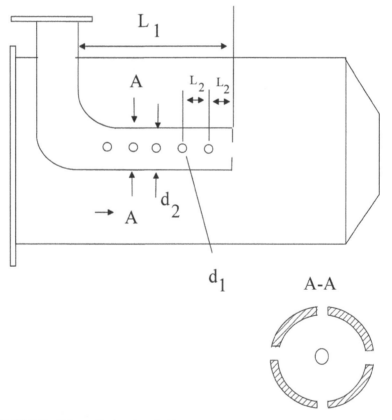

FIGURE 9.5 Scheme for input socket bases.

by 20-one hole with a diameter $d_1 = 5$ mm, where 20 holes are in the walls for the radial outlet to the flow of oil supply bases, and closed front end of the pipe is perforated by hole coaxial with the direction of oil entering the solution of a neutralizing agent. Perforations are arranged symmetrically on the cross section (four holes on one section A-A). Partially dehydrated and desalted oil comes under pressure in the second stage of electric dehydrators E2 (6). Before this, electric dehydrators oil mixed in the diaphragm mixer (7) with preheated to 65–70°C pumped (8) fresh water (IV). Electric dehydrators E1 and E2 (3 and 6) by automatic reset valve saltwater (9) and (10) disperse the water to the sump (11). The extracted water streams containing oil are received for recycling by pumps (12) and (13). Desalted and dehydrated oil from the top V electric dehydrators E2 (6) – play with the installation.

9.2 CONCLUSIONS

1. Tubular turbulent apparatus confusor-diffuser design allows for effective neutralization of aqueous alkali oil and organic amines in equimolar ratio.
2. The proposed low metal device confusor-diffuser design determines the differential pressure at the ends of the device of the five sections of no more than 0.52 atmospheres and is installed as part of the pipeline flow of oil on the node CDU.

This work was supported by the grant of the President of Russian Federation MD-3178.2011.8, RFBR (№ 11–03–97017).

KEYWORDS

- **confusor-diffuser design**
- **crude oil preparation**
- **electric desalting plant**
- **oil neutralization**
- **tubular turbulent apparatus**

REFERENCES

1. V.F. Sorochenko, A.P. Shutko, N. Pavlenko, T.P. Bukolova. The effectiveness of corrosion inhibitors in water recycling systems. Chemistry and technology of fuels and oils. 7, 37 (1984).
2. G.A.Yushmanov, N. Starostin, V.G. Dyakov: Current status of anticorrosion protection techniques and material selection for equipment installation training of primary oil refining. CNIIT petrochemestry, Moscow, 1985.
3. N.V. Bergstein, F.M. Khutoryansky, D.N. Levchenko. Improving the process for desalting EDU. Chemistry and technology of fuels and oils. 1, 8 (1983).
4. F.M. Khutoryansky, G.D. Zalischevsky, N.A. Voronin, G.M. Urivskii. The use of a static mixer for enhanced mixing desalinated oil with an aqueous solution of alkali. Refining and neftehimiya, 1, 11, (2005).
5. G.D. Zalischevsky, F.M. Khutoryansky, O.M.Warsaw, G.M. URIVSKII, N.A. Voronin. Pilot plant evaluation of the static mixer SMV type of firm "Sulzer Hemiteh" with desalting by CDU. Refining and Petrochemicals, 5, 16, (2000).
6. V.P. Zakharov, F.B. Shevlyakov Longitudinal mixing in the flow of fast liquid-phase chemical reactions in the two-phase mixture. Journal of Applied Chemistry, 79 (3), 410, (2006).
7. V.P. Zakharov, A.G. Mukhametzyanova, R.G. Takhavutdinov, G.S. Dyakonov, K.S. Minsker. Creating homogeneous emulsions tubular turbulent apparatus diffuser confusor structure. Journal of Applied Chemistry, 75 (9), 1462, (2002).
8. K.S. Minsker, V.P. Zakharov, R.G. Takhavutdinov, G.S. Dyakonov, A.A. Berlin. Increase in the coefficient of turbulent diffusion in the reaction zone as a way to improve the technical and economic performance in the production of polymers. Journal of Applied Chemistry, 74 (1), 87, (2001).

CHAPTER 10

STRUCTURE-PROPERTY CORRELATION AND FORECAST OF CORROSION OF THE ALKENYLARILAMINES ACTIVITY WITH THE HELP OF DENSITY FUNCTIONAL THEORY

R. N. KHUSNITDINOV, S. L. KHURSAN, K. R. KHUSNITDINOV, A. G. MOUSTAFIN, and I. B. ABDRAKHMANOV

Establishment of Russian Science Academy Institute of Organic Chemistry of Ufa Science Center of RSA, 450054,71 Prospect Octyabrya str, Ufa, Russian Federation; Tel.: + (347)235-38-15; E-mail: Chemhet@anrb.ru

CONTENTS

ABSTRACT

Correlation between inhibitory property 1-methyl-2-butenyl-aniline (MBA) and the electronic structure was investigated using such indices as the reactivity of higher energy occupied by molecular orbitals of lower energy free of molecular orbitals, the negative charge on the nitrogen atom, and the dipole moment of the electrophilicity index. The best correlation was observed between the braking rate of corrosion and the negative charge on the nitrogen atom. Using the regression equation and the forecast of anticorrosive properties of the MBA series, we synthesized a series of compounds with high inhibitory activity.

10.1　INTRODUCTION

Correlation study "structure–anticorrosive activity" number of papers [1–3]. Currently, the use of quantum chemical methods to study the mechanism and prediction of corrosion inhibition is widely used [4–25]. Of particular interest in recent years is the use of density functional theory for the prediction of inhibitory activity of various organic compounds [26, 41].

Alkene-anilines have emerged as potential corrosion inhibitors, easily accessible through the condensation reaction of anilines with 1,4-adducts hydrohalogenation pentadiene or cyclopentadiene [42]. A review of published data showed that the investigation of the dependence of the inhibitory activity of the structural parameters in the series of these compounds was carried out.

The purpose of our research is to study the relationship between the electronic structure of the molecules and their inhibitory activity in a series of derivatives of aniline, N-and C-substituted mono-or poly pentadiene-anilines and identification of optimal structures with the maximum inhibitory efficacy.

10.2　MATERIALS AND THE METHODOLOGY OF THE EXPERIMENT

Pentadiene-arylamines were synthesized by known methods, published in Ref. [42]. In addition to the known compounds new compounds synthesized spectral characteristics are listed below:

2,4,6-tri-(1'-methylbutyl) aniline (**16**) was obtained by hydrogenating compound (19) with Raney nickel in an alcoholic solvent. nd20 = 1.5856.

Found, %: C 83.05; H 12.24; N 4.60. C21H37N. Calculated (%): C 83.10; H 12.29; N 4.61. IR spectrum (ν, см-1): 3420, 3490 (NH2). 1H NMR spectrum (CDCl3, δ, J/Hz): 0.72–0.97 (м, 9H, 3CH3); 1.17–1.28 (м, 9H, CH3); 1.45–1.55 (м, 6H, 3CH2); 1.64–1.75 (м, 6H, 3CH2); 2.72–2.80 (3H, 3CH); 3.56 (уш.с., 2H, NH2); 6.77 (с, 2H, ArH). NMR spectrum 13C: (CDCl3, δ): 14.4 (3CH3); 20.7 (3CH3); 21.8 (3CH2); 33.2 (3CH2); 40.0 (3CH); 121.7, 132.0, 137.4, 138.6 (C-аром.).

2,6-di-(3-methylbutyl)-4-(1'-methylbutyl) aniline (17) was prepared by alkylation by Friedel-Crafts reaction of 4-(1-methylbutyl) aniline, prepared according to known methods [42] izoamilhloridom in the presence A1C13. Found, %: C 83.05; H 12.24; N 4.60. C21H37N. Calculated (%): C 83.10; H 12.29; N 4.61. IR spectrum (ν, см-1): 3425, 3493 (NH2). 1H (CDCl3, δ, J/Hz): 0.88–0.89 (м, 12H, 4CH3); 0.96–0.98 (м, 3H, CH3); 1.16–1.18 (м, 3H, CH3); 1.23–1.48 (м, 8H, 4CH2); 1.58 (м, CH, 2H); 2.20 – 2.24 (м, 4H, CH2); 2.59 (1H, CH); 4.38 (уш.с., 2H, NH2). NMR Spectrum 13C: (CDCl3, δ): 14.19 (CH3); 20.7 (CH2); 22.67 (CH3); 27.90 (2CH); 28.36 (2CH2); 28.70 (CH2); 41.34 (2CH2); 43.90 (CH); 128.09, 132.16, 144.86, 146.11 (C-аром.).

2,6-di-(1-methyl-2-buten-1-yl)-4-methoxy-aniline (20) was obtained by a known method [42] reacting methoxyaniline with 4-chloro-2-pentene at a temperature of 150°C. The compound is a viscous brown mass. T. heated. 143°C (3 mm Hg. Tbsp.). Found (%): C 78.70; H 9.70; N 5.40. C17H25NO. Calculated (%): C 78.70; H 9.71; N 5.40. IR spectrum (ν, см-1): 3433, 3367 (NH2). NMR spectrum 1H (CDCl3, δ): 1.28 (м, 6H, 2CH3, J = 6.86 Hz); 1.69 (м, 6H, 2CH3, J 5.69 Hz); 3.46 (м, 2H, 2CH); 3.49 (уш. с. 2H, NH2); 3.77 (с, 3H, OCH3); 5.52–5.56 (м, 4H,CH = CH); 6.65 (с, 2H, Ar-H). NMR spectrum 13C: (CDCl3, δ): 17.8 (2CH3); 19.7 (2CH3); 30.0 (2CH); 55.5 (OCH3); 110.4 (C2,' C2''); 134.8 (C3, C5); 132.3 (C3,' C3''); 124.3, 135.6, 152.5 (C- arom.)

2,6-di-(1-methyl-2-buten-1-yl)-4-ethoxyaniline (21). Prepared analogously to compound (20) from ethoxyaniline and 4-chloro-2-pentene at a temperature of 150°C. Compound (21) is a viscous dark-brown mass. B.p.. 151°C (3 mm Hg. Tbsp.). Found (%): C 79.00; H 9.90; N 5.10. C18H27NO. Calculated: C 79.07; H 9.95; N 5.12. IR spectrum (ν, см-1): 3433, 3372, 1287 (NH2); 1603, 970 (CH = CH); NMR spectrum 1H (CDCl3, δ): 1.28 (м, 6H, 2CH3, J 6.8 Hz); 1.41 (т, CH3, 3H, J 14.0 Hz);

1.70 (м, 6H, 2CH3, J 5.7 Hz); 3.43 (уш.с. 2H, NH2); 3.99 (м, 2H, CH); 4.01 (м, 2H, CH2); 5.53–5.55 (4H, CH = CH); 6.65 (с, 2H, Ar-H). NMR spectrum 13C: (CDCl3, δ): 15.0 (CH3); 17.8 (2CH3); 19.8 (2CH3); 29.7 (2CH); 63.7 (CH2); 111.2 (C2,' C2"); 133.2 (C3, C5); 135.6 (C3,' C3"); 14.2, 134.7, 157.8 C-arom.)

Quantum-chemical parameters (energy of the highest occupied molecular orbital (EHOMO, AU), the energy of the lowest unoccupied molecular orbitals (ELUMO, AU), the negative charge on the nitrogen atom (Qmin, AU), electrophilicity index (W) and the dipole moment (μ, a)) are calculated using the FIREFLY (PC GAMESS 7.15) [43] in the approximation V3LYP/6 31G (d, p) [44, 45]. Visualization and primary processing of results of calculation was performed using the ChemCraft 1.5 [46].

All structures calculated in this study were subjected to the optimization procedure and are stationary points on the potential energy surface (PES), which is proven by solution of a vibrational problem: for minima on the PES diagonalized Hessian matrix contains only positive terms.

Anticorrosion studies were performed in accordance with industry standard (OST) OST 39–099–79 [47]. As a corrosive hydrochloric acid solutions were used, and as the specimens of the plate steel grade 3.

10.3   RESULTS AND DISCUSSION

With the purpose of obtaining corrosive models adequately responding to changing the structure of selected compounds (introduction of an alkyl substituent, change in position of the double bond, etc.) we performed studies of the effect of acid concentration on their inhibitory activity. Corrosive environments using solutions of hydrochloric acid witness samples of steel grade-20. Preliminary corrosion studies revealed that the optimal concentration of hydrochloric acid meets the required specification, selection of compounds is a 14% solution of hydrochloric acid. At low concentrations of hydrochloric acid (10–12%) poor reproducibility of the experiment is observed, at high concentrations of hydrochloric acid corrosion system (20–22% r HCl, samples of steel grade 20, temperature 25°C) does not respond to changes in the structure of the simplest aromatic amines (aniline, toluidines, xylidines).

As a result of preliminary investigations there were selected corrosion system satisfying the above requirements, the solution is 14% hydrochloric acid, the samples of steel grade 3, dosages of compounds 0.033 mol/L and the temperature 25°C. In order to obtain primary information we tried to find correlation analysis of inhibitory activity in the series of elementary arylamines (aniline, toluidines, xylidines) as a function of the calculated parameters of the above molecules. Table 10.1 shows the coefficients for the corrosion inhibition of these compounds and various reaction indices, and Table 10.2 shows the binding of the regression equation and correlation coefficients.

Analysis of the information showed that the trend for corrosion activity arylamines observed for parameters, such as, the negative charge on the nitrogen atom, the energy of the highest molecular orbitals, electrophilicity index and the dipole moment for the last two indices is less pronounced.

It should be noted that o-substituted aryl amines deviate from the correlation, and p-substituted aryl amines have maximum corrosion activity. We investigated the properties selected pentadien-arylamine mono-substituted pentadiene-arylamines and synthesized by known methods [42]. Based on the findings obtained for the simplest of arylamines, we stopped for corrosion activity indices such as the negative charge on the nitrogen atom, the electrophilicity index and the dipole moment. Table 10.3 shows calculations by the electronic properties of the mono-pentadien-arylamine, and certain compounds for the inhibition of corrosion rates.

Table 10.2 shows the regression equations and correlation coefficients for the sample compounds.

TABLE 10.1 Quantum-Chemical Parameters and Coefficients of Simple Arylamines Corrosion Inhibition

№	Coefficients	E_{HOMO}	E_{LUMO}	W	Q_{min}	μ	γ
I	Aniline	−0.1981	0.0086	0.0217	−0.319	1.710	1.29
II	o-Toluidine	−0.1956	0.0130	0.0200	−0.316	1.747	1.38
III	м-Toluidine	−0.1959	0.0089	0.0213	−0.657	1.617	1.51
IV	p-Toluidine	−0.1920	0.0093	0.0207	−0.657	1.527	1.69
V	2,5-Xilydine	−0.1922	0.0171	0.0183	−0.663	1.568	1.75
VI	2,4-Xylidine	−0.1897	0.0135	0.0191	−0.663	1.557	1.82

TABLE 10.2   Results of the Regressive Analyzes (Tables 11.1–11.3)

№	Index	Options	R^2	Regression equation
1	E_{HOMO}	1–6	0.93	$\gamma = 15.03 -$
		1, 2, 4–6	0.95	$69.46 \times E_{HOMO}$
1	E_{LUMO}	1–6	0.28	$\gamma = 1.15 + 40.03 \times E_{LU}$
		1, 3–6	0.48	$_{MO}$
1	W	1–6	0.52	$\gamma = 4.1 - 122.8W$
		1, 3–6	0.70	
		1, 3, 4, 6	0.91	
1	μ	1–6	0.81	$\gamma = 5.64 - 2.5\mu$
		1–3, 5, 6	0.86	
1	Q_{min}	1–6	0.76	$\gamma = 0.95 - 1.22Q_{min}$
		1–2, 4–6	0.95	
2	E_{HOMO}	7–15	0.27	$\gamma = -561.4 -$
		7–12, 14, 15	0.76	$3033.8 \times E_{HOMO}$
		7–10, 12, 15	0.89	
2	W	7–15	0.04	$\gamma = -108.7 - 6076.2W$
		7–13, 15	0.33	
		7–12, 15	0.59	
		7–11, 15	0.76	
2	μ	7–15	0.40	$\gamma = 91.3 - 47.9\mu$
		7–10, 12, 13, 15	0.50	
		7–10, 13,15	0.56	
2	Q_{min}	7–15	0.47	$\gamma = -232.22$
		7–11, 13, 15	0.91	$-392.34Q_{min}$
		7, 8, 10, 11, 13, 15	0.95	
		7, 8, 10, 13, 15	0.97	
2, 3	Q_{min}	7–21	0.86	$\gamma = -173.4 - 298.7Q_{min}$
		7, 8, 10, 11, 13, 15–21	0.97	

Analysis of the results showed that the best trend in the number of parameters is the negative charge on the nitrogen atom. For the entire sample (**7–15**), the correlation coefficient R2 is 0.47. With the exception of ortho-substituted-pentadien arylamines of the total sample (**12, 14**) there is an

TABLE 10.3 Quantum Chemical Indices of Pentadien-Arylamines and Corrosion Inhibition Factors

№	Coefficients	E_{HOMO}	E_{LUMO}	W	Q_{min}	μ	γ	$\gamma_{pacч}$
1	N-(1-methyl-2-buthenyl) anilyne	0.1869	0.0124	0.0191	0.600	1.852	2.02	-
2	o-(1-methyl-2-buthenyl) anilyne	0.1869	0.0124	0.0191	0.600	1.852	2.29	-
3	N-(1-methyl-2-buthenyl) metaanilyne	0.185	0.0165	0.0176	0.610	1.916	2.8	-
4	N-(1-methyl-2-buthenyl)3 metaanilyne	0.1852	0.0137	0.0185	0.601	1.662	3.0	-
5	N-(1-methyl-2-buthenyl)4 meta anilyne	0.1819	0.0128	0.0184	0.600	1.552	6.0	5.82
6	N-(1-methyl-2-buthenyl)4 metaanilyne	0.1879	0.0069	0.021	0.667	1.616	8.1	-
7	N-(1-methyl-2-buthenyl)2–4 demetaanilyne	0.1802	0.0164	0.0171	0.610	1.627	10.3	8.81
8	o-(1-methyl-2-buthenyl) demetaanilyne	0.1865	0.0113	0.0145	0.676	1.505	13.0	-
9	P-(1-methyl-2-buthenyl) anilyne	0.1927	0.0075	0.0214	0.657	1.576	25.0	22.8

increase of the correlation coefficient to 0.95, and to the complete exclusion of the total sample ortho-substituted pentadien-arylamine (**9, 12** and **14**) R^2 reaches 0.97. This is consistent with the phenomenon known as "ortho-effect" [1] associated with the influence of steric bulk ortho-substituents in these compounds.

Correlations of energy of the highest occupied molecular orbitals, dipole moment and the electrophilicity index is less pronounced compared to the minimum negative charge, for which the correlation coefficients are within 0.56–0.89.

Proceeding from these considerations, we selected the most reliable regression equation with the maximum correlation coefficient: γ = −232.22–392.34 h Q min, R2 = 0.97.

Using the regression equation for the minimum negative charge we came to the forecast of the anticorrosive properties of some compounds. Based on this prediction we synthesized 2,4,6 tri-(1' methylbutyl) aniline (16), 2,6 di-(3' methylbutyl)-4 (1' methylbutyl) aniline (17), 2,6 di-(1' methyl- 2' butenyl)-4 methylaniline (18) and three 2,4,6-(1' methyl- 2' butenyl)-aniline (19), 2,6 di-(1' methyl-2' butenyl)-4 methoxyaniline (20), 2,6 di-(1' methyl-2' butenyl)-4 ethoxyaniline (21). Compounds (18–19) have been previously described in the literature [42] and compound (16,17,20,21), first synthesized by us.

Table 10.4 shows the electronic codes coefficients of corrosion inhibition effect of the above compounds.

Using the regression equation for the negative charge on the nitrogen atom we calculated values of the coefficients of the compounds studied (7–21). As seen from Tables 10.2 and 10.3, the experimental and calculated

TABLE 10.4   Quantum Chemical Indices of Synthesized Pentenilarilamines and Corrosion Inhibition Factors

№	Coefficients	E_{HOMO}	E_{LUMO}	W	Q_{min}	μ	γ	γ_{pac4}
1	Three-2,4,6-(1'-methylbuthyl)anilyne	0.1882	0.0118	0.0194	0.666	1.428	28.0	30.8
2	De-2,6-(3'-methyl-buthyl)-4-(1'-methyl-buthyl)anilyne	0.1877	0.0139	0.0187	0.672	1.325	29.1	31.2
3	De-2,6-(1'-meth-yl-2'-buthrnyl)-4-methylanilyne	0.1837	0.0063	0.0207	0.678	1.689	31.9	34.45
4	Three-2,4,6-(1'-methyl-2'-buthenyl)-anilyne	0.1858	0.0057	0.0212	0.678	1.898	35.0	34.45
5	De-2,6-(1'-meth-yl-2'-buthenyl)-4-metoxenianilyne	0.2726	0.1445	0.0049	0.768	2.825	52.1	52.9
6	De-2,6-(1'-meth-yl-2'-buthenyl)-4-etoxeanelyne	0.2734	0.1439	0.0050	0.768	2.97	53.2	52.9

ratios for the inhibition of the synthesized compounds are the same within experimental error and standard calculations.

10.4 CONCLUSIONS

1. Studies of the dependence of the inhibitory activity MBA from their electronic structure showed that the correlation in this series of compounds exhibit indices such as the energy of the highest occupied molecular orbitals, electrophilicity index, the charge on the nitrogen atom and the dipole moment, correlate with minimal negative charge characterized by the maximum value of the coefficient correlation.

2. Using the regression equation for the minimum negative charge a series of compounds with a maximum inhibitory activity predicted and synthesized.

3. The regression equation for the charge on the nitrogen atom of the calculated braking coefficient (γ) coincide with the experimental values within the experimental error and standard calculations.

KEYWORDS

- **correlation**
- **correlation coefficients**
- **density functional theory**
- **forecast corrosion activity**
- **quantum-chemical parameters**
- **reactivity indices**
- **regression equation**

REFERENCES

1. Grigoriev V. P., Ekilik V. V. The chemical structure and the protective effect of corrosion inhibitors. Rostov-on-Don. Publishing House RSU. 1978. p. 301.
2. Durnie W., Demarco R., Jefferson A., Kinsella B. J. Elektrochem. Soc. 1999. V. 146 N 5. P. 1751.

3. Yu. I. Kusnetsov. – New. York: Plenum Press, 1996. 283 p.
4. Obraztsov V., Danilov F. Physicochemical Mechanics of Materials, 2004. – Special Issue number 4. – V.2. – S. 757–762.
5. Awad G. Kh., Adel -Assad N. Abdel Gaber A. M,. Massoud S.S. Protection of Metals. 1997. V.33, № 6, s.565.
6. Buhay J.E., Gabitov A.I., Bresler I.G., Rahmankulov D.L., Report of Bashkirian Academy of Sciences, v. 314., № 2, p.887.
7. Buhay D.E., Golubev V., Laptev A.B., Liapina N., Golubev I.V., Rahmankulov D.L. Bashkirian Journal of Chemistry. 1996. V.3, no.4. p. 48.
8. Doroshchenko T.F., Ckrypnik Y.G., Lyaschuk S.N. Protection of Metals Journal, 1996. V.32. № 5, 543–549.
9. Costa J.M., Lich J.M. Corros, Sci. 1984. V. 24. P.929
10. Growcock F.B., Frenier W.W., Andreozzi P.A. Corros, Sci. 1989. V. 45. P.1007
11. Gece G. and Bilgig S, Corros Sci., 2009, 51, 1876.
12. El Sayed H. El Ashry, Ahmed El Nemr, Samy A. Essawy and Safaa Ragab. ARKIVOS 2006 (xi) 205–220.
13. Bentiss, F.; Lebrini, M.; Lagrenee, Metall Corros Science, 2005, 47, 2915.
14. Emreguel, K. C.; Hayvali, Metall Corros Science. 2006, 48, 797
15. Yurt, A.; Bereket, G.; Ogretir, C. J Mol Struct (THEOCHEM) 2005, 725, 215.
16. Talati, J.D.; Desai, M.N.,; Shah, N.K. Anti-corros. Meth. Mater. 2005, 52, 108.
17. Dadgarnezhad, A.; Sheikhshoie, I.; Baghae, F; Anti-corros. Meth. Mater.2004, 51, 266.
18. Lukovits, I., Kalman, E.; Zucchi, F. Corrosion 2001, 57, 3
19. Obot, I.B., Obi-Egbedi, N.O., Umoren, S.A. Int. J. Elektrochem. Sci., 4 (2009) 863–877.
20. Lewis G. Corrosion, 1982; 38: 60.
21. Ma, H.; Chen, S.; Liu, Z.; Sun, Y. J Mol Struct (THEOCHEM) 2006, 774, 19.
22. Kandemirli, F.; Sagdinc, S. Corros Sci 2007, 49, 2118.
23. Nnabuk O. Eddy, Benedict I. Ita, Nkechi E. Ibisi, Eno E. Ebenso. Int. J. Elektrochem. Sci., 6 (2011) 1027–1044.
24. Obot, I.B., Obi-Egbedi, N.O. Der Pharma Chemica; 2009, 1 (1): 106–123.
25. F. Bentiss, M. Lebrini, M. Lagrene, M. Lagrenee, M. Traisnell, A. Elfarouk, H. Vesin, Elektrochim Acta, 2007, 52, 6865.
26. S.G.Zhang, W. Lei, M.Z. Xia, F.Y.Wang, J.Mol. Struct. ((Theochem.), 2005, 732, 173.
27. Lopez, N., Illas, F. J.Phus, Chem., 1998, B 102, 1430.
28. Eno E. Ebenso, Taner Arslan, Fatma Kandemirli, Ian Love, Cemil Ogretir, Saracoglu, Saviour A. Umoren. Int. J. of Quant. Chem., vol 110, 2614–2636.
29. Arslan, T.; Kandemirli, F.; Ebenso, E. E.; Love, L; Alemu, H. Corros Sci 2009, 51, 35.
30. Cai, X.; Zhang, Y.; Zhang, X.; Jiang, J. J Mol Struct (THE-OCHEM) 2006, 801, 71.
31. Larabi, L.; Benali, O.; Mekelleche, S. M.; Harek, Y. Appl Surf Sci 2006, 253.
32. Rodriguez-Valdez, L.; Villamisar, W.; Casales, M.; Gonzalez-Rodriguez, J. G.; Martinez-Villafane, A.; Martinez, L.; Glossman-Mitnik, D. Corros Sci 2006, 48, 4053.
33. Lashkari, M.; Arshadi, M. R. Chem Phys 2004, 299,131.
34. Sein, L. T., Jr.; Wei, Y.; Jansen, S. A. Comput Theor Polym Sci 2001, 11,

35. Sein, L. T.; Wei, Y.; Jansen, S. A. Synth Met 2004,143,1.
36. Blajiev, O.; Hubin, A. Electrochim Acta 2004, 49, 2761.
37. Lebrini, M.; Lagrenee, M.; Vezin, H.; Traisnel, M.; Bentiss, F. Corros Sci 2007, 49, 2254.
38. Gao, G.; Liang, C. Electrochim Acta 2007, 52, 4554.
39. I.B. Abdrakhmanov, Thesis, Bashkirian State University and Institute of Organic Chemistry, Bashkirian Branch of Russian Academy of Sciences, 1989. 323c.
40. Granovsky A.A., http://assik/chem.msu.su/gran/amess/ndex.html
41. Stephens P.J., Delvin F.J., Chabalovcky C.F., Frisch M.J. J. Phys. Chem. 1994. V. 98. P. 11623.
42. Rassolov V., Pople J.A., Ratner M., Windus T.L. J. Chem. Phys. 1998. 109(4), p.1223.
43. http://www.chemcraftprog.com.
44. OST 39.099–79. Corrosion inhibitors. Methods for evaluating the effectiveness of the protective effect of corrosion inhibitors in oilfield wastewater. Ufa: 1979.

CHAPTER 11

PROTECTION OF RUBBERS AGAINST AGING WITH THE USE OF STRUCTURAL, DIFFUSION AND KINETIC EFFECTS

V. F. KABLOV[1] and G. E. ZAIKOV[2]

[1]Volzhsky Polytechnical Institute (branch) Volgograd State Technical University, Russia; E-mail: kablov@volpi.ru, www.volpi.ru

[2]N.M. Emanuel Institute of Biochemical Physics Russian Academy of Sciences, Russia, E-mail: chembio@sky.chph.ras.ru, www.ibcp.chph.ras.ru

CONTENTS

ABSTRACT

Processes of rubber aging in the different kinetic modes and mathematical models of the processes including aging and destruction processes under extreme conditions are studied. The aging process for rubbers as thermodynamically open nonlinear systems is considered. It was revealed that the aging process can be controlled by means of the internal physical and chemical processes organization and by creation of external influences (by the organization of thermodynamic forces and streams). According to the Onsager principle, in certain conditions of aging the conjugation of thermodynamic forces and streams is possible. The diffusion and structural aspects of aging of elastomeric rubbers and articles thereof are considered. The composite antiaging systems of prolonged action based on the use of microparticles with a saturated solution of the antiagers migrating into an elastomeric matrix at the exploitation are proposed.

11.1 INTRODUCTION

In this chapter, we shall consider the problem of developing the system technology as a new scientific approach to producing elastomer materials. This technology combines general and special theory of systems, thermodynamics, kinetics, physics and chemistry of polymers, and computer methods of data processing.

The distinct features of the additives are:

- Launched product in a form of particles of micro and macrosizes.
- Composing different functionally active components into the synergetic systems with mutual activation.
- Active groups blocking on a stage of technological processing and discharging the groups on a stage of vulcanization and exploitation.
- Using eutectic alloys for decreasing the melting temperature of poorly dispersing components in an elastomeric matrix.

- Activation of components reactivity at the expense of surface effects.
- Creating the chains of physical and mechanical conversions of components both in volume of a rubber compound and in time (at technological processes and exploitation).
- Use of structural and diffusion effects and use of interfacing the thermodynamic forces and streams in nonequilibrium systems.
- Influence on the energy efficiency of technological processes.

The problem of polymers aging and their stabilization consists a large chapter of polymer material engineering – a science about creation of polymer materials, their processing, preserving, and regulating their performance characteristics [1, 2].

The term "destruction" (sometimes "degradation") is often used instead of the term "aging."

Nowadays, more than 120 million tons of synthetic polymers and almost the same amount of natural polymers (natural rubber, polymers based on polysaccharides) are produced in the world. The production dynamics is also important: the rates of growing polymer production are ahead of the rates of metals production by 25–30%.

If it is possible to prolong the lifetime of polymeric articles, for example, two times as large, then it will be equivalent to an equal increase of their production.

The other objective is a quantitative prediction of resistance of polymeric products. If the time of reliable operation is underestimated, the polymeric articles will be excluded of operation earlier than their resources are depleted, and it is not of economic benefit. Given the overestimated working lifespan, a polymeric product will fail during operation, which may lead to an accident or even more serious consequences.

One more goal is to apply the destruction process as a method for modification of polymeric articles.

The next objective is related to reuse of worked out (aged) polymeric products. Today, when the amount of wormed out polymeric wastes is large, the problems of polymer reuse, regeneration of monomers, and pyrolysis of polymeric wastes to obtain the gaseous fuel become urgent.

The environmental issues and economic considerations are important as well: petroleum (the raw material for polymers) goes up in price, therefore, polymers will be more expensive. Waste plastics have not to be burned at dumps, since in this case a large number of toxic substances, including hydrogen cyanide and dioxin, are formed.

The following problem can be named as creation and search for ways of application self-degrading disposable polymers. The idea is to design polymers made of units connected with swivel groups. In plus, units have to be stable constantly, and swivel groups have to be stable in one cases (while operating) and unstable in other cases (after operating). As a rule, such swivel groups are not resistant to light and hydrolytic processes. That is why, once such polymers get to wastes after use, they easily degrade upon the groups to units-oligomers that dissolve in soil and then eaten by bacteria.

Kinetics is powerful and fine instrument in investigations of chemical, biochemical and biological processes, as well as in medicine, agriculture, etc.

1. Making of artificial diamonds from compounds having no carbon.
2. Development of highly sensitive methods of investigation of chemical reactions (chemiluminescence) and analysis of reaction products (e.g., determination of a one double bond in a carbochain polymer and its location against 50,000 single bonds).
3. Multitonnage chemical production design (production of acetic acid and methylethylketone by oxidation of n-butane in liquid phase, oxidation of propylene in propyleneoxyde in solution).
4. Degradation as a method of modification of polymer materials. Creation of new materials (artificial silk by ozonolysis of poly-ethylenetherephtalate, fibers with natural polymer molecules on the surface and artificial polymer molecules inside of filaments by hydrolysis of triacetate of cellulose, purification of smoke sausage by action of ozone, creation of roughness on surface of polymer materials by ozonolysis, production of membranes with good properties by radiation of films by α-particles and hydrolysis after radiation).
5. Stabilization of polymers.
6. Combustion of polymers.
7. Relation between diffusion properties and rate of chemical reactions in polymer matrix.
8. Prediction (prognoses) of life spent of polymers and composites.
9. Polymers with precise life spent.
10. Recycling.

11.2   A SYSTEM APPROACH TO THE PROBLEM OF RUBBER AGING

Rubber aging is a complicated physical and chemical process running in the nonequilibrium open heterogeneous system [3, 4]. Ageing leads to destruction and/or macromolecules structuring, changes in physical structure, running diffusion processes, and other structure and chemical changes in a material.

A large variety of operating environments, a wide range of effects on polymer materials and the related multitude of technical materials (TM) requires a common conceptual approach to development of new efficient materials. Such a common conceptual approach is a system approach. The system technology combines various complementary methods and approaches to creation of elastomeric materials. The development strategy based on complementary approaches is seen as the most effective. The main theoretical approaches that underlie development of materials based on elastomeric systems are given below.

The task of the system description is a record of all diversity of problems facing both the design and TM as well as finding an optimal pair TM – product. The system description has to be based on the fundamental objective of Materials Science, which is to establish relationships between composition, technology, structure and properties of materials.

The basis of the system approach is representation of an object as a system, that is, as an integral set of interrelated elements of any nature. The main criterion for establishing the need for this element in the system is its participation in the system operation resulting in obtaining the desired result. The system approach allows for functionally supported dissection of any system into subsystems which volume and number are determined by the composition of the system and the consideration scope. One might add that the same object can be represented as different systems while the number of ways to view an object using the system approach has no limitations.

Representing an object as a system, we only get the opportunity to approach the structure of the object; a further step is to search for patterns of whole object relations in a system.

One of the system approach methods is also functional and physical analysis, in which a multicomponent material is considered as a technical system carrying out operations to convert some of the input actions on the output ones. At that, the main attention is paid not to the material structure of the object, but to functional transformations of matter and energy flows [2–4].

External flows of matter and energy through functional systems affect a TM and cause its various physical and chemical transformations (PCT). They lead to a change in the molecular, supramolecular, phase micro and macrostructure of the TM. Since PCT and structural changes occur at different rates, it is necessary to consider the heterochronic behavior of system changes. As a result of all the processes occurring in the TM during operation change, their structural and functional characteristics, so we can talk about a system change in the TM or TM systemogenesis.

11.3   A THERMODYNAMIC APPROACH

Modern thermodynamic methods can be applied not only to identify the possibility or impossibility of the running processes, but also as methods to find new technical solutions [5–7].

Ageing of elastomeric materials from a thermodynamic point of view can be seen as processes of self-organization and disorganization in open nonlinear systems. Processes occurring in such systems can be operated as an organization of chains of physical and chemical transformations in the systems themselves, highlighting the various functional subsystems as well as by controlled external exposures (thermodynamic forces and flows). In case of exposures, in accordance with the Onsager principle, pairing of thermodynamic forces and flows may take place. This allows for additional opportunities to influence the aging process.

A polymer operated at high temperatures, in reactive environments, under intense friction, and other intensive effects can be regarded as a nonequilibrium open system, that is, a system exchanging the medium substance (or energy). Energy and substance exchange is carried out by heat conduction, diffusion of the medium in the material, low molecular additives and products of thermal and chemical destruction of the polymer matrix. Under these conditions, the material as a system is not in equilibrium state, and the further it from equilibrium, the greater the intensity of exposure to the material and, consequently, the more intensive mass and energy exchange with the medium is. The presence of reactive components in a material increases an equilibrium deviation and leads to greater intensification of mass and energy exchange.

For an open system total entropy change dS is given by:

$$dS = d_iS + d_eS \tag{1}$$

where d_iS is a change (production) in entropy in a system due to irreversible processes, d_eS *is the* entropy flow due to energy and matter exchange with the medium.

It is important that the entropy of an open system can decrease due to the entropy output in the medium ($d_eS < 0$), and provided that $|d_eS| <> |d_iS|$. Then, despite the fact that $d_iS > 0$ (in accordance with the second law of thermodynamics), $dS < 0$.

The decrease in entropy of an open system means that in such system self-organization processes start as well as forming specific spatio-temporal dissipative structures supporting sustainable state of the system.

To ensure the flow of negative external entropy can be implemented as a specially organized by external influences such as the artificially created concentration gradient, temperature, potential, and due to the chemical reactions that lead to high entropy products output of the system and its enrich with low-entropy products (e.g., polycondensation system with by-product output).

The main conditions that lead to ordering in an open system are as follows:

1. A system is thermodynamically open.
2. Dynamic equations are nonlinear.
3. A deviation from equilibrium exceeds a critical value.
4. Microscopic processes are coordinated.

Analysis of behavior of polymers and processes flowing in them shows the ability to perform these conditions, and, hence, the occurrence of ordered spatial or temporal structures that stabilize a system. It is only necessary to provide special external and internal conditions for this.

When in a nonequilibrium system several irreversible processes run, they are superimposed on one another and cause a new effect. Irreversible effects may occur due to a gradient of temperature, concentration, the electrical or chemical potential, etc. In thermodynamics all these quantities are called "thermodynamic" or "summarized forces," and denoted by X. These forces cause irreversible phenomena: heat flow, diffusion current, chemical reactions, and other flows, called summarized and denoted by J.

Generally, any force can cause any flow that is reflected in the famous Onsager equation:

$$J_i = \sum_{k=1}^{n} L_{ik} X_k \ (i=1,2,...n) \tag{2}$$

where the coefficients L_{ik} are called phenomenological coefficients.

If the system deviates from the stationary state, then, due to the desire of the entropy production to the minimum according to Prigogine's theorem (extreme principle) about the minimum speed of the entropy production in a stationary state, there come such internal changes that will tend to bring the system to that state. Such autostabilization phenomenon represents an extension to stationary nonequilibrium systems on Le Chatelier's principle applicable to chemical equilibrium.

A technical task for a developer of physical and chemical systems is to ensure the maximum deviation dA_p (the work of changes in a system) from zero in the negative direction, if his purpose is to use the external influences to put the system in a stationary state.

Obtaining of elastomeric materials based on systems with functionally active components aimed to work in high temperatures, pressure and reactive environments is essential when rubber technical products are produced for oil drilling equipment, geophysical instruments and articles for chemical industry.

Theoretical analysis of capabilities to apply principles of the nonequilibrium thermodynamics and theory of open systems for production of elastomeric materials with physical and chemical transformations under operational exposures represents the prospects of the direction.

Functionally active components include the components capable of chemical reactions and physical and chemical transformations in bulk or on a surface of a material under external exposures (heat, mechanical, reactive environments). In addition, either protective agents are formed or physical effects are realized that allow for increasing service durability of rubbers.

The unstable systems are especially sensitive to external exposures which may lead not only to the system degradation, but, by contrast, transform them to a new stationary state that, despite its instability, is characterized by sufficient stability. There are certain conditions for flowing such processes.

When a system is deviated to the highly nonlinear area, as well as, if the system deviation exceeds some critical values, we have its high-grade reconfiguration including possible occurrence of self-ordering processes.

Applied to elastomeric materials, the specified conditions can be created by the introduction to a system of the deliberately chosen functionally active components, which undergo physical and chemical changes under the operational impacts or some external "control" actions that can be interfaced with the operational impacts so to maximize the system stabilization.

For generating the stabilizing physical and chemical transformations, it was proposed in the work to use polycondensation capable monomers (PCCM) and other compounds with reactive groups. PCCM in an elastomeric matrix can react with formation of a new polymeric phase and heat absorption.

Thermodynamic analysis of open polycondensation systems and substantiation of various PCCM application as functionally active components of elastomeric materials have been conducted in the work; research results of polycondensation in an elastomeric matrix have been represented and a possibility of improving heat and corrosion resistance of elastomeric materials with introduction PCCM has been shown; different ways of applying PCCM have been proposed and experimentally proved [8–10].

Polycondensation processes allow for obtaining a large variety of chemical structures and, consequently, a possible wide range of properties when PCCM are applied as modifying agents. A distinctive feature of the developing direction is the thermodynamic nonequilibrium of polycondensation systems. In this case, nonequilibrium means that polycondensation systems have functionally active groups capable of further transformations (both the growth of macromolecules and a reversible reaction) nearly at any stage of a transformation. Low molecular product recovery, exo- and endothermic effects enable to classify polycondensation systems as open systems. Since the heat effects of the polycondensation are not too large, and the obtained low molecular product (usually water) takes much heat away, then, when an open system is considered, the process runs as endothermic. Besides, when the low molecular product is removed, an additional negative entropy flow arises resulting in thermodynamic assumptions of self-organization within the system.

According to Le Chatelier's principle, an increase in temperature shifts endothermic reactions towards heat absorption, that is, towards the polymer formation. In this regard, PCCM introduction to rubbers is of interest in terms of creating a kind of self-cooling rubbers for heat protective materials.

The following reactions have been studied in the research: polyetherification, polytransesterification, polyamidation, the formation of polyhydrazides and their polycyclization into polyoxidiazoles, polyurethane formation and production of poly 1-acylhydrazides, three-dimensional polymerization, and catalytic polycondensation.

Diffusive and kinetic features of polycondensation in the high viscosity rubber matrix are very specific because of the difficult removal of the obtained product, thermodynamic incompatibility of a rubber to monomers and the formed polymer. All this required a study of polycondensation kinetics in a matrix of different elastomers.

One of the most interesting phenomena found at polycondensation in elastomeric matrices is acceleration of the reaction compared to the reaction in a melt. Kinetic constants of reactions running in rubbers exceed greatly the rates of reactions running in a melt, and the rates of reactions in rubbers of different nature are very different. It has been shown that such dependence of the reaction rate on the rubber nature is connected with a type of the phase structure in systems "rubber–monomer" and "rubber–reaction products." Electron-microscopic analysis has revealed that the reaction rate is maximum, if system "rubber – monomer" is a system with emulsion of one monomer in a rubber with the second monomer dissolved in the rubber. In this case, polycondensation runs on the interfacial mechanism with the product recovery into an individual phase. The rate of the interfacial polycondensation goes up with an increase in the interfacial surface and increase in activating effect of free surface energy, respectively.

The essential thing is that polycondensation runs also in vulcanizates at time-temperature aging modes.

That allows for realizing the proposed concept concerned to the production of materials with physical and chemical structure, which is nonequilibrium in operating conditions. Before the certain values of conversion degree of a polycondensation system, the stabilization of rubber characteristics takes place, which is related to inhibition of residual functional groups and endothermic effects of polycondensation. The results of differential thermal analysis represent a flowing of polycondensation and a slow-down of thermal-oxidative processes in rubbers modified with PCCM.

The features of flowing polycondensation in bulk and surface layers have been investigated while application of mechanical loads and high pressure.

The observed activation of polycondensation under mechanical loading of the elastomeric matrix is an occurrence of coupling chemical and mechanical processes in accordance with the Onsager principle. The possibility of using high pressure (up to 1000 MPa) to selectively target polycondensation in an elastomeric matrix has been shown.

The opportunity to improve heat and corrosion elastomeric materials by using physical and chemical effects running at operational exposures into monomer containing nonequillibrium and open elastomeric systems has been displayed (Table 11.1). PCCM application is especially effective for some rubbers exploited in extreme conditions (high temperatures, aggressive oxidative environments based on mixes of inorganic and organic acids). It has been demonstrated that PCCM and polycondensation products form some unique buffer system inside a rubber.

An increase in corrosion resistance of rubbers is also possible in elastomeric compositions based on PVC. In the designed compositions for work in nitrating environments a number of proposed concepts are implemented, in particular the use of aggressive nitrating mixture to seal and structure a surface layer and creation of internal functional subsystems of physical and chemical transformations based PCCM and other components. All this avoids surface gumming.

TABLE 11.1 Coefficient of Properties Variation for Monomer Containing Rubbers at Aging

PCCM	Δfp	ΔEp	PCCM	Δfp	ΔEp
Aging at 398 K x 72 h			Aging at 423 K x 72 h		
Vulcanized rubbers based on butadiene-nitrile rubber SKN-40					
without PCCM	0.30	0.44	without PCCM	0.29	0.07
PA – Glyc.	0.45	0.30	HFDP – AA	0.69	0.59
PA – Glyc. – DEG	0.70	0.40	PFAA – AA	0.88	0.82
PA – DEG	0.60	0.30	Aging at 473 K x 60 min		
AA– DEG	0.51	0.35	without PCM	0.52	0.44
DHAA – AA	0.64	0.68	PA– Glyc.	0.84	0.78
unfilled PA– Glyc.	1.20	0.70	PA– Glyc. – DEG	1.50	0.83
unfilled PA– Glyc. – DEG	0.90	0.80	PA– DEG	1.70	0.66
unfilled PA– DEG	1.25	0.80			
Vulcanized rubbers based on ethylene propylene rubber SKEP					
Aging at 423 K x 72 h			Aging at 443 K x 72 h		
without PCCM	0.70	0.62	without PCCM	0.15	0.13
AA– DEG	0.87	0.88	DCA–MPDA	0.52	0.40
1,3 ADA– DAA	1.40	1.50			

HFDP – hexafluorodiphenylolpropane; Glyc. – glycerin; PA – phthalic anhydride; AA – adipic acid, DHAA – dihydrazine of adipic acid, 1,3 ADA-1,3 adamantanedicarboxylic acid, PFAA-bis(monoethanolamide)-perfluoroadipic acid, DCA – 1,3-dimethyl-5,7-dicarboxyladamantane, MPDA – m-phenylenediamine, f – the aging coefficient on strength, E – the aging coefficient on elongation.

11.4 KINETICS OF AGING

The aging kinetics is rather complicated [14–17]. The Fig. 11.1 represents different types of kinetic curves and their models.

As it is seen from Fig. 11.1, aging of polymers runs on various kinetic curves. It is connected with different processes running at aging including simultaneous running of destruction and structuring, aging by the mechanisms of autoacceleration and autobraking, etc.

In some cases it is possible to use kinetic effects to prolong a working period of elastomeric materials.

With aging materials near the degradation temperature, the mechanism of aging can vary significantly and, thus, methods of protection.

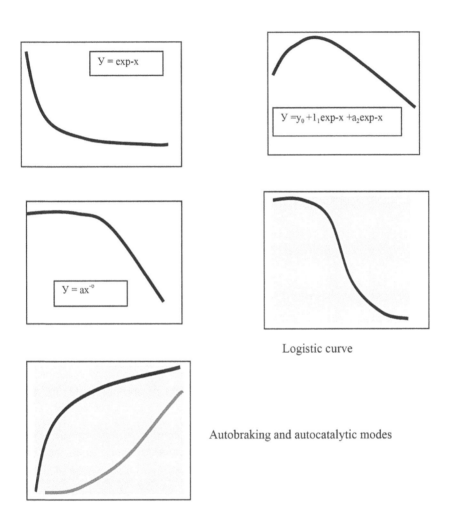

Kinetics. Types of kinetic curves.
Not only mechanics, but also chemistry of polymer systems
is highly nonlinear

FIGURE 11.1   Types of kinetic curves.

The processes are especially difficult while intensive thermal destruc-tion-aging can shift to a "thermal explosion" at that.

When exposed to operational factors on a polymer material, we can distinguish two modes of operation: a normal mode and operation in extreme conditions. A normal mode is characterized by a relatively slow and gradual change of parameters during operation. Such changes in the

Dependence of strength change coefficient on aging at 423 K for rubbers based on SKN-40

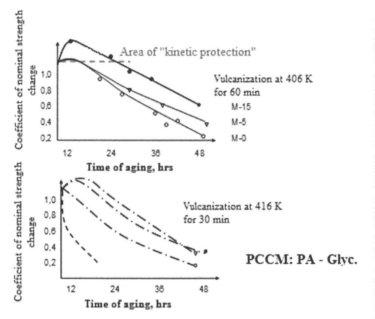

FIGURE 11.2   Dependence of a change in properties of rubbers containing PCCM at aging.

parameters are usually described by linear dependencies, linear sections of nonlinear dependencies or Arrhenius equations. During normal operation, the effective values of the parameters (temperature, concentration of a corrosive environment, etc.) are far from the limits of the material performance. Under extreme conditions, the effective values approach the limits of performance, or even fall in the "nonoperation" area. In this mode a material fails dramatically. A change in parameters is described, as a rule, by nonlinear dependencies including extreme ones or dependencies described by equations of the catastrophe theory. Figure 11.3 shows an example of aging in time-temperature coordinates. Under certain conditions – in the area of instability or bifurcation – the investigated parameter of a polymer x_1 can abruptly change its values.

Durability, however, may be sufficient in some cases – at disposable use, large stocks in thickness (mass, volume, etc.), as well as in the case

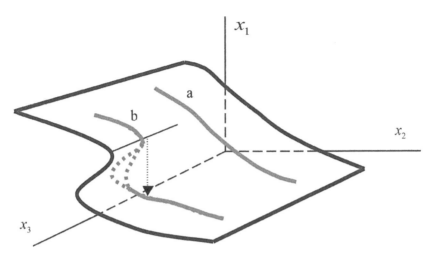

FIGURE 11.3   Behavior of a nonlinear system in normal (a) and extreme (b) modes.

of constructive features of an assembly, the application of protective coatings and other protection methods. Finally, durability of a material may be significantly increased by introduction of functionally active components in its composition that provide the material protection while exploitation. Nonlinear behavior of polymeric materials increases with the approximation of operational parameters to the critical ones. Concurrently, instability of materials increases as well, in particular, structural stability and dynamic stability (according to Lyapunov). At the same time, nonlinear systems may also have the stability areas under certain values of the parameters. So, a system with autocatalysis or autoinhibition can be described by the following equation:

$$\dot{X} = -kX - k_1 X \tag{3}$$

where X – a change in a certain characteristic, k and k_1 – parameters typical for the system.

When $k > 0$, the system has the only real root $X = 0$ corresponding to the stable state, and when $k < 0$ it has three real roots, from which $X = 0$ becomes instable, and two last roots $X_{2,3} = \pm\sqrt{(-k/k_1)}$ turn out stable. It is seen that when the parameter passes through 0 to the negative side, resistance of the initial state is lost, but instead two new stable states arise:

there is a complication (differentiation) of the system by bifurcation (splitting) of the initial stable state. Mathematical analysis indicates that for sufficiently strong nonlinearity a complication of the structure with the formation of stable states may occur. Complexity of the system structure is even more multivariate in systems with two or more variables.

Systems in a state of instability are particularly sensitive to small external influences that can lead not only to the destruction of the system, but, on the contrary, transfer it to a new steady state, which, in spite of its nonequilibrium, has sufficient stability. A typical example is the effect of ultralow friction manifested at relatively low irradiation of rubbing pair "polymer-metal." Another example is rubber hardening on the initial stages of aging, when the rubber composition contains polymer forming or structuring additives. Thus, the polymer systems, which are in the nonlinear area typical for extreme conditions, can be stabilized by internal or external "control" effects [2, 5]. Extreme conditions for heat protective materials are working conditions in high temperature gas streams, operation of rubber products in a dynamic mode in combination with other exposures, etc. Figure 11.4 displays the dependence of the rubber durability depending on the strain. At strains over some critical parameters,

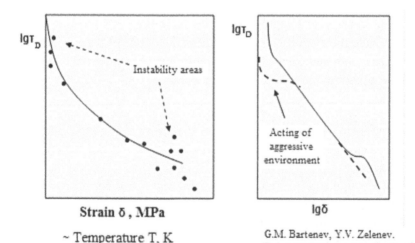

FIGURE 11.4   Dependence of durability on strain.

there is not only the violated usual dependence, but also the dramatically increased variation of parameters. This indicates about instability in the system.

The work of Ref. [14] demonstrates the occurrence of nonlinearity and instability of the aging process in ozone and oil environments.

So, when materials are aged near the destruction temperature, the aging mechanism and, respectively, protection methods can vary considerably [3].

11.5   CRITICAL PHENOMENA IN ELASTOMERS DURING THERMAL EXPOSURE [11–12, 16]

It is known that between the mode of stationary process and thermal explosion there is a clear boundary defined by the value of Frank-Kamenetsky parameter δ:

$$\delta = (Q/\lambda)\,(E/RT^2)\,r^2 Z e^{-E/RT} \tag{4}$$

where Q – enthalpy of a reaction; E, Z – kinetic parameters of a reaction; r – relevant size of reacting system (sample size); λ – heat conductivity coefficient.

When $\delta > \delta_{cr}$ heat balance is disrupted, that provides critical conditions of explosion.

Figure 11.5 shows the effect of sample weight on the kinetics of thermal destruction of polybutadiene during a thermogravimetric study. An increase in the sample mass leads to a distortion of the destruction kinetics, and a thermogravimetric curve becomes stepwise.

In the dynamic heating the heating rate becomes essential. The existence of critical conditions for the dynamic heating is determined by the critical heating rate – critical factor analog δ_{cr}:

$$dT/dt_{\text{кр}} = \text{const}\,(aS/QV)(RT^2/E) \tag{5}$$

where S, V – surface and volume of a reacting system; a – heat exchange coefficient.

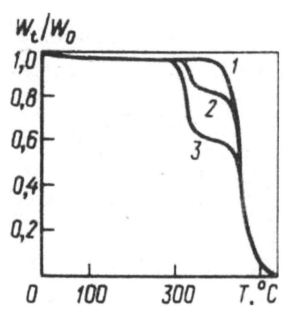

FIGURE 11.5 Impact of sample weight on thermal destruction kinetics of polybutadiene: 1 – 10 mg; 2 – 40 mg; 3 – 70 mg [16].

TABLE 11.2 Effect of Heating Rate on the Weight of Polybutadiene Samples

Heating rate, °C/min	40	20	10	5	2
Sample weight (critical), mg	3	7	10	15	25

The effect of heating rate on the weight of the spherical polybutadiene samples, at which the distortion of thermal destruction kinetics takes place, is shown in Table 11.2.

11.6 AN IMPORTANT FEATURE OF ELASTOMERIC MATERIALS IS FLOWING OF COOXIDATION REACTIONS INTO THEM

A feature of oxidation of organic compounds is that only in the very initial stage it can be regarded as the conversion of an individual substance. As a rule, there are reactive oxygen-containing derivatives during oxidation, which react themselves changing significantly the reaction speed and sometimes its direction [13]. Therefore, in the later stages the process proceeds as the conjugated oxidation of the initial material with the primary oxidation products. It is complicated by the continuous change in the composition of the reacting mixture, and the rate of initiation and chain termination during the reaction. In turn, the mixture composition determines the composition and ratio of peroxy radicals involved in oxidation, which can cause a significant change in the reaction rate as a whole, as well as accumulation and spending of individual products. This generally serves as the primary reason for reduction of selectivity to the desired product. To regulate the process of oxidation, increase its speed and selectivity, it is crucial to know the basic regularities of cooxidation featuring different classes of oxygen-containing organic substances.

Oxidation of mixtures, or, as it is also called, cooxidation or conjugated oxidation attracted the attention of researchers for a long time. First of all, because the rate of mixture oxidation is often different from the total oxidation rate of components taken at the same concentrations, and, in some cases, we get even other reaction products. Here, a number of interesting and important results have been obtained. These results include detection of such systems in which the intermediate radicals of a substance during oxidation act as reactants in the reaction with the other substance, which leads to the formation of the desired product, which in oxidation of the individual substance is formed with low yields or not formed at all. This reaction is cooxidation of aldehydes with unsaturated hydrocarbons:

$$R\text{-}CH{=}CH_2 + R_1\text{-}\underset{\underset{H}{|}}{C}{=}O + O_2 \longrightarrow R\text{-}\underset{\underset{O}{\diagdown\diagup}}{CH}\text{-}CH_2 + R_1\underset{\underset{OH}{|}}{C}{=}O$$

This reaction was proposed as an effective method for producing olefin oxides, particularly, propylene oxides.

The yield of the desired product in these reactions is dependent on the activity of the aldehyde and alkene, the process conditions, presence of a catalyst and other substances. In particular, it was shown that for regulating the composition of the products in oxidation of mixtures of organic compounds, additives of nitroxides may be used as they are selective inhibitors of multicenter processes. The essence the selective inhibition consists in follows. In the "aldehyde–alkene" system the chains lead two types of radicals: acyl-(AO_2) and poly peroxy radicals (MO_2) resulting to monomer and polymer products, respectively. There is a definite concentration of the nitroxide, in which it does not inhibit the aldehyde oxidation but completely inhibits the formation of polymer products, that is, preferably reacts with the radicals MO_2 reducing notably their concentration. A significant change ratio in concentration of radicals towards AO_2 leads to improved selectivity of the reaction on the monomer product.

Features of chain cooxidation in mixtures of several substances are determined by the presence of different peroxy radicals formed from the molecules of each component having different reactivity. However, there are some cases where a substance is oxidized by various bonds C-H (e.g., branched paraffins, alkylaryls containing primary, secondary and tertiary C-H bonds in α-position to the ring; alkenes having C-H bonds in α-position to the double bond, etc.). In this case, an individual substance is oxidized involving heterogeneous peroxy radicals and the reaction mechanism corresponds to cooxidation of a mixture of several individual compounds with a constant fraction of their initial concentrations.

Knowledge of the regularities of cooxidation is necessary to solve such urgent problem as finding ways of its terminating or slowing. There are two important aspects. The first one is finding compositions that are resistant to oxidation. Thus, in earlier investigations of cooxidation it was shown that certain mixtures, for example, tetralin – cumene, were oxidized much worse than individual components. The second one is the suppression of oxidation of substances of natural or synthetic origin having a complex structure and able to oxidize by several reaction centers with different chemical nature, as well as mixtures using specially selected inhibitory systems.

Over the years the attention of researchers is attracted with not quite usual reaction of cooxidation of alkenes and thiols, which can be described as follows:

$$R\text{-}CH{=}CH_2 \ + \ RSH \ + \ O_2 \ \longrightarrow \ R\text{-}CH\text{-}CH_2\text{-}S\text{-}R$$
$$\underset{OH}{\vert} \qquad \overset{\parallel}{\underset{O}{}}$$

The reaction proceeds rapidly even at low temperatures leads to the formation of β-hydroxydisulfoxides and disulfoxides. The mechanism of these reactions involves both radical and nonradical stages.

Kinetic features of radical reactions in multicomponent polymer systems can lead to accelerated aging of a polymer with a more movable hydrogen atoms and slowing oxidation reactions in a more inert polymer. Consideration of this factor requires an adjustment of antioxidant group.

11.7   DIFFUSION PROCESSES

Since elastomeric materials are microheterogeneous multiphase materials, diffusion processes have the great influence on the aging process including migration components [13, 14]. With an increase in the operating temperature, the volatility of components enhances including the volatility of antioxidants. In this regard, the urgent task is reduction of the volatility of antioxidants while keeping their sufficient ability (ability to "hit" in areas with occurring free radicals on destructing macromolecules). The problem is solved effectively enough by using composite systems of antioxidants. We propose the antioxidant systems capable of continuous "release" of active labile antioxidants from micro and nanoparticles. Antioxidants are in a matrix of the particles in the supersaturated state. During operation, they continuously migrate into the dispersion phase of an elastomer. The various implementations of this method have been considered in the study.

The use of PCCM for surface modification and obtainment of gradient nonequilibrium systems with optimal organized spatial structure when the concentration of functionally active components is increased toward the surface of the article have been proposed and experimentally substantiated.

The method for producing gradient systems is that the first condensation monomer is introduced in rubber, then the rubber is cured, and, finally, processing in the second monomer at a temperature providing polycondensation is carried out. Herewith, due to the diffusion of the second monomer into the product the generation of a gradient structure is provided. An important advantage of the gradient systems of is the absence of the phase boundary, and, thus, the layering during operation.

11.8   INHIBITION OF HIGH TEMPERATURE AGING

Ordinary antioxidants successfully operate at temperatures below 100°C, and also can temporarily restrain the oxidation process at temperatures up to 200°C [17–23]. Above 200°C such inhibitors as phenols, amines and other start to oxidize:

$$InH + O_2 \rightarrow In + H_2O$$

Their strong point (rather weak bond *In-H*), which allows them to easily give a hydrogen atom for bringing the radical out of operation (turning it into inactive products), turns at high temperatures in the weak point of an inhibitor: at high temperatures the inhibitor due to weak chemical bond *In-H* easily gives hydrogen up turning into the radical, which at high temperatures actively reacts continuing oxidation chains. Thus, at high temperatures the inhibitor provides not only inactive radicals *In*, but also active HO_2 radicals, which react actively continuing oxidation chains.

For high temperature stabilization, substances are now developed (or mixtures thereof) which are themselves stable at low temperatures, but at temperature increasing, for example, above 200–300°C, begin to decompose into radicals which, in turn, readily react with alkyl and peroxy radicals breaking oxidation chains.

Another example of possibilities of high-temperature stabilization is the introduction of metal salts of variable valency or organic acids into a polymer. At high temperatures, they are decomposed to form an atomized metal or its oxide into a lower degree valence. Such metals (and oxides) intercept oxygen reacting with it almost at every encounter (activation energy is close to 0) and, consequently, prevent organic substance from oxidation (we will return to this issue in the next paragraph).

Regarding variable valency metals, their ions may initiate the process of oxidation, and can inhibit. In most cases, they initiate the oxidation process.

During oxidation of elastomers, the most active initiators are compounds of iron, cobalt, manganese, copper, nickel and titanium. They fall into a polymer as a polymerization catalyst residue, or washed out the walls of reactors where the polymerization takes place. In practice, the accelerating effect of these catalysts is shown at their trace concentrations in rubbers (and plastics) – in the hundredth and thousandth parts of a percent.

The catalytic action of metal compounds is concerned with their ability to participate in redox processes. They can participate in reactions (nucleation and degenerate branching) of chain initiation, chain propagation and termination.

In particular, variable valency metals (Me) easily participate in the reaction by decomposition of hydroperoxide radicals increasing the rate of their formation.

$$ROOH + Me^{++} \rightarrow ROO^* + Me^+ + H^+$$

$$ROOH + Me^+ \rightarrow RO^{*+} Me^{++} + OH^-$$

Effect of variable valency metals on the rate of polymer destruction may occur without the presence of oxygen. Metal compounds may be a kind of radical which reacts with the labile hydrogen molecule of a polymer (elastomer):

$$R - H + Me^{++} \rightarrow R^* + H^+ + Me^+$$

The resulting radical R is then reacted with the oxygen molecule. The catalytic effect of polyvalent metals can be suppressed by introducing into the reaction mixture of materials which forming soluble and inactive compounds with metals. From this point of view, there are especially useful substances forming undissociated complexes and chelate structures. An example is phthalocyanine and salts of dithiocarbamic acids generating such complexes with copper, and thereby largely suppressing the action of the catalyst.

The most effective way of eliminating catalytic action of variable valency metals is the production of polymers, which do not contain these metals. Thus, the presence of the metals can inhibit the high-temperature oxidation process, but the same metals can initiate the oxidation process.

At studying the effect of dispersed metal particles (Cu, Mn, Fe, Co, Ni, Zn, Pb, Al) on the oxidative destruction of natural rubber it was found that Cu, Mn, Co are effective catalysts of thermal oxidation, and Ni, Zn and Pb do not influence on this process. The most active catalyst of destruction is copper, in the presence of which the apparent activation energy decreases from 112.8 to 69.8 kJ/mol. It is assumed that copper and its oxides are involved in redox reactions of thermal oxidation of natural rubber.

The few carried out studies show the high sensitivity of polybutadiene, polyisoprene and natural rubber to the physical and chemical properties of dispersed fillers, which catalyze the thermal-oxidative destruction of elastomers. Some studies reported the opposite effect. An increase in thermal resistance of rubbers was pointed when variable valency metals including rare earth metals, were used. They are introduced either in the form of oxides or in the form of easily decomposing compounds generating a free metal, or in the form of organic, organosiloxane, phosphorus-siloxane compounds. Thus, the introduction of iron compounds generating highly dispersed a free iron under heating leads to the interaction of the latter with SiR_2O radicals and tenfold increase in performance of siloxane rubbers at 250–450°C [18, 19].

In addition, superfine metal particles may be active radical acceptors initiating thermal decomposition. There are known reactions of Pb, Zn, Bi and other variable valency metals with alkyl radicals, leading to the formation of organometallic compounds as follows [17]:

$$Pb + 4\ R\bullet \rightarrow PbR_4$$

In works of Refs. [20–23] the effect of metal particles released in the rubber matrix during thermal destruction of nickel, copper and other metals carbonyls was investigated, and it was showed the stabilizing effect of superfine particles of copper, nickel, lead and bismuth in the thermal destruction of elastomeric materials based on polyolefins and a possibility of developing rubbers for high-temperature use based on ethylene-propylene rubber modified by ultrafine particles of variable valency metals. It was noted that metal particles are the most effective for development of rubbers operating at high-temperature exposures under the conditions of limited access of atmospheric oxygen (LAAO). In Fig. 11.6, the comparative data of investigating the effect of particles of different metals at high temperature aging of EPDM are shown.

As it is seen from Fig. 11.6, the vulcanizates with metal particles sized 5–20 nm superior a base composition in aging under the LAAO conditions 1.8–2.4 fold on the aging coefficient by strength and 1.3–1.7 fold on the aging coefficient by elongation at break.

Thus, in the course of studies the stabilizing effect of superfine copper and nickel particles under thermal destruction of elastomer materials

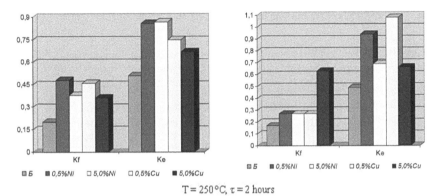

$T = 250\,^{\circ}C,\ \tau = 2$ hours

FIGURE 11.6 Impact of metal nanoparticles on the aging coefficients of rubbers on nominal strength (K_f) and elongation (K_e).

based on carbon-chain rubbers has been shown. An increase in thermal stability of a polymer matrix modified by superfine corpuscles of variable valency metals is firstly determined by the formation of the spatial network structure consisting of metal particles and the polymer macromolecules. Secondly, the metal particles act as an oxygen acceptor that presents in the polymer system. Thirdly, they enter into a chemical reaction with alkyl radicals inhibiting thermal destruction processes and forming the secondary network structures.

11.9 EFFECT OF MICRO AND MACROSTRUCTURE OF COMPOSITIONS

Structure of elastomeric materials at different levels (supramolecular structures, the structure in filled systems, the presence of several polymer phases, polymer-filler, polymer-fiber and polymer-polymer interfaces) largely predestinates the kinetics of aging and ways to improve lifetime while creating the optimal structure materials and products [4, 24, 25]. In particular, the optimal solution is to provide a continuous phase of aging resistant polymer in the dispersion phase of another one, development of barrier layers of fillers as also protective surface layers and coatings able to withstand static and dynamic loads.

11.10   EFFECT OF FUNCTIONAL GROUPS IN ADDITIVES

It is known that modifiers with active functional groups (epoxy, amino, hydroxyl groups) interact with reactive capable oxygen-containing groups of rubbers formed during plastication, oxidation, and preparation of compositions further operation of the rubbers [26].

Besides, the modifiers with the functional groups may also chemically react with peroxy radicals formed during oxidation of rubbers, and thereby take part in stabilization of the reactions preventing oxidation that, finally, prevents the premature rubber destruction.

We have also developed new oligomer compounds based on epoxydiane resin ED-20 and aniline derivatives – products of aniline production wastes (that is known as KPA).

Approbation of the obtained amine containing oligomers was carried out in rubber mixtures based on isoprene, butadiene, nitrile butadiene and chloroprene rubbers.

On the effectiveness of the stabilizing effect, KPA compound is located along with well-known antioxidants and even surpasses them by some measures.

Compound KPA-50 has the volatility which is almost next order lower than the volatility of such effective stabilizer as Diaphene FP (IPPD) and almost twice lower than the volatility of Acetonanil having the oligomeric nature as KPA-50.

Thus, the conducted investigation has shown that there is a possibility to create heat and aggressive resistant elastomeric materials based on systems with functionally active ingredients for different operating conditions.

11.11   NEW TRENDS IN THE STABILIZATION OF POLYMERS

The development of polymer materials with improved characteristics and solving the technological problems and resource-saving questions at the expense of application of nano- and microheterogeneous modifiers, nanofillers, adhesive additives, dispersants and other nano- and microheterogeneous functionally active components has been conducted [27–45].

At use of obtained dispersants-activators the next things are achieved in rubbers:

1. acceleration of the vulcanization process at keeping the inductive period;
2. increment in dispersion degree of filling agents;
3. decreasing the energy expenses on vulcanization at the expense of decreasing the generic vulcanization cycle;
4 decreasing the energy expenses on the rubber compounds preparation processes;
5. improvement of the quality of ingredients dispersion.

The technology of obtaining adhesion promoters has been developed as well as the testing in rubbers for tires has been conducted (blocked diisocyanates or complex compounds of the blocked diisocyanates with metals of variable valency, complex salts of metals of variable valency with organic ligands). The obtained compounds can be capsulated or involved in clay sheets. A possibility of the cobalt stearate replacement on the new promoters providing the reduction of prices at keeping the high adhesion characteristics and their stability has been shown.

The adhesion promoters for polymer materials and rubber-cord composites with fire-retardant properties and adhesion active additives with the properties of antiagers for glue compositions and coatings based on element organic compounds have been developed as well.

As elastomeric materials are microheterogeneous multiphase materials, diffusion processes have the large influence on aging including migration of components. By increasing the operating temperature, the volatility of components, comprising the volatility of antiagers, enhances, and the volatility of antioxidants is possible while keeping them sufficient lability (ability to hit the areas on destructing macromolecules where free radicals appear). The problem is solved effectively enough by application of composite antiaging systems.

In Russian rubber industry, a range of antioxidants is mainly represented by N-isopropyl-N-phenyl-n-phenylenediamine (IPPD) and 2,2,4-trimethyl 1,2-dihydroquinoline (Acetonanil H) in spite of their certain drawbacks. In particular, one of them has inherent wastage due to high diffusion activity, and the second is effective under dynamic loads, but does not provide sufficient protection against heat, oxygen and ozone.

We have carried out the work associated with a possibility of protection of elastomers from oxidative aging using molecular complexes. In turn, molecular complexes were eutectic melts of antioxidants. The studied antioxidants have relatively low diffusion activity that enables to protect an elastomer against oxidative aging for a longer period of time. A decrease in the diffusion capacity is caused by the fact that such antioxidants are in the form of relatively large sized hydrogen-bonded complexes.

Among all the variety of the investigated molecular complexes, the greatest role is assigned to caprolactam containing molecular complexes, and the latter has a number of exceptional properties. Exclusive properties are expressed in conformational transformations. In the binary melt with IPPD lactam conformational transformations allow to obtain a molecular complex with very small activation energy (Fig. 11.7).

Using ε-caprolactam feature that is the ability to form eutectic melts at a relatively low melting point, the synthesis is conducted at relatively low temperatures of 80–90°C. So, the obtained molecular complexes are characterized by a low melting point, high lability and compatibility with a polymer. For example, a molecular complex (of 2 molecules IPPD and 4 – ε-caprolactam), is a liquid with a crystallization temperature −19°C.

Production of molecular complexes is not the only possibility to provide impact of prolonged antiaging effect. There are other techniques such as the use of clays – sorbents, encapsulation. We have also used an extraordinary feature of ε-caprolactam that is capability to form low-melting alloys

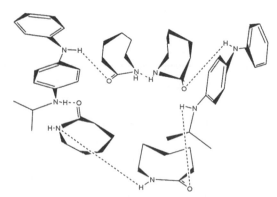

FIGURE 11.7 Hydrogen-bonded conformer consisting of two IPPD and four ε-caprolactam molecules with activation energy of 38.1 kJ/mol.

with many aliphatic and aromatic substances and use them as a dispersion phase for the preparation of complex compounds. Complex compounds were prepared according to the scheme, which involves interaction of a complex agent with acid, and neutral molecules of caprolactam and IPPD are able to enter into the inner sphere of the complex in coordination with the complex agent. Acids were selected among the used in the formulation of rubber or rubber-cord compositions that bring a contribution to the additional functional properties. The peculiarity of obtaining complex compounds in a melt with caprolactam is relatively rapid salt formation, usually accompanied by rapid release of water vapor and relatively slow formation of a complex with a ligand in the ligand field of caprolactam and IPPD. Thus, substances performing the functions of antioxidants and combined in molecular complexes or which are in the ligand field of the complex compound increase resistance of elastomers to oxidative aging.

11.12   STABILIZERS BASED ON NATURAL PRODUCTS [46–49]

Professor Cosimo Carfagna from the Institute of Chemistry and Technology of Polymers (Naples, Italy) in its report to the XIX Mendeleev Congress on General and Applied Chemistry, which was held in September 2012, Volgograd, spoke of two directions in research: the first one is the use of natural antioxidants for stabilization of polymers; the second one is materials intended for food packaging, with antibacterial natural additives.

Where are these substances in fruits and vegetables from? During photosynthesis, quite a lot of free radicals are produced from which the plant must somehow defend itself. However, this is not the only reason for the formation of secondary metabolites – substances which are not necessary as proteins, fats and carbohydrates, but useful for plants and for us. For example, some plants synthesize them to defend against parasites and fungi (that is why resveratrol is produced in red grapes).

These are large classes of very different compounds [46–49]: red carotenoids – long hydrocarbon chains with double bonds; orange, yellow, red and purple flavonoids (including anthocyanins and flavonols). The general formula of flavonoids is two benzene nuclei linked by a three-carbon fragment. By the way, this group also includes many useful components of green tea and soy beans. And there are vitamins, tocopherols and many other classes of antioxidants. Red wine contains a set of flavonoids and other nutrients passing in it not so much of the juice, but as peel and seeds of red grapes. Active natural substances are also found in red pepper, ginger, tomatoes, and cabbage.

Some of the most common antioxidants is "Irganox 1010" and "Irgafos 168." For stabilization of polyethylene and polypropylene, they are used simultaneously because they enhance each other action. Both the one and the other are not environmentally friendly. But it may be a natural antioxidant. Then, on the one hand, a polymer will be safer, and, on the other side, it will be possible to use wastes of raw vegetable and food industries.

In Italy in 2010, 126,000 tons of tomato production wastes were thrown out. One kilogram of lycopene – the main pigment of tomatoes – can be obtained from ten tons of tomato wastes. In the Mediterranean region 300,000 tons of bagasse remains each year after the processing of grapes. Each kilogram of it contains about 10 g of bisphenols, the market value of which is not less than 500 euros per kilo.

Several samples of polypropylene (PP) were studied. A standard was the usual unstabilized PP and other samples with extracts from red or white grapes and tomatoes added to them. They were compared with a sample, which had a couple of standard industrial antioxidants "Irganox1010" and "Irgafos168" in its composition.

The samples of extracts obtained from white grapes and tomatoes were more stable than the pure PP with no additives, but nevertheless were inferior to PP containing industrial stabilizers. But polypropylene with stabilizers (1% by weight) obtained from red grapes was not behind the production prototype. However, polypropylene with the addition of Mediterranean pine bark extract was the most stable. The pine bark extract is a complex mixture of bioflavonoids including proanthocyanidin oligomers (chains containing different amounts of units). There is also a mixture of proanthocyanidins in the seeds of grapes, wine, cranberries, apples, pears and pomegranates.

The second direction of modification of synthetic materials is application of natural antibacterial additives. When spices were brought to Europe in the Middle Ages, they started gaining widespread acceptance not only because it made food taste spicy, but also because it did not allow bacteria to grow up.

Carvacrol is a derivative of phenol (Fig. 11.8), which is rich in essential oils of oregano, thyme, wild bergamot and some other plants. It not only smells good, but also has antibacterial properties.

People try to produce antibacterial food packaging for a long time. After all, the ability to keep food fresh as long as it possible is necessary quality. Today, complex composite materials are actively developed. For this purpose various options are offered based on low density of polyethylene and other polymers with fillers and additives such as nanoparticles of silver, zinc and titanium oxide. Many of them give a very good effect, for example, zinc oxide nanoparticles in polyethylene prolong storage of fresh orange juice in the pack up to 28 days.

Italian researchers as the basis of a nanocomposite also took a low-density polyethylene – it has irregular structure in which micropores of a fine filler good "fit." As the last one "Nanomer" – clay mineral called montmorillonite with the modified surface, was used. In nanofiller carvacrol produced from oregano was added to impart antimicrobial properties throughout the material.

The samples of food packages made of conventional low density polyethylene (LDPE) with the addition of 5% by weight of "Nanomer" were investigated as well as samples based on LDPE containing 5% "Nanomer" and 10% carvacrol. Additive carvacrol increased thermal stability and reduced gas permeability of a material. The composite with carvacrol effectively inhibited the growth of pathogenic bacteria such as *Brochotrix*

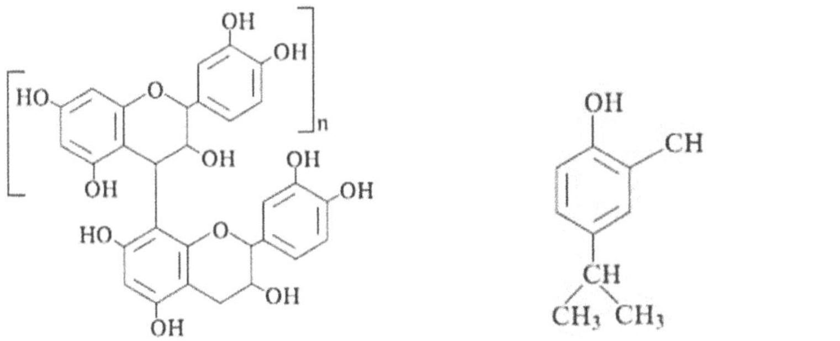

Proanthocyanidins – flavonoids that are found in pine bark

Carvacrol – antiseptic from oregano and thyme

FIGURE 11.8   Natural antioxidants and antiseptics.

thermosphacta, Listeria innocua, Carnobacterium compared to the rest samples. It was assumed that carvacrol as a phenol derivative disturbed a lipid membrane structure of bacteria.

So, protection of elastomeric materials can be performed using various structural, diffusion and kinetic effects.

KEYWORDS

- **aging**
- **antiaging**
- **degradation**
- **destruction**
- **elastomeric rubbers**
- **physical and chemical transformations**
- **polycondensation capable monomers**
- **rubber aging**
- **technical materials**

REFERENCES

1. G.E. Zaikov. Why do polymers age? Soros journal, 2000.
2. V.F. Kablov. Protection of rubbers against aging in the different application conditions with the use of structural, diffusion and kinetic effects. Proceedings of XVII International scientific-practical conference "Rubber industry: raw materials, technology," Moscow, STC NIISHP [et al.], 21–25 May, 2012, pp. 5–7.
3. V.F. Kablov. Protection of rubbers against aging in the different application conditions with the use of structural, diffusion and kinetic effects. Rubber industry. Raw materials, technology: Proceedings of XVIII International scientific-practical conference, STC"NIISHP" et al., 2013. pp. 4–6.
4. V.F. Kablov. System technology of rubber-oligomer compositions. Proceedings of X international conference on chemistry and physical chemistry of oligomers "Oligomers-2009," Volgograd, 2009, pp. 162–191.
5. V.F. Kablov. Regulation of properties of elastomeric materials with functionally active components by conjugation of thermodynamic forces and flows. II Russian scientific-practical conference "Raw materials and materials for rubber industry: Today and Tomorrow": Proceedings. – Moscow, STC"NIISHP," 1995.

6. Kablov V.F. Sychev, N.V., Ogrel A.M. Using thermodynamic forces and flows to create composite materials. International conference on rubber, Beijing, 1992.

7. Kablov V.F. Creation of elastomeric materials based on nonequilibrium systems with functionally active components. Raw materials and materials for rubber industry. Today and Tomorrow (SM. RKR-99): Proceedings of 5 Anniversary Russian scientific-practical conference, May 1998. Moscow, 1998, pp. 385–386.

8. Effect dicarboxylic acid and their dihydrazides on curing characteristics of EPDM / V.E. Derbisher, V.F. Kablov, A.M. Koroteeva, A.M. Ogrel. Kauchuk i Rezina, 1983, №1, pp. 24–26.

9. Kablov V.F. Verwendung polymerizationsfahiger Verbindungen zur Regulierung der Eigenschaften von Gummi / V.F. Kablov, V.E. Derbisher, A.M. Ogrel. Plaste und Kautschuk. 1985, №5, pp. 163–167.

10. Kablov V.F. Kinetic features of polycondensation of monomers introduced in elastomeric matrix / V.F. Kablov, A.M. Ogrel, A.M. Koroteeva. Bulletin of Higher Professional Institutions. Series: Chemistry and Chemical Technology. 1985. Vol. 28, Issue 5, pp. 96–98.

11. Kablov V.F. Mathematical modeling of unsteady heating and combustion of elastomeric materials with physical and chemical changes in different temperature zones / V.F. Kablov, V.L. Strakhov. II International conference on polymer materials of lower combustibility: Proceedings of the RAS, VSTU [et al]. Volgograd, 1992, pp. 24–26.

12. Kablov V.F. Computer modeling of extreme heat phenomena in elastomeric materials / V.F. Kablov. Kauchuk i Rezina. 1997, №1, pp. 8–10.

13. N.M. Emanuel, G.E. Zaikov, Z.K. Maizus. The role of environment in radical-chain oxidation reactions of organic substances. Moscow, "Nauka" Publishing House, 1973, 279 p.

14. Yu.S. Zuev. Resistance of elastomers in operation conditions / Yu.S. Zuev, T.G. Degteva. Moscow, "Chemistry" Publishing House, 1986, 264 p.

15. N.M. Emanuel, A.L. Buchachenko. Chemical physics of aging polymers. Moscow, "Nauka" Publishing House, 1984, 342 p.

16. A.S. Kuzminskii, V.V. Sedov. Chemical changes in polymers. Moscow, "Chemistry" Publishing House, 1984, 192 p.

17. G.E. Zaikov. Degradation and stabilization of polymers. Moscow, MITHT, 1990, 154 p.

18. Gladyshev G.P. Stabilization of heat resistant polymers / G.P. Gladyshev, Yu.A. Ershov, O.A. Shustova. Moscow, "Chemistry" Publishing House, 1979, 272 p.

19. Bryk M.T. Degradation of filled polymers / M.T. Bryk. Moscow, "Chemistry" Publishing House, 1989, 192 p.

20. Novakov I.A., Kablov V.F., Petryuk I.P., Somova A.E. Modifying the elastomer matrix with particles of variable-valency metals for rubbers used at high temperature. International Polymer Science and Technology, 2009. Vol. 36, № 11. P. 15–18.

21. Novakov I.A. Effect of ultrafine particles of transition metals on thermal stability of ethylene-propylene copolymer / I.A. Novakov, V.F. Kablov, I.P. Petryuk, A.E. Somova. Bulletin of VSTU. Series "Chemistry and Technology of Element Organic Monomers and Polymer Materials," 2008, Vol. 5, Issue 1. P. 154–157.

22. Novakov I.A., Kablov V.F., Petryuk I.P., Mikhailyuk A.E. High-temperature aging of vulcanized rubber products based on ethylene-propylene rubber modified using

variable-valency metal particles. International Polymer Science and Technology, 2011. Vol. 38, № 4. P. 13–14.

23. High-temperature aging of vulcanizates based on ethylene-propylene rubber modified with variable-valency metal particles / I.A. Novakov, V.F. Kablov, I.P. Petrjuk, A.E. Mikhailjuk. Kauchuk i Rezina. 2010. № 5. P. 27–32.

24. Ogrel A.M., Kiryukhin N.N., Kablov V.F. To a dependence of properties of vulcanized rubbers based on thermodynamically incompatible rubbers on their morphology and extent of relative affinity. Kauchuk i Rezina, 1978, №11, pp.16–20.

25. Khaimovich A.M., Ogrel A.M., Kablov V.F. Wear resistant composite coatings for rubber articles of friction joints. Bulletin of Higher Professional Institutions. Series: Chemistry and Chemical Technology. 1988, Vol. 31.

26. Kablov V.F. Amine containing polyfunctional modifier for rubbers and glue compositions / V.F. Kablov, N.A. Keibal, S.N. Bondarenko. Proceedings of XV International scientific-practical conference "Rubber industry: raw materials, materials and technology," Moscow, STC NIISHP [et al.], 25–29 May, 2009, pp. 120–122.

27. Modeling and prediction of diffusion and solubility of antiozonants derived from NN-substituted p-phenylenediamines in industrial rubbers / S.M. Kavun, Yu.M. Genkina, V.S. Fillipov. Kauchuk i Rezina. 1995. №6. P. 10–14.

28. Some recommendations on a technology of producing antiaging dispersions / A.F. Puchkov, M.P. Spiridonova, V.F. Kablov. Kauchuk i Rezina. 2009. № 5. P. 21–24.

29. A new technique for obtaining antiagers of prolonged action / A.F. Puchkov, M.P. Spiridonova, V.F. Kablov. Kauchuk i Rezina, №3, 2012. P. 12–15.

30. To a problem of ozone aging of rubbers in the presence of antiozonants of different physical nature / A.F. Puchkov, M.P. Spiridonova, V.F. Kablov. Kauchuk i Rezina, №3, 2011. P. 15–18.

31. Complementary views on a process of ozone aging of rubbers and protection techniques in this basis / A.F. Puchkov, M.P. Spiridonova, V.F. Kablov. "Industrial production and use of elastomers" Journal, №2, 2011. P. 32–35.

32. Properties of binary alloy "ε-caprolactam–stearic acid" / A.F. Puchkov, E.V. Talbi. Kauchuk i Rezina. 2006. №6. P. 21–24.

33. Pat. №4532285 US, IPC C08K009/04. Additives to a rubber for ozonation. Publ. 08.07.86

34. The polymeric composition / Evans L.R., Benko D.A.// Rubb.Chem.Technol.-1992.-№5.-p.1333.

35. Boron containing composite antiager PRS-1B / A.F. Puchkov, M.P. Spiridonova, V.F. Kablov. Kauchuk i Rezina. 2010. №6. P. 10–14.

36. Properties of "ε-caprolactam–boracic acid" alloy and transformations in it / A.F. Puchkov, M.P. Spiridonova, S.V. Lapin, V.F. Kablov. Kauchuk i Rezina. 2013. № 6. P. 10–14.

37. New method for producing prolonged-action antiagers. Puchkov, A.F., Spiridonova, M.P., Kablov, V.F., Kaznacheeva, V.A. / *International Polymer Science and Technology*, 2013, 40 (1), pp. 231–234.

38. Ozone aging of rubbers in the presence of antiozonants of different physical nature. Puchkov, A.F., Spiridonova, M.P., Kablov, V.F., Trusova, E.V., *International Polymer Science and Technology*, 2012.

39. Puchkov A.F. Application of antiagers in form of their alloys for elastomer protection / A.F. Puchkov, V.F. Kablov, S.V. Turenko. Modern science-based technologies. 2005. №8. P. 17–20.
40. Certain recommendations concerning technology for creating antioxidant dispersions / A.F. Puchkov, M.P. Spiridonova, V.F. Kablov, S.V. Turenko. International Polymer Science and Technology. 2010, Vol. 37, № 5, P. 29–31.
41. Capsuling of Liquid Components for Rubbers / A.F. Puchkov, V.F. Kablov, V.B. Svetlichnaya, M.P. Spiridonova, S.V. Turenko. Encyclopedia of chemical engineer, 2011. № 7. P. 47–50.
42. Compositions of chemical and physical antiagers / A.F. Puchkov, M.P. Spiridonova, V.F. Kablov, A.V. Golub. XIX Mendeleev Congress on General and Applied Chemistry (Volgograd, 25–30 September, 2011). Vol. 2 / RAS, D.I. Mendeleev RCS, [et al]. Volgograd, 2011. P. 522.
43. Kablov V.F. Elastomeric systems with nanomicroheterogeneous functionally active technological and modifying additives / V.F. Kablov. Proceedings of XVI International scientific-practical conference "Rubber industry: raw materials, technology," Moscow, STC NIISHP, 23–27 May 2011. P. 28–29.
44. Development of technology and scientific foundations of composite antiagers for rubbers / A.F. Puchkov, M.P. Spiridonova, V.A. Kaznacheeva, V.F. Kablov. Proceedings of XVII International scientific-practical conference "Rubber industry: raw materials, technology," Moscow, STC NIISHP, 21–25 May, 2012, pp. 5–7.
45. Liquid Components Capsuling / A.F. Puchkov, V.F. Kablov, V.B. Svetlichnaya, M.P. Spiridonova, S.V. Turenko. Novel Materials / ed. by Rafiqul Islam. N.Y. (USA): Nova Science Publishers, 2013. P. 115–120.
46. Leshina W. Mediterranean diet for polymers. Chemistry and Life, 2012, № 2.
47. P. Cerruti, M. Malinconico, J. Rychly, L. Matisova-Rychla, C. Carfagna. Effect of natural antioxidants on the stability of polypropylene films. "Polymer Degradation and Stability," 2009, Vol. 94, p. 2095.
48. P. Persico, V. Ambrogi, C. Carfagna, P. Cerruti, I. Ferrocino, G. Mauriello. Nanocomposite polymer film containing carvacrol for Antimicrobial active packaging. "Polymer Engineering & Science," 2009, Vol. 497, p. 1447.
49. http://www.hij.ru/read/articles/technologies-and-materials/90/

INDEX

Milton Keynes UK
Ingram Content Group UK Ltd.
UKHW022057141024
449569UK00031B/1668